Lecture Notes in Mathematics

A collection of informal reports and seminars
Edited by A. Dold, Heidelberg and B. Eckmann, Zürich

188

Symposium on Semantics of Algorithmic Languages

Edited by E. Engeler, University of Minnesota

Springer-Verlag
Berlin · Heidelberg · New York 1971

AMS Subject Classifications (1970): 02E10, 02F43, 68A05, 68A10, 68A30

ISBN 3-540-05377-8 Springer-Verlag Berlin · Heidelberg · New York
ISBN 0-387-05377-8 Springer-Verlag New York · Heidelberg · Berlin

This work is subject to copyright. All rights are reserved, whether the whole or part of the material is concerned, specifically those of translation, reprinting, re-use of illustrations, broadcasting, reproduction by photocopying machine or similar means, and storage in data banks.

Under § 54 of the German Copyright Law where copies are made for other than private use, a fee is payable to the publisher, the amount of the fee to be determined by agreement with the publisher.

© by Springer-Verlag Berlin · Heidelberg 1971. Library of Congress Catalog Card Number 78-151406. Printed in Germany.

Offsetdruck: Julius Beltz, Weinheim/Bergstr.

PREFACE

During the last few years, a number of interesting results and promising ideas have been generated in the area of semantical aspects of programming languages. We felt that we would do a real service to this emerging field by calling a write-in symposium and collecting a number of representative contributions covering the various aspects.

We are happy to present here the results of this endeavor in the form of Lecture Notes. It is understood that many of the contributions represent work in progress and that some may appear elsewhere in final form. We take this opportunity to thank the contributors and Springer Verlag for their exceptionally prompt collaboration.

Minneapolis, October 1970

Erwin Engeler

P.S. The papers are arranged alphabetically according to authors, a joint bibliography is located at the end of the volume.

TABLE OF CONTENTS

Axiom Systems for Simple Assignment Statements 1
Dr. J. W. de Bakker
Mathematisch Centrum
2e Boerhaavestraat 49
Amsterdam (0), Netherlands

A Property of Linear Conditionals. 23
Dr. J. W. de Bakker

Formalization of Storage Properties. 28
Drs. H. Bekic and K. Walk
IBM Laboratory Vienna
Parkring 10
1010 Wien, Austria

Program Schemes, Programs and Logic. 62
Prof. D. C. Cooper
Department of Computer Science
University College of Swansea
Singleton Park, Swansea SA2 8PP
Glamorgan, Wales, U. K.

Algebraic Theories and Program Schemes 71
Dr. C. C. Elgot
IBM Thomas B. Watson Research Center
P. O. Box 218
Yorktown Heights, New York 10598 U.S.A.

Structure and Meaning of Elementary Programs 89
Prof. E. Engeler
School of Mathematics
University of Minnesota
Minneapolis, Minnesota 55455 U.S.A.

Procedures and Parameters: An Axiomatic Approach. 102
Prof. C. A. R. Hoare
Department of Computer Science
The Queen's University of Belfast
Belfast BT7 1NN, Northern Ireland, U. K.

Semantics of ALGOL-Like Statements 117
Prof. S. Igarashi
Computer Science Department
Stanford University
Stanford, California 94305 U.S.A.

Proving Correctness of Implementation Techniques 178
Drs. C. B. Jones and P. Lucas
IBM Laboratory Vienna
Parkring 10
1010 Wien, Austria

Examples of Formal Semantics 212
Prof. D. E. Knuth
Computer Science Department
Stanford University
Stanford, California 94305 U.S.A.

Experience with Inductive Assertions for Proving Programs Correct. 236
Prof. R. L. London
Computer Science Department
University of Wisconsin
Madison, Wisconsin 53706 U.S.A.

Mathematical Theory of Partial Correctness 252
Prof. Z. Manna
Computer Science Department
Stanford University
Stanford, Calfironia 94305 U.S.A.

Towards Automatic Program Synthesis. 270
Prof. Z. Manna and Dr. R. J. Waldinger
Artificial Intelligence Group
Stanford Research Institute
Menlo Park, California 94025 U.S.A.

The Lattice of Flow Diagrams 311
Prof. D. S. Scott
Department of Philosophy
1879 Hall
Princeton University
Princeton, New Jersey 08540 U.S.A.

Joint Bibliography 367

AXIOM SYSTEMS FOR SIMPLE ASSIGNMENT STATEMENTS*

by

J. W. de Bakker

1. Introduction

In this paper we present a number of axiom systems for simple assignment statements and investigate some of their properties and mutual relations.

Simple assignment statements are statements of the form $a:=b$, $x:=y$, etc. We give a formal definition (in section 2) of the effect of a sequence of simple assignment statements upon a variable. Two such sequences are called equivalent if they have the same effect upon all variables. In section 3, we single out four equivalences as axioms, and give a number of rules of inference which allow us to derive other equivalences. We then show that the axiom system is complete in the sense that if two sequences have the same effect upon all variables, then their equivalence is derivable in the system, and vice versa. The axioms are also shown to be independent. In section 5, we investigate the possibility of replacing the four axioms by a smaller number. First we show that three axioms suffice, and then we introduce an infinity of pairs of axioms, all equipollent with the original four. Some variations of these systems are discussed in section 6. In section 7, we give a one-axiom system, but with an extension of the main rule of inference.

Axiomatic characterizations of programming concepts in terms of equivalences have been given by McCarthy (1962, 1963a) as extended by Kaplan (1968b), and Igarashi (1964, 1968). Our paper is closest to Igarashi's, where (general) assignment, conditionals, and goto statements are treated. Igarashi also gives several

---

*This paper is a somewhat modified, and slightly extended, version of de Bakker (1968).

completeness theorems, of which ours is a special case (our proof is different, however). An axiomatic approach to programming concepts, including assignment, is also taken by Hoare (1969), but he does not take equivalence of statements as his starting point. References to various other approaches to assignment, not directly of an axiomatic nature, can be found in our survey paper (de Bakker, 1969) and in the bibliography of London (1970b) which also contains more recent material.

2. Definitions

Let V be an infinite set, V^2 the set of all ordered pairs of elements of V, and V^{2*} the set of all finite non-empty sequences of elements of V^2. The elements of V are denoted by lower case letters, possibly with indices, e.g., $a,b,\ldots,s_1,t_2,\ldots,x,y,z$; the elements of V^2 are denoted by pairs such as ab, s_1t_2, or xy, and the elements of V^{2*} are denoted by sequences such as $ab\ cd$, s_1t_2, or $xy\ yz\ zx$. $S,S_1,S_2,\ldots$ stand for arbitrary elements of V^{2*}.

Definition 2.1 1. The elements of V are called variables.

2. The elements of V^2 are called assignment statements.

3. The elements of V^{2*} are called sequences of assignment statements.

The elements of V correspond to the (simple) variables of, e.g., ALGOL 60; the elements of V^2 to assignment statements such as $a:=b$, $s_1:=t_2$, or $x:=y$; and the elements of V^{2*} to sequences of assignment statements such as $a:=b;c:=d$, or $s_1:=t_2$, or $x:=y;y:=z;z:=x$.

Definition 2.2 Let $S\in V^2$. $p_i(S)$, $i=1,2$, is the ith element of the ordered pair S.

Definition 2.3 Let $S\in V^{2*}$. The set of left parts, $\lambda(S)$, and the set of right parts $\rho(S)$, are defined by:

1. If $S\in V^2$, then $\lambda(S)=\{p_1(S)\}$ and $\rho(S)=\{p_2(S)\}$.
2. If $S=S_1S_2$, $S_1\in V^2$, $S_2\in V^{2*}$, then

$$\lambda(S)=\lambda(S_1)\cup\lambda(S_2),$$

and

$$\rho(S)=\rho(S_1)\cup\rho(S_2).$$

<u>Definition 2.4</u> Let $S \in V^{2*}$. The length $\ell(S)$ of S is defined by:

1. If $S \in V^2$, then $\ell(S) = 1$.
2. If $S = S_1 S_2$, $S_1 \in V^2$, $S_2 \in V^{2*}$, then $\ell(S) = 1 + \ell(S_2)$.

For the simple assignment statements we are concerned with in our paper, it is easy to give a formal description of their usual meaning, by defining the effect E of a sequence S upon a variable a.

<u>Definition 2.5</u> The function $E: V \times V^{2*} \to V$ is defined by:

1. Let $a \in V$ and $S \in V^2$. Then

$$E(a,S) = p_2(S) \text{ , if } a = p_1(S) \text{ ,}$$
$$= a \text{ , if } a \neq p_1(S) \text{ .}$$

2. Let $a \in V$ and $S = S_1 S_2$, with $S_1 \in V^{2*}$ and $S_2 \in V^2$. Then

$$E(a,S) = E(E(a,S_2),S_1) \text{ .}$$

Examples: $E(a\,,ab) = b$, $E(a\,,bc) = a(b \neq c)$, and $E(c\,,ab\;bc\;ca) = E(E(c\,,ca)\,,ab\;bc)$ $= E(a\,,ab\;bc) = E(E(a\,,bc)\,,ab) = E(a\,,ab) = b$.

Next, the notion of equivalence of two sequences of assignment statements is defined in terms of the function E.

<u>Definition 2.6</u> Let $S_1, S_2 \in V^{2*}$. S_1 and S_2 are called equivalent if for all $a \in V$, $E(a,S_1) = E(a,S_2)$.

Example 1: ab bc ca ab and ac cb ba are equivalent, since

$$E(a\,,ab\;bc\;ca\;ab) = c = E(a\,,ac\;cb\;ba) \text{ ,}$$
$$E(b\,,ab\;bc\;ca\;ab) = c = E(b\,,ac\;cb\;ba) \text{ ,}$$
$$E(c\,,ab\;bc\;ca\;ab) = b = E(c\,,ac\;cb\;ba) \text{ ,}$$

and for all $d \neq a,b,c$,

$$E(d\,,ab\;bc\;ca\;ab) = d = E(d\,,ac\;cb\;ba) \text{ .}$$

Example 2: ab bc and bc ab are not equivalent, since

$$E(a\,,ab\;bc) = b \text{ , and } E(a\,,bc\;ab) = c \text{ .}$$

3. An Axiomatic Theory for Equivalence

We now introduce a formal axiomatic theory $\mathfrak{J}$ for the equivalence of simple assignment statements. Well-formed formulas of the theory are expressions of the form $S_1 = S_2$, with $S_1, S_2 \in V^{2*}$. The axioms of the system are:

A_1: For all $a, b \in V$, $ab\ ba = ab$.

A_2: For all $a, b, c \in V$ with $a \neq c$, $ab\ ac = ac$.

A_3: For all $a, b, c \in V$, $ab\ ca = ab\ cb$.

A_4: For all $a, b, c \in V$, $ab\ cb = cb\ ab$.

The rules of inference are:

R_1: If $S_1 ac = S_2 ac$, and $S_1 bd = S_2 bd (a \neq b)$, then $S_1 = S_2$.

R_2: If $S_1 = S_2$, then $S_2 = S_1$. If $S_1 = S_2$ and $S_2 = S_3$, then $S_1 = S_3$.

R_3: If $S_1 = S_2$, then $SS_1 = SS_2$ and $S_1 S = S_2 S$.

Remarks:

1. The set of axioms $\{A_1, A_2, A_3, A_4\}$ is denoted by $\mathfrak{a}$.
2. Rule R_1 may be understood intuitively as follows: If S_1 and S_2 have the same effect upon all variables with the possible exception of a, if they have the same effect upon all variables with the possible exception of b, and if $a \neq b$, then S_1 and S_2 have the same effect upon all variables, i.e., $S_1 = S_2$.
3. We shall use formulae of the form $S_1 = S_2 = S_3 = \ldots$ as an abbreviation for $S_1 = S_2$ and $S_2 = S_3$ and $S_3 = \ldots$.

Lemma 3.1. $S = S$.

Proof.

(1) $ab\ ba = ab$, A_1;

(2) $ab = ab\ ba$, (1), R_2;

(3) $ab = ab$, (2), (1), R_2;

(4) $S = S$, (3), R_3.

From now on, the rules R_2 and R_3 will be used without explicit mentioning.

Lemma 3.2. $ab\ ab = ab$.

Proof. $ab\ ab = ab\ ba\ ab = ab\ ba = ab$, by A_1, A_1, A_1.

Lemma 3.3. If $a \neq c$, $a \neq d$, and $b \neq c$, then $ab\ cd = cd\ ab$.

Proof.

(1) $ab\ cd\ cb = ab\ cb\ (b \neq c)$, A_2;

(2) $cd\ ab\ cb = cd\ cb\ ab = cb\ ab\ (b \neq c)$, A_4, A_2;

(3) $ab\ cd\ cb = cd\ ab\ cb\ (b \neq c)$, (1), (2), A_4;

(4) $ab\ cd\ ad = cd\ ab\ ad\ (a \neq d)$, similar to (3);

(5) $ab\ cd = cd\ ab\ (a \neq c, a \neq d, b \neq c)$, (3), (4), R_1.

Lemma 3.4. If $\lambda(S_1) \cap \lambda(S_2) = \lambda(S_1) \cap \rho(S_2) = \lambda(S_2) \cap \rho(S_1) = \emptyset$, then $S_1S_2 = S_2S_1$.

Proof. By repeated application of lemma 3.3.

(Using the completeness theorem of section 4, it can be proved that the assertion of the lemma also holds with "if" replaced by "only if".)

Lemma 3.5. $aa\ bc = bc\ aa = bc$.

Proof. 1. First we show that $aa\ bc = bc$.

(1) $aa\ ba = ba\ aa = ba\ ab = ba$, A_4, A_3, A_1;

(2) $aa\ ac = ac$, A_2, A_1;

(3) $aa\ bc\ ac = aa\ ac\ bc = ac\ bc = bc\ ac$, A_4, (2), A_4;

(4) $aa\ bc\ ba = aa\ ba = ba = bc\ ba\ (a \neq b)$, A_2, (1), A_2;

(5) $aa\ bc = bc\ (a \neq b)$, (3), (4), R_1;

(6) $aa\ bc = bc$, (2), (5).

2. Now we prove that $bc\ aa = bc$.

(7) $bc\ aa = aa\ bc = bc\ (a \neq b, a \neq c)$, lemma 3.3 and part 1;

(8) $ac\ aa = ac\ ac = ac$, A_3 and lemma 3.2;

(9) $ba\ aa = ba\ ab = ba$, A_3, A_1;

(10) $bc\ aa = bc$, (7), (8), (9).

Lemma 3.6. $aa\ S = S\ aa = S$.

Proof. Follows by lemma 3.5.

The next lemma is included because its proof illustrates the method used in the proof of the completeness theorem in the next section. The lemma shows the effect of two successive interchanges of the two variables b and c .

Lemma 3.7. ab bc ca ab bc ca = ac $(a \neq c)$.

Proof. It is easy to verify, using the previous lemmas, that the assertion holds if a = b or b = c . Now suppose that a , b , c differ from each other. Let x , y , z be three variables, different from a , b , c . Then

$$
\begin{aligned}
&\text{ab bc ca ab bc ca ax by} = \text{ab bc ca ab bc by ca ax} = \\
&\text{ab bc ca ab by ca ax} \quad = \text{ab bc ca ab ca by ax} \quad = \\
&\text{ab bc ca ab cb by ax} \quad = \text{ab bc ca cb ab by ax} \quad = \\
&\text{ab bc cb ab by ax} \quad = \text{ab bc ab by ax} \quad = \\
&\text{ab bc ac by ax} \quad = \text{ab ac bc by ax} \quad = \\
&\text{ac by ax} \quad = \text{ac ax by}
\end{aligned}
$$

by the axioms and lemma 3.3. Hence,

(1) ab bc ca ab bc ca ax by = ac ax by .

Similarly, it is proved that

(2) ab bc ca ab bc ca by cz = ac by cz ,

and

(3) ab bc ca ab bc ca ax cz = ac ax cz .

By (1), (2), and R_1,

(4) ab bc ca ab bc ca by = ac by .

By (1), (3), and R_1,

(5) ab bc ca ab bc ca ax = ac ax .

By (4), (5), and R_1, ab bc ca ab bc ca = ac .

Remark. Lemma 3.7 is a fundamental property of assignment. In fact, it may replace axiom A_2:

(1) ab ab = ab , lemma 3.2;

(2) ab ac = ab ab bc ca ab bc ca = ab bc ca ab bc ca = ac $(a \neq c)$, lemma 3.7, (1), and lemma 3.7.

Hence, A_2 can be proved from A_1 and lemma 3.7.

4. Completeness and Independence of the Axiom System

Theorem 4.1. (Completeness theorem) Two sequences of assignment statements S_1 and S_2 are equivalent (in the sense of definition 2.6) if and only if $S_1 = S_2$ is a theorem of J .

For the proof, we need an auxiliary theorem. We use the notation $\prod_{j=1}^{m} S_j$ as an abbreviation for $S_1 S_2 \ldots S_m$.

__Theorem 4.2.__ Let $S \in V^{2*}$, $\lambda(S) = \{a_1, a_2, \ldots, a_m\}$, $m \geq 1$. Let X be a subset of V such that $X \cap \lambda(S) = \phi$. Then for each i, $1 \leq i \leq m$, and each $x_1, x_2, \ldots, x_m \in X$,

$$S \prod_{\substack{j=1 \\ j \neq i}}^{m} a_j x_j = a_i E(a_i, S) \prod_{\substack{j=1 \\ j \neq i}}^{m} a_j x_j .$$

__Proof.__ We use induction on the length of S

1. $\ell(S) = 1$, i.e., $S = ab$, for some $a, b \in V$. Then, clearly $ab = aE(a, ab)$.

2. Let the assertion be proved for all $S' \in V^{2*}$ with $\ell(S') = n$. Let S be an element of V^{2*}, with $\ell(S) = n+1$. Then $S = S'\, ab$, for some $ab \in V^2$ and $S' \in V^{2*}$ with $\ell(S') = n$. Let $\lambda(S') = \{a_1, a_2, \ldots, a_m\}$, for some $m \leq n$. We distinguish two cases, $a \in \lambda(S')$ and $a \notin \lambda(S')$.

2.1. $a \in \lambda(S')$, i.e., $a = a_k$ for some k, $1 \leq k \leq m$. We have to prove that for each i, $1 \leq i \leq m$,

$$S'\, a_k b \prod_{\substack{j=1 \\ j \neq i}}^{m} a_j x_j = a_i\, E(a_i, S'\, a_k b) \prod_{\substack{j=1 \\ j \neq i}}^{m} a_j x_j .$$

We distinguish the cases $i = k$ and $i \neq k$.

2.1.1. $i = k$

Case (α). $b \notin \lambda(S')$. Then

$$S'\, a_k b \prod_{j \neq k} a_j x_j = S' \prod_{j \neq k} a_j x_j\, a_k b = a_k E(a_k, S') \prod_{j \neq k} a_j x_j\, a_k b =$$

$$a_k b \prod_{j \neq k} a_j x_j = a_k E(b, S') \prod_{j \neq k} a_j x_j = a_k E(a_k, S'\, a_k b) \prod_{j \neq k} a_j x_j ,$$

by repeated use of lemma 3.3, by the induction hypothesis, and since $b \notin \lambda(S')$ implies $E(b, S') = b$.

Case (β). $b = a_k$. This case follows directly from the induction hypothesis and lemma 3.6.

Case (γ). $b = a_h$, for some $h \neq k$. Let x_k be an arbitrary element of X. Then

$$S' a_k a_h \prod_{j\neq k} a_j x_j = S' a_k x_k a_k a_h \prod_{j\neq k} a_j x_j =$$

$$S' \prod_{j\neq h} a_j x_j a_k a_h a_h x_h = a_h E(a_h, S') \prod_{j\neq h} a_j x_j a_k a_h a_h x_h =$$

$$a_h E(a_h, S') a_k a_h \prod_{j\neq k} a_j x_j = a_h E(a_h, S') a_k E(a_h, S') a_h x_h \prod_{j\neq h,k} a_j x_j =$$

$$a_k E(a_h, S') \prod_{j\neq k} a_j x_j = a_k E(a_k, S' a_k a_h) \prod_{j\neq k} a_j x_j$$

by A_2, A_3, lemma 3.3, and the induction hypothesis.

2.1.2. $i \neq k$. Follows easily from the induction hypothesis.

2.2. $a \notin \lambda(S')$, i.e., $\lambda(S) = \{a_1, a_2, \ldots, a_m, a_{m+1}\}$, with $a = a_{m+1}$. Here we have to prove that for each i , $1 \leq i \leq m+1$,

$$S' a_{m+1} b \prod_{\substack{j=1 \\ j\neq i}}^{m+1} a_j x_j = a_i E(a_i, S' a_{m+1} b) \prod_{\substack{j=1 \\ j\neq i}}^{m+1} a_j x_j .$$

2.2.1. $i = m+1$. Again the cases $b \notin \lambda(S)$, $b = a_{m+1}$, and $b = a_h$, $1 \leq h \leq m$, must be distinguished. The proofs are then similar to cases (α), (β), and (γ) above.

2.2.2. $i \neq m+1$. Follows easily from the induction hypothesis. This completes the proof of theorem 4.2.

We can now give the proof of theorem 4.1.

Proof of theorem 4.1.

1. First we prove: If $S_1 = S_2$ is a theorem of $\mathfrak{J}$, then for all $a \in V$, $E(a, S_1) = E(a, S_2)$. Let $A_{\ell i}(A_{ri})$, $i = 1,2,3,4$, denote the left-hand side (right-hand side) of the axiom A_i. It is easy to verify that for all $a \in V$, $E(a, A_{\ell i}) = E(a, A_{ri})$. Moreover, it is easily established that the property of having the same effect upon all variables is preserved by application of the rules of inference.

2. Let $E(a, S_1) = E(a, S_2)$ for all $a \in V$. We prove that then $S_1 = S_2$ is a theorem of $\mathfrak{J}$. We may assume that $\lambda(S_1) = \lambda(S_2)$, say $\lambda(S_1) = \lambda(S_2) = \{a_1, a_2, \ldots, a_m\}$. (If, e.g., $a_i \in \lambda(S_1) \setminus \lambda(S_2)$, then replace S_2 by $S_2 a_i a_i$, etc.) Let $X \subseteq V$ be such that $X \cap \lambda(S_i) = \emptyset$, $i = 1,2$. Then, by theorem 4.1, for each i , $1 \leq i \leq m$,

$$S_1 \prod_{\substack{j=1\\ j\neq i}}^{m} a_j x_j = a_i \; E(a_i , S_1) \prod_{\substack{j=1\\ j\neq i}}^{m} a_j x_j ,$$

and

$$S_2 \prod_{\substack{j=1\\ j\neq i}}^{m} a_j x_j = a_i \; E(a_i , S_2) \prod_{\substack{j=1\\ j\neq i}}^{m} a_j x_j .$$

Since $E(a_i , S_1) = E(a_i , S_2)$, we have

$$S_1 \prod_{\substack{j=1\\ j\neq i}}^{m} a_j x_j = S_2 \prod_{\substack{j=1\\ j\neq i}}^{m} a_j x_j , \quad i = 1,2,\dots,m .$$

Suitable repeated application of R_1 now gives $S_1 = S_2$. This completes the proof of theorem 4.1.

For the proof of the independence of our axiom system, we need a new concept and some notations. We introduce an auxiliary function F . Let N be the set of non-negative integers.

Definition 4.1. The function $F\colon V \times V^{2*} \to N$ is defined by

1. Let $a \in V$ and $S \in V^2$. Then

$$F(a,S) = 1 , \text{ if } a = p_1(S) \text{ and } a \neq p_2(S) ;$$
$$= 0 , \text{ otherwise.}$$

2. Let $a \in V$ and $S = S_1 S_2$, with $S_1 \in V^{2*}$ and $S_2 \in V^2$. Then

$$F(a , S) = F(a , S_2) + F(E(a , S_2) , S_1) .$$

$F(a , S)$ may be described as the number of non-trivial steps which are made in calculating the effect of S upon a .

Example: Let a , b , c be three different variables.

$$F(b , ab\ ca\ bc\ bb) = F(b , bb) + F(E(b , bb) , ab\ ca\ bc) =$$
$$0 + F(b , ab\ ca\ bc) = F(b , bc) + F(E(b , bc) , ab\ ca) =$$
$$1 + F(c , ab\ ca) = 1 + F(c , ca) + F(E(c , ca) , ab) =$$
$$1 + 1 + F(a , ab) = 3 .$$

Definition 4.2. The sets of axioms $\mathcal{A} \setminus \{A_i\}$, $i = 1,2,3,4$, are denoted by $\mathcal{A}_i$.

In the remainder of this section and in the following sections, we shall consider sets of axioms which differ from $\mathfrak{a}$. Therefore, the following notation is introduced:

Definition 4.3. Let $\mathcal{F}$ be a set of axioms, and let $S_1, S_2 \in V^{2*}$. $\mathcal{F} \vdash S_1 = S_2$ means that $S_1 = S_2$ can be derived from the set of axioms $\mathcal{F}$ by application of R_1, R_2, R_3.

Usually, it is clear from the context which set of axioms is meant. Explicit mentioning of it is then omitted. E.g., up to now, $S_1 = S_2$ always meant $\mathfrak{a} \vdash S_1 = S_2$. We now prove the independence theorem.

Theorem 4.3. The set of axioms $\mathfrak{a}$ is independent.

Proof. We exhibit four properties $P_i = P_i(S_1, S_2)$, $i = 1,2,3,4$, such that if $\mathfrak{a}_i \vdash S_1 = S_2$, then $P_i(S_1, S_2)$ holds, but $P_i(A_{\ell i}, A_{ri})$ does not hold. ($A_{\ell i}$ and A_{ri} are the left- and right-hand sides of A_i.) These properties are:

P_1: $\lambda(S_1) = \lambda(S_2)$.

P_2: $s(S_1) = s(S_2)$, where, for $S \in V^{2*}$, $s(S)$ is the second variable of the first assignment statement in the sequence S.

P_3: For all $a \in V$, $F(a, S_1) + F(a, S_2)$ is an even number.

P_4: $f(S_1) = f(S_2)$, where, for $S \in V^{2*}$, $f(S)$ is the first variable of the first assignment statement in the sequence S.

5. Equipollent Axiom Systems

In this section, we introduce several (in fact, an infinity of) smaller sets of axioms for assignment statements, and we prove that from these systems the same equivalences can be derived as from $\mathfrak{a}$. (We do not change the rules of inference, R_1, R_2, and R_3.)

Definition 5.1. Let $\mathcal{F}_1, \mathcal{F}_2$ be two sets of axioms for assignment statements. $\mathcal{F}_1 \Rightarrow \mathcal{F}_2$ is used as an abbreviation for: For all $S_1, S_2 \in V^{2*}$, if $\mathcal{F}_1 \vdash S_1 = S_2$, then $\mathcal{F}_2 \vdash S_1 = S_2$. $\mathcal{F}_1$ and $\mathcal{F}_2$ are called equipollent, denoted by $\mathcal{F}_1 \Leftrightarrow \mathcal{F}_2$, if $\mathcal{F}_1 \Rightarrow \mathcal{F}_2$ and $\mathcal{F}_2 \Rightarrow \mathcal{F}_1$.

In order to reduce the number of axioms, one looks for equivalences which, in some sense, combine the properties of some of the axioms A_1, A_2, A_3, and A_4. Two equivalences which combine A_3 and A_4 are

B: ab ca = cb ab and B': ab ca = cb ac .

Combinations of A_1, A_3, and A_4 are given by

C_1: ab ca bc = cb ab and C_1': ab ca bc = cb ac .

The structure of C_1 and C_1' suggests that one also considers

D_1: ab ca bc ab = cb ab and D_1': ab ca bc ab = cb ac ,

E_1: ab ca bc ab ca = cb ab and E_1': ab ca bc ab ca = cb ac ,

C_2: $(\text{ab ca bc})^2$ = cb ab and C_2': $(\text{ab ca bc})^2$ = cb ac ,

etc. ($(S)^n$ is a sequence of n times S .) The general form of these equivalences is:

C_n: $(\text{ab ca bc})^n$ = cb ab and C_n': $(\text{ab ca bc})^n$ = cb ac ,

D_n: $(\text{ab ca bc})^n$ab = cb ab and D_n': $(\text{ab ca bc})^n$ab = cb ac ,

E_n: $(\text{ab ca bc})^n$ab ca = cb ab and E_n': $(\text{ab ca bc})^n$ab ca = cb ac .

It is easily verified that all these equivalences are indeed provable from $\mathfrak{A}$.

Let $\mathfrak{B} = \{A_1, A_2, B\}$ and $\mathfrak{B}' = \{A_1, A_2, B'\}$,

$\mathfrak{C}_n = \{A_2, C_n\}$ and $\mathfrak{C}_n' = \{A_2, C_n'\}$,

$\mathfrak{D}_n = \{A_2, D_n\}$ and $\mathfrak{D}_n' = \{A_2, D_n'\}$,

$\mathfrak{E}_n = \{A_2, E_n\}$ and $\mathfrak{E}_n' = \{A_2, E_n'\}$.

We shall prove that $\mathfrak{B}$, $\mathfrak{B}'$, and, for each $n \geq 1$, $\mathfrak{C}_n$, $\mathfrak{D}_n'$, $\mathfrak{E}_n$, $\mathfrak{E}_n'$ are all equipollent with $\mathfrak{A}$. As we shall see in section 6, this does not hold, in general, for $\mathfrak{C}_n'$ and $\mathfrak{D}_n$.

<u>Theorem 5.1</u>. $\mathfrak{A} \Leftrightarrow \mathfrak{B}$.

<u>Proof</u>. $\mathfrak{B} \Rightarrow \mathfrak{A}$ is clear. In order to prove $\mathfrak{A} \Rightarrow \mathfrak{B}$, it is sufficient to show that $\mathfrak{B} \vdash A_3$ and $\mathfrak{B} \vdash A_4$:

(1) ab ca = ab ca ac = cb ab ac = cb ac = ab cb $(a \neq c)$, A_1, B, A_2, B ;

(2) ab aa = ab ab , B ;

(3) ab ca = ab cb , (1), (2).

Hence, $\mathfrak{B} \vdash A_3$. From A_3 and B , A_4 follows directly.

<u>Theorem 5.2</u>. $\mathfrak{A} \Leftrightarrow \mathfrak{B}'$.

<u>Proof</u>. $\mathfrak{B}' \Rightarrow \mathfrak{A}$ is clear. Proof of $\mathfrak{A} \Rightarrow \mathfrak{B}'$:

(1) $ab\ cb = ab\ ba\ cb = ab\ ca\ bc = cb\ ac\ bc$, A_1, B', B' ;

(2) $ac\ bc = bc\ ab\ cb = bc\ cb\ ac\ bc = bc\ ac\ bc$, (1), (1), A_1 ;

(3) $ac\ bc\ bc = bc\ ac\ bc$, A_1, (2) ;

(4) $bc\ ac\ ac = ac\ bc\ ac$, (3) ;

(5) $ac\ bc = bc\ ac$, (3), (4), R_1 .

Hence, $\mathcal{B}' \vdash A_4$.

(6) $ab\ cb = cb\ ac\ bc = cb\ bc\ ac = cb\ ac$, (1), (5), A_1 .

$\mathcal{B}' \vdash A_3$ now follows as in theorem 5.1.

Remark. It is not true that $\{A_2, B\} \Leftrightarrow \mathcal{A}$ or $\{A_2, B'\} \Leftrightarrow \mathcal{A}$. This follows from:

a. If $\{A_2, B\} \vdash S_1 = S_2$ or $\{A_2, B'\} \vdash S_1 = S_2$, then $\lambda(S_1) = \lambda(S_2)$.

b. $\lambda(ab\ ba) = \{a,b\} \neq \{a\} = \lambda(ab)$.

Theorem 5.3. For each $n \geq 1$, $\mathcal{A} \Leftrightarrow C_n$.

Proof. $C_n \Rightarrow \mathcal{A}$ is clear. Proof of $\mathcal{A} \Rightarrow C_n$:

(1) $cb\ ab\ bc = (ab\ ca\ bc)^n\ bc = (ab\ ca\ bc)^n = cb\ ab\ (b \neq c)$, C_n, A_2, C_n ;

(2) $cb\ ab\ ab = (ab\ ca\ bc)^n\ ab = ab(ca\ bc\ ab)^n = ab\ ba\ ca$, C_n, C_n ;

(3) $ab\ ab\ ba = ab\ ab\ (a \neq b)$, (1) ;

(4) $ab\ ba = ab\ (a \neq b)$, (3), A_2 ;

(5) $aa\ aa\ ba = aa\ aa\ ba\ ba = aa\ ba\ ab\ ab = aa\ ba\ (a \neq b)$, A_2, (2), (4), A_2 ;

(6) $aa\ aa\ ab = aa\ ab\ (a \neq b)$, A_2 ;

(7) $aa\ aa = aa$, (5), (6), R_1 ;

(8) $ab\ ba = ab$, (4), (7) .

Hence, $C_n \vdash A_1$.

(9) $cb\ ab = ab\ ca$, (2), (8) .

(9) and theorem 5.1 yield $C_n \vdash A_3$ and $C_n \vdash A_4$.

Theorem 5.4. For each $n \geq 1$, $\mathcal{A} \Leftrightarrow \mathcal{D}'_n$.

Proof. $\mathcal{D}'_n \Rightarrow \mathcal{A}$ is clear. Proof of $\mathcal{A} \Rightarrow \mathcal{D}'_n$:

(1) $cb\ ab = cb\ ac\ (a \neq b)$, similar to (1) of theorem 5.3;

(2) $cb\ ac\ ca = ab\ ba\ cb$, similar to (2) of theorem 5.3;

(3) $ab\ ba\ ab = ab\ aa\ aa = ab\ ab\ aa = ab\ ab\ ab = ab\ ab\ (a \neq b)$, (2), (1), (1), A_2 ;

(4) ab ba ba = ab ba $(a \neq b)$, A_2 ;

(5) ab ba = ab $(a \neq b)$, (3), (4), R_1 ;

(6) aa aa ba = aa aa ba ab = aa ba ab aa = aa ba ab ab = aa ba $(a \neq b)$, (5), (2), (1), A_2, (5);

(7) ab ba = ab , (5), (6), and similar to theorem 5.4, (5) to (8).

Hence, $\mathcal{D}'_n \vdash A_1$.

(8) cb ac = ab cb , (2), (7) .

(8) and theorem 5.1 yield $\mathcal{D}'_n \vdash A_3$ and $\mathcal{D}'_n \vdash A_4$.

__Theorem 5.5.__ For each $n \geq 1$, $\mathcal{A} \Leftrightarrow \mathcal{E}_n$.

__Proof.__ $\mathcal{E}_n \Rightarrow \mathcal{A}$ is clear. Proof of $\mathcal{A} \Rightarrow \mathcal{E}_n$:

(1) cb ab bc = ab ba ca , cf. (2) of theorem 5.3;

(2) $(\text{ab aa ba})^n$ ab aa = ab ab , E_n ;

(3) $(\text{ab aa ba})^n$ ab aa ab = ab ab ab , (2);

(4) $(\text{ab aa ba})^n$ ab = ab $(a \neq b)$, (3), A_2 ;

(5) aa ba = $(\text{ba ab aa})^n$ ba ab = ba $(\text{ab aa ba})^n$ ab = ba ab $(a \neq b)$, E_n, (4);

(6) ba ab ba = aa ba ba = aa ba = ba ab $(a \neq b)$, (5), A_2 , (5);

(7) ab ba = $(\text{ab aa ba})^n$ ab ba = $(\text{ab aa ba})^n$ ab = ab $(a \neq b)$, (4), (6), (4);

(8) aa ba = ba $(a \neq b)$, (5), (7);

(9) ab ba = ab , (7), and similar to theorem 5.4, (5) to (8).

Hence, $\mathcal{E}_n \vdash A_1$.

(10) ab cb ab = ab $(\text{ab ca bc})^n$ ab ca =
$(\text{ab ca bc})^n$ ab ca = cb ab = cb ab ab , E_n , (9), E_n , (9);

(11) cb ab cb = ab cb cb , (10);

(12) ab cb = cb ab , (10), (11), R_1 .

Hence, $\mathcal{E}_n \vdash A_4$.

(13) ab ca = ab ba ca = cb ab bc = ab cb bc = ab cb , (9), (1), (12), (9) .

Hence, $\mathcal{E}_n \vdash A_3$.

__Theorem 5.6.__ For each $n \geq 1$, $\mathcal{G} \Leftrightarrow \mathcal{E}'_n$.

__Proof.__ $\mathcal{E}'_n \Rightarrow \mathcal{G}$ is clear. Proof of $\mathcal{G} \Rightarrow \mathcal{E}'_n$:

(1) cb ac ca = cb ac $(a \neq c)$, cf. (1) of theorem 5.3;

(2) cb ac bc = ab ba cb , cf. (2) of theorem 5.3;

(3) ba ab ba = ba ab $(a \neq b)$, (1);

(4) $(\text{ab aa ba})^n$ ab = ab $(a \neq b)$, cf. theorem 5.5, (2) to (4);

(5) ab ba = ab , (3), (4), and similar to theorem 5.5.

Hence, $\mathcal{E}'_n \vdash A_1$.

(6) cb ac bc   cb , (2), (5);

(7) ab cb = cb ab , (6), and similar to th. 5.2, (1) to (5).

Hence, $\mathcal{E}'_n \vdash A_4$.

(8) cb ab = ab cb = cb ac bc = cb bc ac = cb ac , (7), (6), (7), (5).

Hence, $\mathcal{E}'_n \vdash A_3$.

6. Non-equipollent Axiom Systems

The first main result of this section is the somewhat surprising fact that, in general, neither of the systems C'_n and $\mathcal{D}_n$, where

$$C'_n = \{A_2, (\text{ab ca bc})^n = \text{cb ac}\} ,$$
$$\mathcal{D}_n = \{A_2, (\text{ab ca bc})^n\ \text{ab} = \text{cb ab}\} ,$$

is equipollent with $\mathcal{A}$. More precisely, the following holds:

1. For all $n \geq 1$, $C'_n \vdash A_1$ and $C'_n \vdash A_4$.
2. For all $n \geq 1$, $\mathcal{D}_n \vdash A_1$ and $\mathcal{D}_n \vdash A_4$.
3. For all $n \geq 1$, $C'_n \Leftrightarrow \mathcal{D}_n$.
4. For all $n \geq 1$, $C'_n \Rightarrow \mathcal{A}$ and, hence, $\mathcal{D}_n \Rightarrow \mathcal{A}$.
5. $\mathcal{A} \Rightarrow C'_1$ and, hence $\mathcal{A} \Rightarrow \mathcal{D}_1$; thus, $\mathcal{A} \Leftrightarrow C'_1 \Leftrightarrow \mathcal{D}_1$.
6. For no even $n \geq 2$, $\mathcal{A} \Rightarrow C'_n$.

Thus, for even n , $\mathcal{A}$ is not equipollent with C'_n (or $\mathcal{D}_n$) . The equipollence of $\mathcal{A}$ and C'_n for odd $n \geq 3$ is still open. We conjecture that in this case as well, $C'_n \Leftrightarrow \mathcal{A}$ does not hold.

It is not clear how to explain the difference in the properties of C_n , $\mathcal{D}'_n$, $\mathcal{E}_n$, and $\mathcal{E}'_n$ on the one hand, and C'_n and $\mathcal{D}_n$ on the other hand. Clarification might arise either from the construction of a model for C'_n (as was given for $\mathcal{A}$

by the E-function of definition 2.5), or by investigating the connection with properties of permutations.

Some first results which can be derived from C_n' are given in theorems 6.3 and 6.4, the second of which concerns a generalization of lemma 3.7.

Theorem 6.1. For each $n \geq 1$,

1. $C_n' \vdash A_1$ and $C_n' \vdash A_4$;
2. $\mathcal{D}_n \vdash A_1$ and $\mathcal{D}_n \vdash A_4$;
3. $C_n' \Leftrightarrow \mathcal{D}_n$.

Proof. The proofs of 1 and 2 are quite similar to the proofs of the previous section and are, therefore, omitted. Once 1 and 2 are established, the proof of 3 offers no difficulties.

Theorem 6.2. 1. For each $n \geq 1$, $C_n' \Rightarrow \mathcal{A}$.

2. $\mathcal{A} \Rightarrow C_1'$.

3. For no even $n \geq 2$, $\mathcal{A} \Rightarrow C_n'$.

Proof. 1. Clear

2. By theorem 6.1, it is only necessary to prove $C_1' \vdash ab\ ca = ab\ cb$. The proofs of the special cases that $a = b$, $a = c$, or $b = c$ are again similar to the proofs of section 5, and omitted. Now suppose that a, b, c are different and that x, y, z are arbitrary variables, different from a, b, c .

(1) $ab\ cd = cd\ ab$ $(a \neq c, a \neq d, b \neq c)$, the proof of lemma 3.3 does not use A_3 ;

(2) $ab\ ca\ ax\ by = cb\ ac\ ba\ ax\ by = cb\ ax\ by = ab\ cb\ ax\ by$, C_1', (1), A_2;

(3) $ab\ ca\ ax\ cz = ab\ cb\ ax\ cz$, (1), A_2;

(4) $ab\ ca\ by\ cz = ab\ cb\ by\ cz$, (1), A_2;

(5) $ab\ ca\ ax = ab\ cb\ ax$, (2), (3), R_1;

(6) $ab\ ca\ by = ab\ cb\ by$, (2), (4), R_1;

(7) $ab\ ca = ab\ cb$, (5), (6), R_1.

3. Let n be even ≥ 2 . Suppose $C_n' \vdash S_1 = S_2$. Then S_1 and S_2 have the following property. (P): For all $a \in V$, $F(a, S_1) + F(a, S_2)$ is an even number. This is clearly true for A_2 . Next we consider C_n' . First suppose that a, b, c are all different. Then

$$F(a, (ab\ ca\ bc)^n) = 3n-2 \quad \text{and} \quad F(a, cb\ ac) = 2,$$
$$F(b, (ab\ ca\ bc)^n) = 3n \quad \text{and} \quad F(b, cb\ ac) = 0,$$
$$F(c, (ab\ ca\ bc)^n) = 3n-1 \quad \text{and} \quad F(c, cb\ ac) = 1,$$
$$F(d, (ab\ ca\ bc)^n) = 0 \quad \text{and} \quad F(d, cb\ ac) = 0,$$

for all $d \neq a,b,c$. Hence, in all cases, $F(e, (ab\ ca\ bc)^n) + F(e, cb\ ac)$ is even, since n is even. It is also easily verified that (P) holds if two or more variables in C'_n are equal. Moreover, it is clear that (P) is preserved by R_1, R_2, and R_3 . Since $F(c, ab\ ca) + F(c, ab\ cb) = 2+1 = 3$, it follows that A_3 does not have property (P) , and hence cannot be derived from C'_n . This means that $\mathfrak{a} \Rightarrow C'_n$ holds for no even n . This completes the proof of theorem 6.2.

According to theorem 6.1, $\{A_2, C'_n\} \vdash A_1, A_4$, and $\{A_2, D_n\} \vdash A_1, A_4$. The next theorem shows that not all of A_2 is needed for these derivations. It is sufficient to assume, together with C'_n :

$$A'_n:\ (ab)^{3n-2} = ab ,$$

and together with D_n :

$$A_n:\ (ab)^{3n-1} = ab .$$

Moreover, we derive two equivalences using only $C'_n(D_n)$; the first of these is used in the proof of theorem 6.4.

<u>Theorem 6.3</u>. For each $n \geq 1$,

1. $\{C'_n\} \vdash (ab)^{3n-1} = ab$;
2. $\{A'_n, C'_n\} \vdash A_1, A_4$;
3. $\{D_n\} \vdash (ab)^{3n} = ab$;
4. $\{A_n, D_n\} \vdash A_1, A_4$.

<u>Proof</u>.

1:

(1) cb ac ab = ab ba cb , cf. (2) of theorem 5.3;

(2) ba ab aa = aa aa ba , (1);

(3) ab aa ab = ab ba ab , (1);

(4) aa ba ba = ba ab aa , (1);

(5)	$(aa\ ba\ ab)^n = ba\ ab$	, C'_n;
(6)	$(ba\ ab\ aa)^n = aa\ ba$	, C'_n;
(7)	$aa\ ba\ ab\ aa\ ba\ ab = ba\ ab\ ba\ ab\ ba\ ab$	, (4), (2), (4), (3);
(8)	$(aa\ ba\ ab)^{2n} = (ba\ ab)^{3n}$	, (7);
(9)	$(ba\ ab)^2 = (ba\ ab)^{3n}$	, (5), (8);
(10)	$aa\ ba\ aa\ ba = (ba\ ab\ aa)^n aa\ ba =$	
	$(ba\ ab\ aa)^n ba\ ba = aa\ ba\ ba\ ba = ba\ ab\ aa\ ba$	, (6), (2), (4), (6), (4);
(11)	$ba\ ab\ aa\ ba\ ab\ aa = aa\ ba\ aa\ ba\ aa\ ba$	, (4), (2), (10), (10);
(12)	$(aa\ ba)^2 = (ba\ ab\ aa\ ba\ ab\ aa)^n = (aa\ ba)^{3n}$	, (6), (11);
(13)	$(ab\ ba)^{3n-1} ab\ ba = ab\ ba\ ab\ ba$	, (9);
(14)	$(ab\ ba)^{3n-1} ab\ aa\ ab = ab\ ba\ ab\ aa\ ab$	, (3), (9);
(15)	$(ab\ ba)^{3n-1} ab\ aa\ ba = ab\ ba\ ab\ aa\ ba$	, (12), (10);
(16)	$(ab\ ba)^{3n-1} ab\ aa = ab\ ba\ ab\ aa\ (a \neq b)$	, (14), (15), R_1;
(17)	$(ab\ ba)^{3n-1} ab = ab\ ba\ ab\ (a \neq b)$	, (13), (16), R_1;
(18)	$(ba\ ab)^{3n-1} aa\ ab = ba\ ab\ aa\ ab$	, (3), (9);
(19)	$(ba\ ab)^{3n-1} aa\ ba = ba\ ab\ aa\ ba$	, (12), (10);
(20)	$(ba\ ab)^{3n-1} aa = ba\ ab\ aa\ (a \neq b)$	, (18), (19), R_1;
(21)	$(ba\ ab)^{3n-1} = ba\ ab\ (a \neq b)$	, (17), (20), R_1;
(22)	$(ba\ ab)^{3n-2}(ba\ ab\ aa)^n = (ba\ ab\ aa)^n\ (a \neq b)$	, (21),
(23)	$(aa\ ba)^{3n-1} = (ba\ ab)^{3n-2} aa\ ba = aa\ ba\ (a \neq b)$	, (10), (6), (22), (6);
(24)	$(ab\ ba)^{3n-2} ab\ aa\ ab = ab\ aa\ ab\ (a \neq b)$	, (21), (3);
(25)	$(ab\ ba)^{3n-2} ab\ aa\ ba = ab\ aa\ ba\ (a \neq b)$	, (23), (10);
(26)	$(ab\ ba)^{3n-2} ab\ aa = ab\ aa\ (a \neq b)$	, (24), (25), R_1;
(27)	$(ab\ ba)^{3n-2} ab = ab\ (a \neq b)$	, (21), (26), R_1;
(28)	$ba\ ab\ aa = aa\ ba\ ba = aa\ ba\ ba\ (ab\ ba)^{3n-2} =$	
	$ba\ ab\ aa\ (ab\ ba)^{3n-2} = ba\ ab\ ba\ (ab\ ba)^{3n-2} =$	
	$ba\ ab\ ba\ (a \neq b)$	, (4), (27), (4), (3), (27);
(29)	$(ba\ ab)^{3n-2} aa = ba\ (a \neq b)$	, (27), (28);
(30)	$(ba\ ab)^{3n-2} aa\ ba = aa\ ba\ (a \neq b)$	, (10), (23);
(31)	$ba\ ba = aa\ ba$	, (29), (30);
(32)	$(ba)^{3n} = (ba)^2\ (a \neq b)$	, (6), (4), (31)

(33) $ba\ ab = (aa\ ba\ ab)^n = aa\ (ba\ ab\ aa)^{n-1} ba\ ab =$

$aa\ (ba)^{3n-3} ba\ ab = (ba)^{3n-1} ab\ (a \neq b)$, (5), (4), (31), (31);

(34) $(ba)^{3n-1} = ba\ (a \neq b)$, (32), (33), R_1;

(35) $(aa)^{3n} = aa\ aa$, C'_n;

(36) $(aa)^{3n-1} ba = aa\ ba\ (a \neq b)$, (31), (32);

(37) $(aa)^{3n-1} = aa$, (35), (36), R_1;

(38) $(ab)^{3n-1} = ab$, (34), (37).

Hence, $\{C'_n\} \vdash (ab)^{3n-1} = ab$.

2:

(39) $(ab)^{3n-2} = ab$, A'_n;

(40) $(ab)^{3n-1} = ab$, (38);

(41) $ab\ ab = ab$, (39), (40);

(42) $ab\ ba\ ab = ab\ ab\ ab$, (28), (31), (4);

(43) $ab\ ba\ ba = ab\ ab\ ba$, (41);

(44) $ab\ ba = ab$, (42), (43), R_1, (41).

Hence, $\{A'_n, C'_n\} \vdash A_1$.

(45) $ab\ cb = cb\ ac\ ab$, (1), (44);

(46) $ab\ cb\ ac = cb\ ac$, (41), C'_n;

(47) $cb\ ac\ bc = cb\ ac$, (41), C'_n;

(48) $ac\ bc = bc\ ab\ ac = bc\ ab\ cb\ ac = bc\ cb\ ac = bc\ ac$, (45), (47), (46), (44).

Hence, $\{A'_n, C'_n\} \vdash A_4$.

3:

(49) $cb\ ab\ ca = ab\ ba\ ca$, cf. (2) of theorem 5.4;

(50) $ba\ aa\ ba = aa\ aa\ ba$, (49);

(51) $ab\ ab\ aa = ab\ ba\ aa$, (49);

(52) $aa\ ba\ ab = ba\ ab\ ab$, (49);

(53) $(aa\ ba\ ab)^n aa = ba\ aa$, D_n;

(54) $(ab\ aa\ ba)^n ab = ab\ ab$, D_n;

(55) $ab\ ab\ ba\ aa = ab\ ab\ (aa\ ba\ ab)^n aa =$

$ab\ ba\ (aa\ ba\ ab)^n aa = ab\ ba\ ba\ aa$, (53), (51), (53);

(56) $ba\ ba\ ab\ ab = ba\ aa\ ba\ ab = aa\ aa\ ba\ ab =$

$aa\ ba\ ab\ ab = ba\ ab\ ab\ ab$, (52), (50), (52), (52);

(57) $ba\ ba\ ab = ba\ ab\ ab$, (55), (56), R_1;

(58) $ba\ ba = ba\ ab$, (51), (57), R_1;

(59) $aa\ ba\ ba = ba\ ba\ ba$, (52), (58);

(60) $aa\ ba\ ab = ba\ ba\ ab$, (59), (58);

(61) $aa\ ba = ba\ ba$, (59), (60), R_1;

(62) $(ab)^{3n+1} = ab\ ab$, (54), (58), (61);

(63) $(ab)^{3n}ba = ab\ ba$, (62), (58);

(64) $(ab)^{3n} = ab\ (a \neq b)$, (62), (63), R_1;

(65) $(aa)^{3n}aa = aa\ aa$, D_n;

(66) $(aa)^{3n}ba = aa\ ba$, (61), (62);

(67) $(aa)^{3n} = aa$, (65), (66), R_1;

(68) $(ab)^{3n} = ab$, (64), (67).

Hence, $\{D_n\} \vdash (ab)^{3n} = ab$.

4. The proof of A_1 is similar to part 2. Moreover, from $ab\ ab = ab$ and D_n, $ab\ cb\ ab = cb\ ab$ follows. The proof of A_4 is then straightforward. This completes the proof of theorem 6.3.

Finally, theorem 6.4 gives the analogon of lemma 3.7 and the remark following it. Consider the equivalence

$$T_n\text{: } (ab\ bc\ ca)^{2n} = ac\ (a \neq c)\ .$$

We show that T_n can be derived from C'_n and A_2, and, conversely, that A_2 can be derived from C'_n and T_n.

<u>Theorem 6.4</u>. For each $n \geq 1$,

1. $\{A_2, C'_n\} \vdash T_n$;
2. $\{T_n, C'_n\} \vdash A_2$.

<u>Proof</u>. 1. We prove that $(ab\ bc\ ca)^{2n} = ac\ (a \neq c)$ can be derived from C'_n and A_2. It is easy to verify this for $a = b$ or $b = c$. From now on we suppose that a, b, c are different, and that x, y, z are arbitrary variables, different from a, b, c.

(1) A_1, theorem 6.1;

(2) A_4, theorem 6.1;

(3) $(ab\ bc\ ca)^{2n-2}(ba\ cb\ ac)^{n-1} =$

$(ab\ bc\ ca)^{2n-4}ab\ bc\ ca\ ab\ bc\ ca\ ba\ cb\ ac\ (ba\ cb\ ac)^{n-2} =$

$(ab\ bc\ ca)^{2n-4}(ba\ cb\ ac)^{n-2} = \ldots =$

$(ab\ bc\ ca)^{2}ba\ cb\ ac = bc\ ac$, A_1, A_2, A_4;

(4) $(ab\ bc\ ca)^{2n-2}(cb\ ac\ ba)^{n-1} =$

$(ab\ bc\ ca)^{2n-4}ab\ bc\ ca\ ab\ bc\ ca\ cb\ ac\ ba\ (cb\ ac\ ba)^{n-2} =$

$(ab\ bc\ ca)^{2n-4}(cb\ ac\ ba)^{n-2} = \ldots =$

$(ab\ bc\ ca)^{2}cb\ ac\ ba = ab\ ca$, A_1, A_2, A_4;

(5) $ab\ cd = cd\ ab\ (a \neq c,\ a \neq d,\ b \neq c)$, lemma 3.3;

(6) $(ab\ bc\ ca)^{2n}ax\ by = (ab\ bc\ ca)^{2n-1}ab\ bc\ ca\ ax\ by =$

$(ab\ bc\ ca)^{2n-1}ab\ ca\ ax\ by = (ab\ bc\ ca)^{2n-1}(cb\,ac\,ba)^{n}ax\ by =$

$ab\ bc\ ca\ ab\ ca\ cb\ ac\ ba\ ax\ by = ac\ ax\ by$, A_1, A_2, A_4, C'_n, (4), (5);

(7) $(ab\ bc\ ca)^{2n}ax\ cz = (ab\ bc\ ca)^{2n-2}ab\ bc\ ca\ ab\ bc\ ca\ ax\ cz =$

$(ab\ bc\ ca)^{2n-2}ab\ bc\ (ba\ cb\ ac)^{n}ax\ cz =$

$(ab\ bc\ ca)^{2n-2}ab\ ba\ cb\ ac\ (ba\ cb\ ac)^{n-1}ax\ cz =$

$(ab\ bc\ ca)^{2n-2}(cb\ ac\ ba)^{n-1}ax\ cz = ac\ ax\ cz$, A_1, A_2, A_4, C'_n, (4), (5);

(8) $(ab\ bc\ ca)^{2n}by\ cz = \ldots =$

$(ab\ bc\ ca)^{2n-2}(ba\ cb\ ac)^{n-1}by\ cz = ac\ by\ cz$, A_1, A_2, A_4, C'_n, (3), (5);

(9) $(ab\ bc\ ca)^{2n}ax = ac\ ax$, (6), (7), R_1;

(10) $(ab\ bc\ ca)^{2n}by = ac\ by$, (6), (8), R_1;

(11) $(ab\ bc\ ca)^{2n} = ac$, (9), (10), R_1 .

Hence, $\{A_2, C'_n\} \vdash T_n$.

2. We now prove that $\{T_n, C'_n\} \vdash A_2$.

(1) $(ab)^{3n-1} = ab$, theorem 6.3;

(2) $ac\ ab = (ab\ bc\ ca)^{2n}ab = ab\ (bc\ ca\ ab)^{2n} = ab\ ba\ (a \neq c,\ a \neq b)$, T_n, T_n;

(3) $ab\ ab = ab\ ba$, (2);

(4) $ab\ aa\ ab = ab\ ba\ ab$, (3) of theorem 6.3;

(5) $(aa\ ab\ ba)^{2n} = ab\ (a \neq b)$, T_n;

(6) $(ab)^2 = ab\ (aa\ ab\ ba)^{2n} = (ab)^{6n+1} =$

$(ab)^{3n-1}(ab)^{3n-1}(ab)^3 = (ab)^5\ (a \neq b)$, (5), (3), (4), (1);

(7) $(ab)^2 = (ab)^5 = \ldots = (ab)^{3n-1} = ab\ (a \neq b)$, (6), (1);

(8) $aa\ ba = ba\ ba = ba\ (a \neq b)$, (31) of theorem 6.3, (7);

(9) $aa\ aa = aa$, (8), etc.;

(10) $ab\ ab = ab$, (7), (9);

(11) $ab\ ac = ab\ (ab\ bc\ ca)^{2n} = (ab\ bc\ ca)^{2n} = ac\ (a \neq c)$, T_n, (10), T_n.

Hence, $\{T_n, C'_n\} \vdash A_2$. This completes the proof of theorem 6.4.

7. A One-axiom System

We have not succeeded in finding a one-axiom system with R_1 , R_2 , and R_3 . However, with an extension of rule R_1 to R'_1 :

R'_1 : If, for some positive integer i , and for $a \neq b$,

$$S_1(ac)^i = S_2(ac)^i ,$$
$$S_1(bd)^i = S_2(bd)^i ,$$

then $S_1 = S_2$,

it appears that the following axiom, F: $ab\ ca\ bc\ ab = cd\ cb\ ac\ (b \neq c)$, is sufficient. Let $\vdash'$ denote derivation using R'_1, R_2, and R_3 .

Theorem 7. $\{F\} \vdash' S_1 = S_2 \Leftrightarrow \mathcal{A} \vdash S_1 = S_2$.

Proof. $\Rightarrow$ is clear. Proof of $\Leftarrow$:

(1) $ab\ ca\ bc\ ab = cd\ cb\ ac\ (b \neq c)$, F;

(2) $cd_1\ cb\ a_1c = cd_2\ cb\ a_1c\ (b \neq c)$, (1);

(3) $cd_1\ cb\ a_2c = cd_2\ cb\ a_2c\ (b \neq c)$, (1);

(4) $cd_1\ cb = cd_2\ cb\ (b \neq c)$, (2), (3), R'_1;

(5) $ab\ aa\ ba\ ab = ab\ ab\ aa\ (a \neq b)$, (1);

(6) $aa\ ba\ ab\ aa = ba\ ba\ ab\ (a \neq b)$, (1);

(7) $ab\ ab\ aa\ aa = ab\ aa\ ba\ ab\ aa = ab\ ba\ ba\ ab\ (a \neq b)$, (5), (6);

(8) $ab\ ab\ aa\ aa\ ab = ab\ ba\ ba\ ab\ ab\ (a \neq b)$, (7),

(9) $ab^5 = ab\ ba^2\ ab^2\ (a \neq b)$, (8), (4);

(10) $ba^2\ ab^3 = aa\ ba\ ab\ aa\ ab\ ab = aa\ ba\ ab\ ab\ aa\ ab =$
$aa\ ba\ ab\ aa\ ba\ ab\ ab = ba^2\ ab\ ba\ ab^2\ (a \neq b)$, (6), (4), (5), (6);

(11) $ba^2\ ab\ ba\ ab\ ba\ ab^2 = aa\ ba\ ab\ aa\ ba\ ab\ ba\ ab^2 = aa\ ba\ ab\ ab\ aa\ ba\ ab^2 =$
$aa\ ba\ ab^5 = aa\ ba\ ab\ aa\ ab^3 = ba^2\ ab^4\ (a \neq b)$, (6), (5), (5), (4), (6);

(12) $ab\ ba^2\ ab\ ba\ ab^2 = ab\ ba^2\ ab^3 = ab^6 \ (a \neq b)$, (10), (9);

(13) $ab^7 = ab\ ab\ aa\ aa\ ba\ ab^2\ ab =$

$ab\,ab\,aa\ aa\ ba\ ab\ aa\ ab = ab\ ab\ aa\ ba^2\ ab^2 \ (a \neq b)$, (12), (7), (4), (6);

(14) $ab^7\ ba^2 = ab\ ab\ aa\ aa\ ba^2\ ab^2\ ba^2 =$

$ab\ ab\ aa\ ba\ ab^2\ ba^3 = ab\ ab\ ab\ aa\ ab\ ba^3 =$

$ab^5\ ba^3 \ (a \neq b)$, (13), (9), (5), (4);

(15) $ab^5\ ba^2 = ab\ ba^6 \ (a \neq b)$, (9);

(16) $ba^2\ ab^4\ ba^3 = ba^2\ ab\ ba\ ab\ ba\ ab^2\ ba^3 =$

$ba^2\ ab\ ba\ ab\ ba^6 = ba^2\ ab\ ba\ ab^5\ ba^2 =$

$ba^2\ ab^6\ ba^2 = ba^2\ ab^2\ ba^6 = ba^{10} \ (a \neq b)$, (11), (9), (15), (10), (15), (9);

(17) $ab^9\ ba^2 = ab^7\ ba^3 = ab^5\ ba^4 \ (a \neq b)$, (14), (14);

(18) $ab^{14} = ab^9\ ba^2\ ab^3 = ab^5\ ba^4\ ab^3 = ab^{13} \ (a \neq b)$, (9), (17), (6);

(19) $ab\ ba^{14} = ab^{13}\ ba^2 = ab^{14}\ ba^2 = ab^2\ ba^{14} \ (a \neq b)$, (15), (18), (15);

(20) $ab\ ab^{14} = ab^2\ ab^{14} \ (a \neq b)$, (18);

(21) $ab = ab^2 \ (a \neq b)$, (19), (20), R_1';

(22) $cd\ cb = cb\ cb = cb \ (b \neq c)$, (4), (21);

(23) $ab\ ca\ bc\ ab = cb\ ac \ (b \neq c)$, (1), (22);

(24) $cb\ ab = cb\ ac\ ab = cb\ ac \ (a \neq b,\ b \neq c)$, (23), (22);

(25) $ab\ ab = ab\ aa$, (24);

(26) $ab\ ab = ab\ ba\ ab \ (a \neq b)$, (7), (25), (22);

(27) $ab\ ba = ab\ ba\ ba \ (a \neq b)$, (22);

(28) $ab = ab\ ba \ (a \neq b)$, (26), (27), R_1';

(29) $aa\ ba = ba \ (a \neq b)$, (6), (28), (25), (22);

(30) $ba\ ab\ aa\ ba = aa\ ba \ (a \neq b)$, (28), (29);

(31) $aa\ aa = aa$, (29), etc.;

(32) $ab\ ca\ bc\ ab = cb\ ac$, (23), (30), (31).

The theorem now follows from (22), (32), and theorem 5.4.

A PROPERTY OF LINEAR CONDITIONALS

by

J. W. de Bakker

INTRODUCTION

Consider the usual conditional statements, by which we mean here statements which are made up from certain elementary statements by (repeated) composition $(\sigma_1;\sigma_2)$ and selection (<u>if</u> p <u>then</u> σ_1 <u>else</u> σ_2). Let E be the dummy statement and Ω the undefined statement (L: <u>goto</u> L , say).

We introduce a special class of conditional statements (conditionals for short), which we call linear, and which are shown to have the following property: Let $\sigma_1(X)$ and $\sigma_2(X)$ be two conditionals containing zero or more occurrences of the elementary statement X . Let $\sigma_i(E)$ and $\sigma_i(\Omega)$, $i=1,2$, be the result of replacing all occurrences of X in σ_i by E and Ω , respectively. Then the following holds: If $\sigma_1(X)$ and $\sigma_2(X)$ are linear in X , if $\sigma_1(E)=\sigma_2(E)$ and $\sigma_1(\Omega)=\sigma_2(\Omega)$, then $\sigma_1(X)=\sigma_2(X)$. (In fact, we prove the generalization of this for statements which are linear in n variables, and which are equal in the 2^n elements of $\{E,\Omega\}^n$.)

In section 2 (conditionals), we give a formal axiomatic theory for (general) conditionals. Most of this section is a restatement of standard material, as present e.g. in the papers of McCarthy (1962), Igarashi (1964, 1968), and in unpublished work by Scott. The linear conditionals section contains the definition of linearity and the proof of the above mentioned property.

CONDITIONALS

We introduce a language with the following elements:

1. Variables, denoted by upper case letters, A , B , ... , X , Y , Z , ...

2. Constants E and Ω.

3. Boolean variables, denoted by lower case letters, $p, q, r, \ldots$.

A term is defined by

1. Each variable or constant is a term.

2. If σ and τ are terms, then $(\sigma;\tau)$ and $(p\to\sigma,\tau)$ are terms.

We use $\sigma(X)$ to indicate that σ possibly contains the variable X. Well-formed formulas of the system are equations $\sigma=\tau$. An interpretation I of the system is provided by

1. A domain $\mathcal{D}$ is selected.

2. A variable X is interpreted as a partial function X^I from $\mathcal{D}$ to $\mathcal{D}$.

3. E^I is the identity function on $\mathcal{D}$, Ω^I the nowhere defined function.

4. $p^I, q^I, \ldots$ are partial functions from $\mathcal{D}$ to $\{0,1\}$.

5. $(\sigma;\tau)^I$ and $(p\to\sigma,\tau)^I$ are derived from σ^I, τ^I, and p^I as follows: Let $x\in\mathcal{D}$. Then

$$(\sigma;\tau)^I(x)=\tau^I(\sigma^I(x))\ ;$$

$$\begin{aligned}(p\to\sigma,\tau)^I(x)&=\sigma^I(x)\ , \text{ if } p^I(x)=1\ ,\\ &=\tau^I(x)\ , \text{ if } p^I(x)=0\ ,\\ &=\Omega^I(x)\ , \text{ if } p^I(x) \text{ is undefined.}\end{aligned}$$

$\sigma=\tau$ is valid if and only if $\sigma^I=\tau^I$ holds in all interpretations I.

An axiomatic characterization of $=$ is provided by the following axioms and rules (see McCarthy (1962)):

A_1: $(p\to\Omega,\Omega)=\Omega$

A_2: $(p\to(p\to X,Y),Z)=(p\to X,Z)$

A_3: $(p\to X,(p\to Y,Z))=(p\to X,Z)$

A_4: $(p_1\to(p_2\to X,Y),(p_2\to U,V))=(p_2\to(p_1\to X,U),(p_1\to Y,V))$

A_5: $((p\to X,Y);Z)=(p\to(X;Z),(Y;Z))$

A_6: $(\Omega;X)=(X;\Omega)=\Omega$

A_7: $(E;X)=(X;E)=X$

A_8: $((X;Y);Z)=(X;(Y;Z))$

R_1: $=$ is symmetric and transitive.

R_2: If $\sigma_1=\sigma_2$, then $(\sigma;\sigma_1)=(\sigma;\sigma_2)$, $(\sigma_1;\sigma)=(\sigma_2;\sigma)$, $(p\to\sigma_1,\sigma)=(p\to\sigma_2,\sigma)$, and $(p\to\sigma,\sigma_1)=(p\to\sigma,\sigma_2)$.

S: If $\sigma_1(X) = \sigma_2(X)$, then $\sigma_1(\tau) = \sigma_2(\tau)$, where $\sigma_i(\tau)$, $i = 1,2$, is the result of substituting τ for all occurrences of X in σ_i .

Use of R_1 , R_2 , S is not indicated in the sequel. A_8 allows us to abbreviate $(\sigma;\tau)$ to $\sigma\tau$.

For the statement of theorem 1, it is convenient to introduce the notion of <u>partition</u> of a term.

A term has one or more partitions:

If τ is, or begins with, a variable or constant, then its (only) partition is τ .

If τ is of the form $(p \rightarrow \tau_1,\tau_2)$, then a partition of τ is either τ , or a partition of τ_1 followed by a partition of τ_2 .

If τ is of the form $(p \rightarrow \tau_1,\tau_2)\tau_3$, then its partitions are the partitions of $(p \rightarrow \tau_1\tau_3,\tau_2\tau_3)$.

Example: The partitions of $(p_1 \rightarrow X,(p_2 \rightarrow AY,Z))$ are

$$(p_1 \rightarrow X,(p_2 \rightarrow AY,Z))$$

$$X,(p_2 \rightarrow AY,Z)$$

$$X,AY,Z .$$

If $\sigma_1,\sigma_2,\ldots,\sigma_n$ is a partition of τ , we write $\tau = \tau[\sigma_1,\sigma_2,\ldots,\sigma_n]$.

The following theorem generalizes the equivalences of axioms A_1 to A_5 .

<u>Theorem 1</u>. T_1: $\tau[\Omega,\Omega,\ldots,\Omega] = \Omega$

T_2: $\tau[\sigma_1,\ldots,\sigma_{i-1},\tau[\rho_1,\ldots,\rho_n],\sigma_{i+1},\ldots,\sigma_n] = \tau[\sigma_1,\ldots,\sigma_{i-1},\rho_i,\sigma_{i+1},\ldots,\sigma_n]$

T_3: $\tau_1[\tau_2[\sigma_{11},\ldots,\sigma_{1n}],\ldots,\tau_2[\sigma_{m1},\ldots,\sigma_{mn}]] = \tau_2[\tau_1[\sigma_{11},\ldots,\sigma_{m1}],\ldots,\tau_1[\sigma_{1n},\ldots,\sigma_{mn}]]$

T_4: $\tau[\sigma_1,\ldots,\sigma_n]\sigma = \tau[\sigma_1\sigma,\ldots,\sigma_n\sigma]$.

<u>Proof</u>. The proof is not given here. It follows by a straightforward induction on the number of terms in the partitions involved. The basis steps of the induction are provided by A_1 , A_2 and A_3 , A_4 and A_5 .

<u>Remark</u>. Rule T_3 and a variation of rule T_2 are taken as axioms by Caracciolo (1964) in his study of generalized selection.

LINEAR CONDITIONALS

A term $\tau(X)$ may be <u>linear</u> in X :

1. X is linear in X .

2. If τ does not contain X, then τ and τX are linear in X.

3. If τ_1 and τ_2 are linear in X, then $(p \to \tau_1, \tau_2)$ is linear in X.

Example: $(p_1 \to AX,(p_2 \to X,Y))$ is linear in X. XY and $(p_1 \to A(p_2 \to X,Y),Z)$ are not linear in X.

Theorem 2. Let $\tau_1(X_1,X_2,\ldots,X_n)$ and $\tau_2(X_1,X_2,\ldots,X_n)$ be linear in each of $X_1,X_2,\ldots,X_n$. Let $\gamma_1,\ldots,\gamma_{2^n}$ be the elements of $\{E,\Omega\}^n$. If $\tau_1(\gamma_i) = \tau_2(\gamma_i)$, $i = 1,2,\ldots,2^n$, then $\tau_1(X_1,\ldots,X_n) = \tau_2(X_1,\ldots,X_n)$.

Proof. First we give the proof for $n=1$. Let $X = X_1$. By the linearity of $\tau_1(X)$ in X, we can write

$$\tau_1(X) = \tau_1(X,\ldots,X,\sigma_1 X,\ldots,\sigma_m X,\sigma_{m+1},\ldots,\sigma_{m+n}) ,$$

where

a. $\sigma_1,\ldots,\sigma_{m+n}$ do not contain X.

b. $X,\ldots,X,\sigma_1 X,\ldots,\sigma_m X,\sigma_{m+1},\ldots,\sigma_{m+n}$ is a permutation of a partition of τ.

Similarly, we write

$$\tau_2(X) = \tau_2(X,\ldots,X,\rho_1 X,\ldots,\rho_p X,\rho_{p+1},\ldots,\rho_{p+q}) .$$

Let $\alpha = \tau_1(E) = \tau_2(E)$. We have

$$\tau_1(E) = \tau_1(E,\ldots,E,\sigma_1,\ldots,\sigma_m,\sigma_{m+1},\ldots,\sigma_{m+n}) ,$$

and, by T_2,

$$\tau_1(E) = \tau_1(\alpha,\ldots,\alpha,\alpha,\ldots,\alpha,\alpha,\ldots,\alpha) ,$$

for which we write

$$\tau_1(E) = \tau_1(\alpha,\alpha) .$$

Also

$$\tau_1(\Omega) = \tau_1(\Omega,\ldots,\Omega,\Omega,\ldots,\Omega,\sigma_{m+1},\ldots,\sigma_{m+n}) ,$$

and, by T_2,

$$\tau_1(\Omega) = \tau_1(\Omega,\ldots,\Omega,\Omega,\ldots,\Omega,\alpha,\ldots,\alpha) ,$$

for which we write

$$\tau_1(\Omega) = \tau_1(\Omega,\alpha) .$$

we have

$$\tau_1(\Omega,\alpha) = \tau_2(\Omega,\alpha) . \tag{1}$$

By T_2 and T_4, we get $\tau_1(X) = \tau_1(\alpha X,\alpha)$ and $\tau_2(X) = \tau_2(\alpha X,\alpha)$. We also note that

$$\alpha X = \tau_1(\alpha,\alpha)X = \tau_1(\alpha X,\alpha X) = \tau_2(\alpha X,\alpha X) . \tag{2}$$

We shall show that $\tau_1(\alpha X,\alpha) = \tau_2(\alpha X,\alpha)$. First we derive

$$\tau_1(\Omega,\tau_2(\alpha X,\alpha)) = \tau_1(\Omega,\alpha) . \tag{3}$$

This follows from

$$\begin{aligned}
&\tau_1(\Omega,\tau_2(\alpha X,\alpha)) = \tau_1(\tau_2(\Omega,\Omega),\tau_2(\alpha X,\alpha)) = \\
&\tau_2(\tau_1(\Omega,\alpha X),\tau_1(\Omega,\alpha)) = \tau_2(\tau_1(\Omega,\alpha)X,\tau_1(\Omega,\alpha)) = \\
&\tau_2(\tau_2(\Omega,\alpha)X,\tau_2(\Omega,\alpha)) = \tau_2(\tau_2(\Omega,\alpha X),\tau_2(\Omega,\alpha)) = \\
&\tau_2(\Omega,\alpha) = \tau_1(\Omega,\alpha) ,
\end{aligned}$$

by T_1, T_3, T_4, (1), T_4, T_2, and (1). From (3) we get

$$\begin{aligned}
&\tau_1(\alpha X,\tau_2(\alpha X,\alpha)) = \tau_1(\alpha X,\tau_1(\Omega,\tau_2(\alpha X,\alpha)) = \\
&\tau_1(\alpha X,\tau_1(\Omega,\alpha)) = \tau_1(\alpha X,\alpha) ,
\end{aligned}$$

by T_2, (3), and T_2. Thus

$$\tau_1(\alpha X,\tau_2(\alpha X,\alpha)) = \tau_1(\alpha X,\alpha) . \tag{4}$$

Also,

$$\begin{aligned}
&\tau_1(\alpha X,\tau_2(\alpha X,\alpha)) = \tau_1(\tau_2(\alpha X,\alpha X),\tau_2(\alpha X,\alpha)) = \\
&\tau_2(\tau_1(\alpha X,\alpha X),\tau_1(\alpha X,\alpha)) = \tau_2(\alpha X,\tau_1(\alpha X,\alpha)) ,
\end{aligned}$$

by (2), T_3, and (2). Thus

$$\tau_1(\alpha X,\tau_2(\alpha X,\alpha)) = \tau_2(\alpha X,\tau_1(\alpha X,\alpha)) . \tag{5}$$

From (4) and (5), the desired result follows.

The proof for $n > 1$ follows by repeated application of part 1.

FORMALIZATION OF STORAGE PROPERTIES

by

H. Bekić and K. Walk

INTRODUCTION

This paper is a contribution to the design of abstract machines for the interpretation of programs. It is concerned with a particular component, the storage component, of the states of abstract machines interpreting programs written in languages like ALGOL 68 and PL/I.

The aim of the paper is to present a model of storage which avoids any special constructions not implied by the languages to be modelled. The design followed closely the informal descriptions of the languages. Notions present in informal descriptions like ALGOL 68 "names" and "modes", PL/I "pointers", etc., will have their formal counterpart in the model. In some cases, however, certain informal notions, like "instances of values" in ALGOL 68, appeared to be unnecessary. They were avoided in the design of the model, making it conceptually simpler than the informal description.

A general storage model is described in section 1. It serves as the common basis for sections 2 and 3, which present the specific storage properties of ALGOL 68 and PL/I, respectively.

The scope of the paper is confined to the investigation of the properties of the storage component of abstract interpreters, i.e., that component which is changed by assignment and allocation of variables, and which represents a (partial) mapping from "locations" to "values". Locations are entities which, during program interpretation, become associated with identifiers of variables. For the languages under consideration, this intermediate step between identifier and value of a variable is mandatory. It introduces the possibility that different variables share their values in the sense that their identifiers are associated with the same location. In these languages locations, or a derivative of locations like the PL/I pointers, also occur as values of variables.

No regard is given in this paper to the problems of the scope of identifiers and the life-time of allocations. "Identifier" will always mean unique identifier. "Storage" will mean internal storage, excluding any devices addressed by input or output operations.

NOTATION

$\neg$	not
&	and
$\vee$	or
$\supset$	implication
$\equiv$	equivalence
$(\exists! x)(\ldots)$	there is exactly one x such that ...
$(\exists! x \in X)(\ldots)$	there is exactly one $x \in X$ such that ...
$\in$	element of
$\cup$, $\bigcup$	set union
$\subseteq$	subset or equal
$\{\ \}$	empty set
$\{x \mid \ldots\}$	the set of elements x such that ...
$\langle x_1, x_2, \ldots \rangle$	list of elements $x_1, x_2, \ldots$
$\cap$	concatenation of lists
$f: L \to V$, $(f: L \rightleftharpoons V)$	f is a (partial) mapping from L to V, $L =$ domain f, $V =$ range f

1. A GENERAL STORAGE MODEL

The storage model to be described in this section originally arose from a study of value classes and storage in ALGOL 68. It then proved possible to generalize some of the underlying postulates in such a way that the model could also be used for reflecting the more machine-like properties of PL/I storage.

Like other models (e.g., in Elgot and Robinson (1964), Strachey (1966)), this one views storage as a "contents"-function from "locations" (called, in different programming languages, "variables", "left-hand values", "generations" of variables, "names") to "values". The main objective in setting up the model was to cover the following two phenomena (exemplified by fixed and flexible arrays in ALGOL 68):

(a) Composite locations. The contents of a location ("the value referred to by a name", in ALGOL 68 terminology) may be allowed to vary over all values of a fixed given structuring, e.g., all one-dimensional arrays with given bounds and given element attributes. This induces a corresponding structure on the location which is preserved by the contents-function: contents of i-th component = i-th component of contents.

(b) Flexible locations. More generally, the contents of a location may have any of a prescribed set of structurings; e.g., it may vary over arrays with arbitrary upper bound. Then assignment of a value to a location will affect not only the contents of the location but also the existence of component locations.

Thus we model storage as a function from a (changeable) set of "active" or "allocated" locations to the set of values satisfying certain consistency conditions; this function will be partial, to mirror allocated locations that have not yet been assigned to.

Each location has associated with it a "range", the set of values over which its contents may vary; in programming languages, these sets are usually denoted by "attributes". Thus our notion of location contains more information than that of PL/I-"pointer", which can be characterized as location minus attributes.

We proceed by isolating properties of values and value classes that affect storage structure. We then introduce the set of locations, the notion of storage, and finally allocation and assignment as operations on storages.

1.1. Values

We assume a set V of values, with a partition into two disjoint subsets, elementary and composite values. Composite values will have a structure that induces a corresponding structure on locations containing such a value; with elementary values, we are not interested in any further structure.

Elementary values

We do not assume any further properties of elementary values. Examples for elementary values would be reals, or intergers, or pointers (even pointers to composite values).

Composite values

A composite value v has:

$$\text{an index-set} \qquad I = \text{indexset}(v) \ , \qquad (1.1)$$

$$\text{components} \qquad v_i = \text{compval}(v,i), \text{ for } i \in I \ .$$

Here compval(v,i) , for given v , is a partial function of $i \in I$, i.e., v_i may be undefined; if v_i is defined, it shall again be a value.

Examples for composite values in conventional programming languages are arrays, where the index-set is the set of n-tuples of integers within given bounds, and structures, where the index-set is a set of identifiers or selectors. Also the storages to be introduced below, and in particular PL/I-areas, can be regarded as composite values.

The value v need not be uniquely determined by its index-set I and its components v_i (see, however, the postulate (1.6) on composite ranges below). Examples are again provided by PL/I-areas, which are additionally characterized by their "size"; also, by the difference of the two sets

union(struct(int,int,int) , struct(int,int,real))

and

struct(int,int,union(int,real))

in ALGOL 68 (assuming some standard selectors for the structure components), which forces the component ranges to be included in the characterization of multiple and structured values (see section 2.1).

Finite nesting assumption

A component v_i may itself be composite and thus have components v_{ij} . We assume that this process cannot be iterated indefinitely, i.e., that $v_{ij\ldots}$ eventually yields elementary or undefined. This excludes, for example, infinitely nested structures, though not self-referential structures: a pointer to a structure would count as elementary.

1.2. Ranges

We assume a set of ranges, i.e., of sets of values. Ranges arise as ranges of locations, i.e., as the sets of values over which the contents of a location may vary. We assume that each value is in some range

$$V = \bigcup \text{ ranges} . \tag{1.2}$$

Since one of the operations for forming new ranges is set union, ranges will, in general, not be disjoint.

In programming languages, ranges are described by "types", or "modes", or "attributes". These may include dynamic information, like array bounds.

Ranges fall into three disjoint sub-classes: elementary, composite, and flexible ranges.

Elementary ranges

Again, we make no further assumptions on elementary ranges R, except, of course,

$$v \in R \supset v \text{ is elementary.} \tag{1.3}$$

Composite ranges

A composite range R has:

$$\text{an index-set} \qquad I = \text{indexset}(R) , \tag{1.4}$$

$$\text{component ranges} \qquad R_i = \text{comprange}(R,i) , \text{ for } i \in I ;$$

$$\text{the } R_i \text{ are again ranges.}$$

The elements of R shall be composite values v with indexset I and with components in the corresponding component range:

$$v \in R \supset v \text{ is composite, } \text{indexset}(v) = I , \tag{1.5}$$

$$v_i \text{, if defined, } \in R_i .$$

In R, each system of components shall uniquely determine a composite value: for any system $v^{(i)}$, $i \in I$, such that $v^{(i)} \in R_i$ or undefined:

$$(\exists! v)(v_i = v^{(i)}) \qquad (1.6)$$

(strong equality: both sides undefined, or both defined and equal).

Examples of composite ranges are: the set of arrays of given dimension, bounds, and elements in a given range (here, the R_i are all equal); the set of structures with given selectors and elements in given component ranges. The last postulate above is used to express that, e.g., the contents of an array location is uniquely determined by the contents of its components.

Flexible ranges

A flexible range R is uniquely decomposable into a union

$$R = R' \cup R'' \cup \ldots \qquad (1.7)$$

of disjoint, non-flexible ranges $R', R'', \ldots$, the alternatives of R. We also write $R = \bigcup$ alternatives(R), where alternatives(R) is the set of ranges $\{R', R'', \ldots\}$.

Thus, taking examples from ALGOL 68, the alternatives of the set union(int,real) would be the sets int and real. The alternatives of [1:flex]real would be the sets [1:0]real, [1:1]real, [1:2]real, ..., and the alternatives of union(int,[ ]real) would be the set int and the infinite number of sets [i:k]real. In PL/I, an example for a flexible range is the set of areas of given size.

We have assumed that elementary, composite, and flexible ranges are mutually different. Therefore, if we disregard the empty set, a flexible range has at least two alternatives. Our assumption of unique decomposition proved to be satisfied in the cases we wanted to model. The assumption could be weakened by introducing a flexible range as a set together with a decomposition rule.

Given a value $v \in R$, we can define the current alternative of R as the alternative that contains v:

$$\text{curralt}(R,v) = \text{the range } R' \in \text{alternatives}(R) \text{ such that } v \in R'. \qquad (1.8)$$

1.3. Locations

We assume a set $\mathcal{L}$ of locations, with a partial relation l indep l' on $\mathcal{L}$, called independence (which expresses "having no common parts").

Each location $l \in \mathcal{L}$ has a range, $R = \text{range}(l)$, and is correspondingly classified as elementary, composite, or flexible.

Components of locations

Corresponding to the components of composite values, we introduce components of composite and flexible locations, i.e., of locations that can contain composite values. Different components of a composite location, or different components "under the same alternative" of a flexible location, can be assigned values independently and hence are postulated to be independent; on the other hand, nothing is assumed concerning the relation between components of flexible locations under different alternatives.

For composite l, $R = \text{range}(l)$:

l has components

$$l_i = \text{comploc}(l,i) \quad \text{for} \quad i \in \text{indexset}(R) \text{, with:} \tag{1.9}$$

$$l_i \in \mathcal{L}, \quad \text{range}(l_i) = R_i, \quad l_i \text{ indep } l_j \quad \text{for} \quad i \neq j .$$

For flexible l, $R \in \text{alternatives}(\text{range}(l))$, R composite:

l has components

$$l_i^R = \text{compflexloc}(l,R,i) \quad \text{for} \quad i \in \text{indexset}(R) \text{, with:} \tag{1.10}$$

$$l_i^R \in \mathcal{L}, \quad \text{range}(l_i^R) = R_i, \quad l_i^R \text{ indep } l_j^R \quad \text{for} \quad i \neq j .$$

We define sub-location as the reflexive transitive closure of the relation component, i.e., l' is a sub-location of l if it is either l itself or a sub-location of a component of l .

Super-locations

Sometimes it is necessary to construct a composite location from its components. Examples for this situation are provided by "cross-sections" of arrays, e.g., taking a row or a column of a matrix, and by "rowing", i.e., constructing an array location whose only component is a given location, in ALGOL 68. The latter example shows that the "super-location" so obtained need not be a part of a previously introduced location. For simplicity, we are quite liberal in postulating the existence of unique super-locations; our assumption below could be weakened by restricting it to <u>some</u> ranges R and systems of components $l^{(i)}$. On the other hand, there is no need to postulate the construction of <u>flexible</u> super-locations.

For composite R, and a system of locations $l^{(i)}$, $i \in \text{indexset}(R)$, with $\text{range}(l^{(i)}) = R_i$, $l^{(i)}$ indep $l^{(j)}$ for $i \neq j$:

$$(\exists! l)(\text{range}(l) = R\,,\ l_i = l^{(i)} \text{ for } i \in \text{indexset}(R))\,. \qquad (1.11)$$

<u>Independence</u>

According to our informal characterization above, we postulate independence as a partial relation that is irreflexive and symmetric. Independence from l implies independence from its components; for composite locations, the converse also holds (whereas flexible locations might use some "hidden" parts to store information concerning the current alternative).

Thus:

$$\neg(l \text{ indep } l)\,,\quad l \text{ indep } l' \equiv l' \text{ indep } l\,. \qquad (1.12)$$

For composite l':

$$l \text{ indep } l' \equiv l \text{ indep } l'_i \text{ for all } i \in \text{indexset}(\text{range}(l'))\,. \qquad (1.13)$$

For flexible l':

$$l \text{ indep } l' \supset l \text{ indep } l'^{R}_{i} \text{ for all } i \in \text{indexset}(R)\,, \qquad (1.14)$$

$$R \text{ composite},\ R \in \text{alternatives}(\text{range}(l'))\,.$$

1.4. <u>Storages</u>

Assume we are given a set $L_o \subseteq \mathcal{L}$ of independent locations, and a partial function

$$f_o : L_o \simeq V$$

from L_o to the set of values which is <u>range-respecting</u>, i.e., satisfies

$$f_o(l)\,,\ \text{if defined},\ \in \text{range}(l)\,.$$

We can extend f_o to a function $f: L \rightleftharpoons V$ that coincides with f_o on L_o, i.e.,

$$f(l) = f_o(l) \quad \text{for } l \in L_o$$

(strong equality), and is consistently complete under the construction of component and super-locations:

$$f(l) = v \equiv f(l_i) = v_i \quad \text{for all } i \in \text{indexset}(\text{range}(l)), \tag{1.15}$$
$$\text{for composite } l \in \mathcal{L}, \text{ composite } v$$

$$f(l) = v \supset f(l_i^R) = v_i \quad \text{for all } i \in \text{indexset}(\text{range}(l)), \tag{1.16}$$
$$R = \text{curralt}(\text{range}(l), v), \text{ for}$$
$$\text{flexible } l \in \mathcal{L}, \text{ composite } v .$$

Thus, for flexible locations l, we use the contents $f(l)$ of l to determine the "currently active" component locations.

It follows from the independence of the elements of L_o and from our uniqueness assumptions on composite values and composite locations that such an extension is indeed possible, provided we identify, for composite l, the case " $f(l)$ undefined" with the case " $f(l)$ = the $v \in \text{range}(l)$ such that all components v_i undefined". For example, any composite $l \in L_o$ will "reappear", in the course of constructing f, as the super-location built from its components l_i; but then $f(l)$ will be $f_o(l)$, because both are the unique $v \in \text{range}(l)$ with $v_i = f(l_i)$.

In particular, we can consider the function $f = \text{extend}(f_o)$ generated from f_o under the above rules, i.e., the smallest consistently complete extension of f_o, where the partial function $f: L \rightleftharpoons V$ is smaller than the partial function function $f': L' \rightleftharpoons V$, if $L \subseteq L'$, $f(l) = f'(l)$ for $l \in L$.

We define now a storage as a pair

$$S = (\mathcal{L}, f_o : L_o \rightleftharpoons V) \tag{1.17}$$

where $\mathcal{L}$ is a set of locations as introduced above (fixed in the present model, but variable in the applications), and f_o is a partial range-respecting function from an independent subset $L_o \subseteq \mathcal{L}$ to V. We write

$$\mathcal{L} = \text{Locs}(S), \quad f_o = \text{level 1 contents}(S), \quad L_o = \text{level 1 locs}(S)$$

and call $\mathcal{L}$ the locations, f_o the level-one contents-function, L_o the level-one locations of S. Extending f_o to the function $\text{extend}(f_o) = f: L \rightleftharpoons V$, we write

$$f = contents(S) \ , \quad L = actlos(S)$$

and call f the <u>contents-function</u>, $\mathcal{L}$ the <u>active locations</u> of S. For $l \in L$, we define

$$l(S) = f(l) \ ,$$

thus using l as a "selector" of S.

We call f_o a <u>base</u> of f. Clearly f has many bases, e.g., we can replace a composite location by its components. In the definition of S, we could have abstracted from the particular base by using f instead of f_o. However, it seemed advantageous to retain f_o, because level-one locations reflect the way locations get and loose existence (through allocating and freeing, e.g., by crossing block boundaries), and because they have special properties in PL/I.

1.5. Allocation, Freeing, Assignment

We can now define the main operations on storages:

Allocation

For $S = (\mathcal{L}, f_o : L_o \overset{\sim}{\to} V)$ and $l \in \mathcal{L}$, l indep l' for all $l' \in L_o$, we define

$$(allocate\ l)(S) = \quad \text{the storage } S' = (\mathcal{L}, f' : L_o \cup \{l\} \overset{\sim}{\to} V) \qquad (1.18)$$
$$\text{such that } l(S') \text{ undefined, } l'(S') = l'(S) \text{ for } l' \in L_o \ ,$$

i.e., we add the independent location l to the set of level-one locations without initializing it. In a programming language, l is usually given by its range R, so we need a non-determinate function <u>generate</u> such that

$$(generate\ R)(S) = \underline{some}\ \ l \in \mathcal{L} \ , \ l \text{ indep } l' \qquad (1.19)$$
$$\text{for all } l' \in L_o \ , \ range(l) = R \ .$$

Allocation then is performed by

$$allocate((generate\ R)(S))(S) \ .$$

Note that the conditions l indep l' for all $l' \in L_o$ and l indep l' for all $l' \in L$ are equivalent.

Freeing

For $S = (\mathcal{L}, f_o : L_o \overset{\sim}{\to} V)$, $l \in L_o$:

$$(\text{free } l)(S) = \text{ the storage } S' = (\mathcal{L}, f'_o : L_o - \{l\} \rightleftarrows V) \text{ such that} \qquad (1.20)$$
$$l'(S') = l'(S) \text{ for } l' \in L_o - \{l\} ,$$

i.e., we restrict f_o to $L_o - \{l\}$.

Assignment

For $S = (\mathcal{L}, f_o : L_o \rightleftarrows V)$, $l \in L = \text{actlocs}(S)$, $v \in \text{range}(l)$:

$$(l := v)(S) = \text{ the storage } S' = (\mathcal{L}, f'_o : L_o \rightleftarrows V) \text{ such that} \qquad (1.21)$$
$$l(S') = v ,$$
$$l'(S') = l(S') \text{ for } l' \in L,\ l' \text{ indep } l ,$$
$$l'(S') \in \text{curralt}(l', l'(S)) \text{ for flexible } l' \in L ,\ \neg(l' \text{ indep } l) ,$$
$$l' \text{ not a sub-location of } l .$$

Thus, we leave unchanged the set L_o of level-one locations and define a new function $f' : L' \rightleftarrows V$ by prescribing its value for l and for locations independent from l. These conditions would be sufficient, except for flexible locations that have parts with l in common, without being completely contained in l; for those locations, we must explicitly add the condition that the current alternative will not be changed by the assignment.

2. STORAGE PROPERTIES OF ALGOL 68

The storage-related properties of ALGOL 68 can easily be introduced in terms of the notions of our general model. Thus, structures and "fixed" arrays form composite ranges, "flexible" arrays and unions form flexible ranges. "Names" are locations, and the relation "to refer to" is the contents-function.

We note the following points in which our description method differs from the one used in the ALGOL 68 Report, Van Wijngaarden (Ed.), (1969). In fact, these points were the original motivation in setting up the general model.

Declarers denote value classes

We associate with each "declarer" a class of values, so that the role of declarers can be described in the following simple way:

an "identity declaration" $\delta id = v$, where δ is a declarer, id an identifier, v denotes a value, tests v for being in the class δ and makes id denote v;

a "generator" (essentially given by a declarer δ) chooses a new name with range δ ;

an assignment statement tests the value of the right-hand side for being in the range of the name denoted by the left-hand side and replaces the contents of the name.

No linearization of arrays

In the ALGOL 68 Report, arrays ("multiple values") are introduced as linear sequences of their elements, plus some descriptive information, rather than as n-dimensional aggregates of their elements. This linearization turns out to be a purely descriptional feature (unlike to the situation in PL/I, where, e.g., overlay-defining forces particular assumptions on linearization). Restoring the usual notion results, among other things, in a simpler definition of "slicing" (i.e., subscripting, forming cross-sections).

Flexibility a property of names

In the ALGOL 68 Report, part of the descriptive information in a multiple value are the "states" which indicate, for each bound, whether the value can be replaced by one with differing bound. But this really is a property of the range of the name containing the value, i.e., a property of names, not of multiple values.

No "instances"

To distinguish between values in different locations, i.e., referred to by different names, one just has to remember the names--there is no need to introduce "instances" of values.

2.1. Values, Modes, Ranges, Sub-Modes

Values are organized in certain classes, called modes; thus, the set of reals of given length is a mode , and the set of n-dimensional arrays (with arbitrary bounds) of elements in a given mode is again a mode. Declarers can prescribe some or all of the bounds of arrays; therefore, we introduce finer classes which, together with the modes, form the possible ranges of names. By prescriptions on bounds of array names (in identity declarations), we obtain even finer classes which, together with the ranges, we call sub-modes.

There are certain basic modes, e.g., the set of reals mentioned before, and certain rules for constructing new modes from given modes, e.g., constructing the set of procedures ("routines") with given argument and result modes.

For the present purpose, we are interested in the following modes and ranges:

(a) Arrays

An n-dimensional array (or "multiple value") v is characterized by:

an n-tuple of integer pairs, $l_1 : u_1, \ldots, l_n : u_n$, the lower and upper bounds of v;

its element mode M;

its elements $v[i_1, \ldots, i_n] \in M$, for $l_k \le i_k \le u_k$, $1 \le k \le n$.

(We have to include the element mode as part of the information characterizing an array, similarly the field modes for structures below. For an example concerning structures, see the beginning of section 1.1.) Clearly, v can be regarded as a composite value, with

$$\mathrm{indexset}(v) = \{\langle i_1, \ldots, i_n\rangle \mid l_k \le i_k \le u_k\},$$

$$\mathrm{compval}(v, \langle i_1, \ldots, i_n\rangle) = v[i_1, \ldots, i_n].$$

Given a non-array range R and an n-tuple $ol_1 : ou_1, \ldots, ol_n : ou_n$ of pairs of "optional" bounds, i.e., each ol_i, ou_i is either an integer or "unspecified", we define the array range

$$[ol_1 : ou_1, \ldots, ol_n : ou_n]R = \text{the set of n-dimensional arrays with} \qquad (2.1)$$

$$\text{bounds} \begin{cases} \text{as given, where specified} \\ \text{arbitrary, otherwise,} \end{cases}$$

$$\text{element mode } M = \text{mode } R,$$

$$\text{elements} \in R,$$

(where the mode of a range, or more generally of a sub-mode below, is obtained by ignoring restrictions on bounds).

We call $[\ldots]R$ fixed or flexible, and correspondingly classify it as composite or flexible in the sense of section 1.2, depending upon whether all bounds are specified or not. Similarly, we say that $[\ldots]R$ is fixed or flexible at the i-th lower-upper bound.

Given a non-array mode M, the array mode $\text{row}^n\, M$ is defined as the set of all n-dimensional arrays with element mode M and elements in M.

(b) Structures

A structure (or "structured value") v is characterized by:

an n-tuple $(n \geq 1)$ of identifiers $id_1, \ldots, id_n$, its selectors,
an n-tuple of modes $M_1, \ldots, M_n$, its field modes,
its fields id_i of $v \in M_i$ $(1 \leq i \leq n)$,

(thus the ordering of the selectors is relevant).

Again, a structure v can be considered as a component value, with $\text{indexset}(v) = \{id_1, \ldots, id_n\}$, $\text{subval}(v, id_i) = id_i$ of v. We define the composite structure range

$$(id_1 : R_1, \ldots, id_n : R_n) = \text{the set of structures } v \text{ with} \quad (2.2)$$

selectors $id_1, \ldots, id_n$,
element modes $M_i = \text{mode } R_i$,
elements id_i of $v \in R_i$,

where the R_i are arbitrary ranges; if the R_i are modes, then the resulting range is a structure mode.

(c) Unions

If the M_i are modes, $M = M_1 \cup \ldots \cup M_n$ is again a mode, the union of the M_i. If we disregard the empty set, it is sufficient to assume $n \geq 2$ and the M_i all different and not again unions. (There is a further restriction that the M_i in such a decomposition are not "related", i.e., transferable into each other by certain automatic conversions.) Unions are classified as flexible ranges.

(d) Names

Names are locations in the sense of section 1.3, with the predicates, operations, and relations postulated there. Thus, composite or flexible names, i.e., names with composite or flexible range, have sub-locations, two names may be independent, etc.

The new thing here is that names are values, i.e., we include the set of names in the set of values. For each range R, we postulate a countable set of independent names with range R. We define

$$\text{ref } R = \text{the set of names with range } R. \qquad (2.3)$$

Even if R is a mode, ref R is not a mode or range, but only a "sub-mode". We get a <u>name mode</u> if we form the union $\bigcup$ ref R over all ranges R such that mode R is a given mode M; we call this mode ref M (slightly ambiguously: ref has a different meaning whether we expect it to produce a mode or a sub-mode).

From given sub-modes T which are name sets, we can form others by the following two operations, starting with T's of the form ref R :

$[ol_1 os_1 : ou_1 ot_1, \ldots, ol_n os_n : ou_n ot_n]_{lhs} T$ = the set of n-dimensional array names (2.4)

which are fixed or flexible at the i-th lower-upper bound, where so restricted by os_i / ot_i ,

which have bounds as given, where they are fixed and a bound is specified ,

which have element mode M ,

which have element names in T ,

where the os_i , ot_i are optional <u>states</u>, i.e., each is "fix", "flex", or unspecified, the ol_i , ou_i are optional bounds, and M is determined by mode T = ref M ;

$(id_1 : T_1, \ldots, id_n : T_n)_{lhs}$ = the set of structure names l with (2.5)

selectors $id_1, \ldots, id_n$,

field modes M_i ,

id_i of $l \in T_i$,

where again mode $T_i = \text{ref } M_i$. Straight-forward extension to certain notions like "fixed" or " id_i of l ", that have been introduced above for ranges or values, can be made for names.

<u>Classification of ranges into elementary, composite, flexible</u>

We have already classified arrays, structures, and unions. All remaining ranges, e.g., sets of names, are elementary. It follows that flexible ranges are indeed uniquely decomposable into disjoint non-flexible alternatives; for example, a flexible array range is decomposed into fixed array ranges; a union is decomposed into non-union alternatives, with those alternatives further decomposed if they are flexible arrays.

Denoting modes, ranges, sub-modes: declarers

Sub-modes (hence ranges, modes) are denoted by declarers; the distinction between modes, ranges, sub-modes, corresponds to the distinction between virtual, actual, formal declarers. We only give a few examples concerning arrays and array names; ε stands for "unspecified".

(a) virtual array declarer

[,]**real** denotes row^2(reals of length 1) .

(b) actual array declarer

[1:0]**flex** **char** denotes [1:ε]characters .

Thus a bound can, and in fact must, be specified in the **flex** case. This bound does not influence the range denoted by the declarer, but restricts the choice of a value at the automatic initialization coupled with allocation.

(c) formal array declarer in a name declarer

ref[1:**flex**,1:10**either**]**char** denotes

$[1\text{"fix"}:\varepsilon\text{"flex"},1\text{"fix"}:10\varepsilon]_{\text{lhs}}$(ref(characters)) .

Thus, in a state position of a formal declarer, "no state" means "fix", **either** means "unspecified".

2.2. Slicing, Selecting, Composing

We define briefly certain operations for getting parts of, or composing, arrays and structures; similarly for array and structure names.

Slicing

For an array v , and integers l_k, u_k, l'_k (resp. i_k) such that the interval $l_k : u_k$, resp. the integer i_k , is within the respective bounds of v :

$$v[s_1,\ldots,s_n] = \text{the } m\text{-dimensional array } w \text{ with} \tag{2.6}$$

bounds $l'_{k1} : l'_{k1} + (u_{k1} - l_{k1}),\ldots,l'_{km} : l'_{km} + (u_{km} - l_{km})$,

element mode M = element mode of v ,

$w[l'_{k1}+j_{k1},\ldots,l'_{km}+j_{km}] = v[s'_1,\ldots,s'_n]$ for $0 \le j_{ki} \le u_{ki} - l_{ki}$, $1 \le i \le m$,

where each s_k is either $l_k : u_k$ at l'_k or i_k and, correspondingly, s'_k is $l_k + j_k$ or i_k ; $k_1,\ldots,k_m$ are the positions (in ascending order) where an s of the first kind is specified. (This definition holds for the case $m \ge 1$; for $m = 0$, $v[i_1,\ldots,i_n]$ has already been defined.) Thus slicing takes a sub-aggregate (special case: an element) and possibly re-indexes it.

An analogous operation $l[\ldots]$ is available for array names, yielding again a name.

Selecting

This is id_k of v , for structures v , which has already been defined; similarly id_k of l .

Composing

Depending on context, more precisely on a mode M provided by context, $(v_1,\ldots,v_m)$ denotes either an array or a structure:

(a) $M = row^{n+1} M'$

M' a non-array mode, $n \ge 0$: if there is a range $R = [l_1 : u_1,\ldots,l_n : u_n]R'$ such that all $v_i \in R$:

$$(v_1,\ldots,v_m) = \text{the } (n+1)\text{-dimensional array with} \tag{2.7}$$

bounds $1 : m$, $l_1 : u_1,\ldots,l_n : u_n$,

element mode M' = mode R' ,

$v[i,i_1,\ldots,i_n] = v_i[i_1,\ldots,i_n]$, $(1 \le i \le m)$.

(This includes the case $n = 0$, if we ignore subscript and bound-pair lists of length 0 .)

(b) $M = (id_1 : M_1,\ldots,id_m : M_m)$

For $v_i \in M_i$:

$$(v_1,\ldots,v_m) = \text{the structure } v \text{ with selectors } id_k , \tag{2.8}$$

field modes M_k , fields v_k .

"lhs"-composing

Besides slicing of names, only a very special case arises (in form of an automatic conversion known as rowing): for a name $l \in \mathrm{refrow}^{n}M'$ (M' a non-array mode, $n \geq 0$) :

$$(l)_{lhs} = \text{the name } l' \in \mathrm{refrow}^{n+1}M' \text{ defined analogously to } (v_1) \text{ in case (a) above.} \qquad (2.9)$$

2.3. Identity Declarations, Generators, Assignment

We can now use the association of declarers with value classes to describe the effect of identity declarations (which are concerned with storage properties only to the extent that the range-test performed by them includes a test on bounds and states of names), of generators, and of assignment.

There is one case, both with identity declarations and generators, in which our interpretation differs from that in the ALGOL 68 Report, namely the case that a declarer prescribes a bound but admits a corresponding flexible state.

Identity declarations

Identity declarations have the form

$$\tau\,id = E \;,$$

where τ is a formal declarer, id an identifier, E an expression. They test the value v of E for being in the class T denoted by τ and make id denote v . Though identity declarations introduce "constants", they include as special case ALGOL 60's variable declarations (by choosing a name declarer τ and a generator E); they are also used to express the different kinds of parameter passing.

Generators

Generators are actual declarers. Execution of a generator ρ denoting a range R is defined as

$$(\mathrm{allocate}_{A68}R)(S) = \mathrm{let}\ l = (\mathrm{generate}\ R)(S) \quad (l := v)(\mathrm{allocate}\ l(S)) \;, \qquad (2.10)$$

where v is some value in R . Thus allocation as defined in the general model is followed by non-determinate initialization.

Assignment

Assignment is assignment of the general model, preceded by a range test:

$$(l :=_{A68} v)(S) = \underline{\text{if}}\ \ v \in \text{range}(l)\ \ \underline{\text{then}}\ \ (l := v)(S)\ . \qquad (2.11)$$

2.4. Some Further Points

Scopes

Scopes, i.e., the life-time of names and other "dynamic" values, are outside the subject of this paper. We remark only that the scope of a name can be influenced by writing loc or heap in front of a generator; a further test on assignment secures that a value is not assigned to a name with bigger scope (i.e., possibly, transported outside its scope).

Sub-names of flexible names

Sub-names of flexible names have a life-time even shorter than that of local names: they can die through re-assignment to the flexible name. A further test on several occasions (identity declaration, assignment, composing, rowing) secures that no access, other than via the flexible name itself through slicing, is established to a proper sub-name of a flexible name.

Nil

There is one special name, nil (serving as a null-pointer in chained structures), which receives its mode through context. In the ALGOL 68 Report, different "instances" of nil can have different modes. We can meet the situation by introducing different values $\underline{\text{nil}}_M$, one for each mode ref M .

3. STORAGE PROPERTIES OF PL/I

For presenting the storage properties of PL/I, we start with the general storage model of section 1 and proceed by giving it more special properties. The PL/I features related to these properties are: based and defined variables, areas, pointers, and offsets.

After a brief informal discussion of variables in PL/I, we present the ranges of values in PL/I, including area ranges, which are of particular interest. After introducing pointers and offsets, we are in a position to describe the special storage mapping properties of PL/I. Consequences of the definitions are shown in a few examples.

The resulting PL/I storage model differs in various respects from the one used in the complete formal description of PL/I (Walk et al. (1969), Lucas, Walk (1969)). There, storages were introduced as a set of entities characterized by axioms. Storage access gave a "value representation" which, via data attributes, could be partially interpreted as values. This approach comes nearer to an implementation (value representations may be thought of as bit patterns in a real machine), the relationship between this axiomatic model and an actual implementation was investigated in Henhapl (1969). The present more explicit model lends itself better to a common description of ALGOL 68 and PL/I.

3.1. Variables in PL/I

Each variable is associated by its declaration with a data attribute classifying it as array, structure, or scalar variable. The evaluation of the data attribute gives the range of the variable.

We distinguish proper, defined, and based variables, and parameters in PL/I.

By a reference to a non-based variable or a parameter, we mean a specification of the (unique) identifier of the variable and a list of indices (integers and/or identifiers). A reference to a based variable in addition specifies an expression whose evaluation gives a pointer. The evaluation of a valid reference gives a location, which in turn gives access to the current value in storage.

The way in which a variable becomes associated with a location depends upon its type (proper, defined, based, or parameter). The way in which a sub-location of this location is determined by the indices specified in a reference is common to all types.

By allocation of a proper variable, we mean the association of a properly selected new location with the (unique) identifier of the variable, and the allocation of that location in storage. The range of the location to be selected is equal to the range of the variable. At what point during the interpretation of a program a proper variable is allocated, and how long the resulting association between identifier and location is maintained, depends upon the declared storage class (STATIC, AUTOMATIC, CONTROLLED) of the variable and is of no concern for our present investigation.

Defined variables do not own locations. By its declaration, a defined variable is rather associated with a reference to a proper variable. The location associated with this reference is evaluated when a reference to the defined variable is evaluated. From this location, by appropriate steps of deriving component and super-locations, a location is constructed which is used as the current location of the defined variable at the point of reference. Thus, referring to a defined variable always means referring to storage owned by a proper variable.

A based variable serves two purposes. First, it can be used to allocate new locations, with the range of the based variable, either in main storage, or in an area. These locations do not become associated with the based variable, but are identified by the contents of pointer or offset variables assigned when the locations are allocated. The contents of pointer variables are called pointers. A pointer represents a piece of information derived from a location (it is not a location itself). Intuitively, a pointer identifies a point in storage, but does not necessarily contain information about a range of values. An offset has the same significance for an area, as a pointer has with respect to main storage.

The second use of a based variable is the reconstruction of a location from a pointer value. Like defined variables, based variables do not own locations, but locations are constructed at the point of reference. From the pointer evaluated from the reference to a based variable and the range of the based variable, a location is constructed which is used as the current location of the based variable at the point of reference. This construction does not necessarily give an allocated location; if it does not, the reference to the based variable is undefined. The rules governing the construction of locations are implied by the properties of storage mapping discussed in section 3.4.

Parameters own locations like proper variables. For value parameters a new location is allocated before the call of the corresponding procedure. For non-value parameters, the location of the argument (a reference) is associated with the parameter, so that argument and parameter share the same location.

3.2. Ranges of Values in PL/I

There is a set V_P of values which is the union of all possible ranges of PL/I variables. This set will be partially characterized in the following.

There is a set $\mathcal{L}_P$ of PL/I locations. PL/I locations have array, structure, or scalar range. In the sense of section 1.2, array and structure ranges are composite (there are no flexible arrays in PL/I), scalar ranges with the exception of area ranges are elementary, and area ranges are flexible ranges.

(a) Examples for elementary ranges are arithmetic, string, label, entry, pointer, and offset ranges. The significance of pointers and offsets is explained in the next section. The properties of the other elementary values are of no concern for our present purpose.

(b) An array range is characterized by

a pair of integers $1:u$, the bound pair (3.1)

a range R, the component range.

The index-set of an array range is the set of integers

$$\{i \mid 1 \leq i \leq u\} .$$

Note that in PL/I, we can treat ranges of multi-dimensional arrays as ranges of one-dimensional arrays with array components (this was not possible in ALGOL 68, since the property of some dimension to have flexible bounds makes a range of multi-dimensional arrays flexible). We also do not need any further characterization of array or structure values, since values are always processed in scalar units.

A structure range is characterized by an n-tuple of ranges

$$R_1, R_2, \ldots, R_n , \tag{3.2}$$

the component ranges. The index-set of a structure range is the set of integers

$$\{1,2,\ldots,n\} .$$

Note: Syntactically, the indices of structure components are identifiers. It is not necessary, however, to distinguish between ranges that differ only in identifiers of structure components.

The ordering of the index-set of array and structure ranges is significant. It gives special properties to the function which maps composite locations into component locations, which in turn are significant for the use of based and defined variables (see section 3.).

(c) Area ranges, though counting as scalar with respect to expression evaluation, are flexible ranges with composite alternative ranges.
Area variables serve the purpose of identifying storage to be used for the allocation of based variables. An area range, therefore, is a set of storages, i.e., we include a set of storages in the set of values V_P. Since the notion of storage has already been introduced, it will be possible to explain the handling of areas using the general storage model of section 1.
An area is a composite value. It components are the values of the level-one locations that have been allocated in the area. Let an area a be the storage

$$a = (\mathcal{M}, f : M_o \rightrightarrows V_P) .$$

Then the index-set of a is the set of level-one locations M_o in a, and the component values are

$$m(a) \text{ , for } m \in M_o .$$

The set of "potential" locations $\mathcal{M} \subseteq \mathcal{L}_P$ of an area is determined by the declared size characterizing the area range. Let the declared size be the integer value s ; then the corresponding area range A_s is

$$A_s = \{a \mid locs(a) = \mathcal{M}_s\} , \qquad (3.3)$$

where $\mathcal{M}_s$ is the set of locations determined by s . The relationship between size and set of locations is not fixed by the PL/I language. A PL/I implementation is free to establish this relationship, observing, however, the condition

$$s_1 < s_2 \supset \mathcal{M}_{s_1} \subseteq \mathcal{M}_{s_2} . \qquad (3.4)$$

We now can define the functions alternatives, curralt, indexset, and comprange, for area ranges. The alternative of an area range A are those maximal subsets of A whose members have the same level-one locations:

$$\text{alternatives}(A) = \text{the set of equivalence classes of members of } A \text{ with respect to their level-one locations.} \quad (3.5)$$

The current alternative of A determined by a member $a \in A$:

$$\text{curralt}(A,a) = \{v \mid v \in A \ \& \ \text{level}\,1\,\text{locs}(v) = \text{level}\,1\,\text{locs}(a)\} \ . \quad (3.6)$$

Let A^M be an alternative of A ; then

$$\text{indexset}(A^M) = \text{the set of level-one locations of the members of } A^M \ ; \quad (3.7)$$

$$\text{comprange}(A^M,m) = \text{range}(m) \ .$$

Finally, we explain the conversion of areas. Before an area a is assigned to an area variable with size s , a is converted to an area with size s . This conversion is possible if $\mathfrak{M}_s$, the set of locations determined by s , contains the active locations of a :

$$\text{actlocs}(a) \subseteq \mathfrak{M}_s \ .$$

The result of the conversion is the area a' such that

$$\text{locs}(a') = \mathfrak{M}_s \ ,$$

$$\text{level}\,1\,\text{locs}(a') = \text{level}\,1\,\text{locs}(a) \ ,$$

$$\text{level}\,1\,\text{contents}(a') = \text{level}\,1\,\text{contents}(a) \ .$$

3.3. Pointers and Offsets

The set of pointers P and the set of offsets O are sets of elementary values in PL/I. Pointers and offsets are derived from locations and are used, in turn, for the construction of locations.

We first have to introduce the notion of "connected locations". As an auxiliary notion, we use "segment" of a location. Since the index-set of a composite range is ordered, we can determine for any composite location l the ordered list of its scalar (i.e., elementary or area) sub-locations. Let

$$l_1, l_2, \ldots, l_n$$

be this list; then we call any list of locations

$$l_i, l_{i+1}, \ldots, l_j \ , \quad (1 \leq i \leq j \leq n) \ ,$$

a <u>segment</u> of 1 .

A location is <u>connected</u> if it is one of the following:

(1) a level-one location. (3.9)

(2) a location whose list of scalar sub-locations forms a segment of a connected location,

(3) a constructed location (see below).

The function addr is a mapping from connected locations to pointers and offsets:

$$\text{addr} : \mathcal{L}_{\text{conn}} \rightarrow P \cup O \ . \qquad (3.10)$$

$\mathcal{L}_{\text{conn}} \subset \mathcal{L}_P$ is the set of connected locations in main storage and in areas. If 1 is a main storage location, then

$$\text{addr}(l) \in P \ .$$

If it is the location of an area, then

$$\text{addr}(l) \in O \ .$$

For a pointer (or offset) p and a range R , there is exactly one connected location $l \in \mathcal{L}_P$

$$l = \text{construct}(p,R) \quad \text{such that} \qquad (3.11)$$

$$\text{addr}(l) = p \ ,$$

$$\text{range}(l) = R \ .$$

The function construct is used in the evaluation of references to based variables. If a reference combines a pointer which was derived from a location 1 , with a based variable whose range is equal to range(1) , then the evaluation of the based reference gives the location 1 . If only the above relation (3.11) between locations, pointers, and ranges is known, we only know how to reconstruct those allocated locations by based references of which the pointer was previously derived. There is, however, more freedom in PL/I for the combination of pointers and ranges of based variables, which derives from special properties of the mapping from locations to component locations discussed in the next section.

3.4. Special Properties of Storage Mapping

Locations are mapped into component locations by the functions comploc and comflexloc introduced in section 1.3. For PL/I, these functions are given special properties in addition to those imposed already by the general model.

3.4.1. The mapping of connected array and structure locations

For connected array and structure locations, the ordering of the index-set is significant. In the previous section, we introduced the notion of a segment of a location. Analogously, we can determine for each range the ordered list of its scalar sub-ranges.

We require that the function comploc(l,i) for connected, composite l can be reduced to another function compconnloc :

$$\text{comploc}(l,i) = \text{compconnloc}(\text{addr}(l), \text{r-list}_i, r_i) \quad \text{where} \tag{3.12}$$

r-list$_i$ = the list of scalar sub-ranges of range(l) up to, but excluding, the sub-ranges of the i-th component range,

r_i = the list of scalar sub-ranges of the i-th component range of range(l) .

This property is called structure-independent mapping. Furthermore, we require a "linear" property of the mapping function, which establishes a well-defined relationship between sub-locations of connected locations having the same pointer and the same list of scalar sub-ranges.

Let r-list$_1 \cap r_1$ be an initial segment of range(l) , and r-list$_2 \cap r_2$ an initial segment of r_1 (see the figure below). Then

$$\text{compconnloc}(\text{addr}(\text{compconnloc}(\text{addr}(l), \text{r-list}_1, r_1)), \text{r-list}_2, r_2) = \text{compconnloc}(\text{addr}(l), \text{r-list}_1 \cap \text{r-list}_2, r_2) \; . \tag{3.13}$$

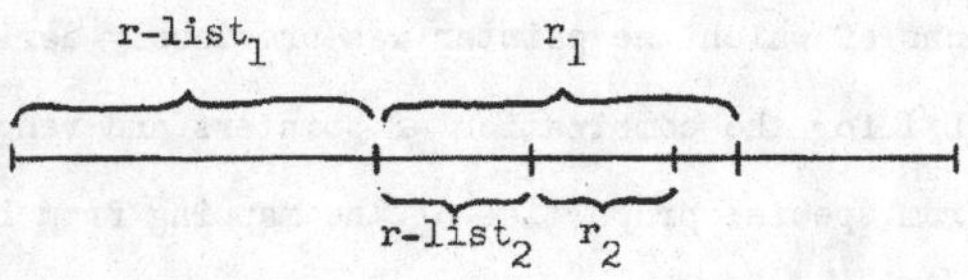

3.4.2. The mapping of area locations

The function compflexloc(l, A_c, m) , for l being an area location, A_c its current range, and $m \in \text{indexset}(A_c)$, reduces to an operation between two arguments:

$$\text{compflexloc}(l, A_c, m) = m \circ l \ . \tag{3.14}$$

The operation $m \circ l$ is called composition of locations. It is easily verified by the properties of compflexloc that

$$m \circ l(S) = m(l(S)) \ ,$$

which explains the notation of functional composition for the operation, and that

$$\text{range}(m \circ l) = \text{range}(m) \ .$$

There is a corresponding operation between pointers and offsets: Let l be an area location and m a connected location of the area; then

$$\text{addr}(l) \oplus \text{addr}(m) = \text{addr}(m \circ l) \ . \tag{3.15}$$

The reverse operation is introduced by

$$(p \oplus 0) \ominus 0 = p \ . \tag{3.16}$$

The operations $\oplus$ and $\ominus$ are called conversion between pointers and offsets. They give the possibility of identifying locations in different areas by the same means, namely offsets, and converting these offsets to pointers (and hence locations) via the pointers of the different area locations. In order to use this technique, we have to know, however, under which circumstances locations in two different areas have the same offset. These follow from properties of allocation in areas.

If a based variable with range R is to be allocated in an area a , a location of the area is selected by the function

$$newloc(level1\,locs(a),R) \ .$$

This function gives a location m such that

$$\begin{aligned} &range(m) = R \ , \text{ and} \\ &m \text{ is independent of all locations in } level1\,locs(a) \ . \end{aligned} \tag{3.17}$$

The new location, therefore, is determined independently of the size of the area. Consequently, if the "history" of two areas is such that they have the same level-one locations, then a new allocation with the same based variable in both areas will identify the same location m . The actual allocation, however, is possible in a only if

$$m \in locs(a) \ ,$$

i.e., if the size of the area allows the allocation.

Some consequences of the above storage mapping properties are shown in section 3.6.

3.5. Reference, Allocation, Assignment

The process of referring to a variable after the evaluation of subscripts, identifiers of structure components, and possibly pointer expressions (in references to based variables) follows from the previous section. It should be noted, however, that this general mechanism gives more freedom than is given syntactically in PL/I. The reader is referred to PL/I Language Specifications (1966) or to the formal description Walk et al. (1969) on that subject.

The central part of allocation of variables is as described in section 1.5 for the general model.

Similarly, assignment of a value to a variable is taken over from section 1.5. The assignment function (1.21) is to be understood as being performed after the evaluation of the left-hand side and the right-hand side of an assignment statements, and <u>after</u> the conversion of the right-hand side value to the characteristics of the left-hand side location.

3.6. Some Consequences of the Storage Mapping Properties

A few examples will show the use of the mapping functions. This section presupposes some familiarity with PL/I notation.

(a) We consider the following program fragment:

```
DECLARE 1 R AUTOMATIC,
          2 A FIXED,
          2 B,
            3 C CHAR(5),
            3 D BIT(5),
          2 E FLOAT,
        1 T BASED,
          2 U FIXED,
          2 V CHAR(5),
          2 W BIT(5),
          2 X FIXED,
        P POINTER;
        P = ADDR(R);
        R.B.D = '10110'B;
        .... P→T.W ....
```

The automatic structure variable R is allocated at declaration time. Let l_R be the location of R ,

$$p_R = \text{addr}(l_R) ;$$

p_R is assigned to the pointer variable P . The location associated with R.B according to (3.12) is

$$l_B = \text{compconnloc}(p, \langle sr_A \rangle, \langle sr_C, sr_D \rangle) ,$$

where sr_A , sr_C , sr_D are the scalar (fixed point arithmetic, character string, and bit string) ranges associated with the scalar components R.A , R.B.C , and R.B.D , respectively.

The location associated with R.B.D is

$$l_D = \text{compconnloc}(\text{addr}(l_B), \langle sr_C \rangle, \langle sr_D \rangle) ,$$

which according to (3.13) is equal to

$$l_D = \text{compconnloc}(p, \langle sr_A, sr_C \rangle, \langle sr_D \rangle) .$$

l_D becomes associated with the bit string value '10110'B through the assignment statement. Let S be the current storage:

$$l_D(S) = \text{'10110'B} .$$

We now ask for the evaluation of the based reference P→T.W . A location l_T is constructed from the pointer p and the range R_T of T :

$$l_T = \text{construct}(p, R_T) .$$

The location associated with T.W is

$$l_W = \text{compconnloc}(p, \langle sr_U, sr_V \rangle, \langle sr_W \rangle) .$$

Since $\langle sr_U, sr_V \rangle = \langle sr_A, sr_C \rangle$ and $\langle sr_W \rangle = \langle sr_D \rangle$ according to the declaration, we have

$$l_w = l_D .$$

The reference P→T.W , therefore, is well-defined and gives the value

$$l_w(S) = \text{'10110'B} .$$

(b) We now show the use of offsets for identifying locations in areas. Let us again consider a program fragment:

```
DECLARE A1 AREA(50), A2 AREA(100),
        O OFFSET,
        B BASED;
        ALLOCATE B SET(O) IN (A1);
        A2 = A1;
        .... POINTER (O,A2) → B ....
        ALLOCATE B SET (O) IN A1;
        ALLOCATE B SET (O) IN A2;
```

Let the locations associated with the area variables be l_{A1} and l_{A2} and its initial values a_1 and a_2 (we denote successive storages by $S, S', S'', \ldots$) :

$$l_{A1}(S) = a_1$$
$$l_{A2}(S) = a_2 \ .$$

Initially, there are no active locations in the areas

$$actlocs(a_1) = \{ \ \}$$
$$actlocs(a_2) = \{ \ \} \ .$$

The allocation of B in A1 adds a new location m_1 to the level-one locations of a_1 . This location according to (3.17) is given by

$$m_1 = newloc(\{ \ \}, R_B) \ ,$$

where R_B is the range of B . The offset o_1 derived from m_1 ,

$$o_1 = addr(m_1) \ ,$$

is assigned to the offset variable O . With S' the current storage, we now have

$$l_{A1}(S') = a_1'$$
$$l_{A2}(S') = a_2$$
$$level\,1\,locs(a_1') = \{m\} \ .$$

Now comes the assignment of a_1' to A2 . a_1' is converted to the size of A2 giving a_2'' :

$$\text{level 1 locs}(a_2'') = \{m_1\}$$

$$\text{contents}(a_2'') = \text{contents}(a_1')$$

$$\text{locs}(a_2'') = \text{locs}(a_2)$$

and assigned to A_2 :

$$l_{A2}(S'') = a_2'' \ .$$

l_{A2} now has one component location $m \circ l_{A2}$.
We now ask for the meaning of the reference POINTER(0,A2) → B . The builtin-in function POINTER performs the conversion of the offset value o of 0 and the pointer of the location l_{A2} of A_2 to a pointer p_1 (see (3.15)):

$$p_1 = \text{addr}(l_{A2}) \oplus o \ .$$

This pointer, together with the range R_B of B , is used to construct a location

$$l_1 = \text{construct}(p_1, R_B) \ .$$

Using (3.15) and (3.14), we can show for the sub-location of l_{A2} that

$$\text{addr}(m_1 \circ l_{A2}) = p_1 \quad \text{and}$$

$$\text{range}(m_1 \circ l_{A2}) = R_B \ ,$$

which by (3.11) means that l_1 is precisely the sub-location of l_{A2}.
POINTER(0,A2) → B , therefore, refers to this sub-location whose value is identical with the value of POINTER(0,A1) → B . A single offset value thus can be used for identifying locations in different areas. Finally, a new location is allocated in both of the areas. This new location is

$$m_2 = \text{newloc}(\{m_1\}, R_B)$$

identically for both areas, despite their different size. These new locations, therefore, can again be identified by the same offset.
(c) The last example shows the working of defined variables. We consider a declaration and an assignment.

```
DECLARE B(2,3),D(2,2) DEFINED B;
        D(1,1) = 0;
```

The defined variable D is a two-dimensional array variable defined on the two-dimensional array variable B. Let the location of B be l_B, its component locations l_1 and l_2, and its scalar sub-locations $l_{11}, l_{12}, l_{13}, l_{21}, l_{22}, l_{23}$. For B we construct a super-location l_D with the range of B from the locations l_{11}, l_{12}, l_{21}, l_{22} such that

$$\text{comploc}(\text{comploc}(l_D,1),1) = l_{11}$$
$$\text{comploc}(\text{comploc}(l_D,1),2) = l_{12}$$
$$\text{comploc}(\text{comploc}(l_D,2),1) = l_{21}$$
$$\text{comploc}(\text{comploc}(l_D,2),2) = l_{22} \; .$$

An assignment to l_{11} via D(1,1) will clearly also define the value of B(1,1). Note that the list of scalar sub-locations of l_D does not form a segment of l_B. l_D, therefore, is not a connected location; $\text{addr}(l_D)$ and hence ADDR(D) would be undefined.

CONCLUDING REMARKS

Storage properties of ALGOL 68 and PL/I have been presented using the same general storage model as the common basis. This allows a thorough comparison of the two languages with respect to storage handling, which is independent of the syntactic appearance of the two languages. Only a few remarks are made in the following.

There is obviously no counterpart of flexible arrays in PL/I, and no counterpart of areas in ALGOL 68. (Somewhat surprisingly, both these features can be modelled using the concept of flexible locations.)

With respect to the building up of more general data structures like lists, rings, etc., the corresponding tools in ALGOL 68 and PL/I are names and pointers. Both serve the purpose of identifying locations in storage, which themselves are processable data. Name variables, however, are restricted with respect to the types of data their values may refer to. This allows compile time checking of the validity of references. PL/I pointer variables are not restricted in the types of data to which their values may point. Pointers also do not necessarily contain range information which makes even run-time checking a problem. Getting defined references in PL/I is the responsibility of the programmer, who has to take into account a number of special properties of storage mapping he need not know for ALGOL 68. On the other hand, PL/I pointers are certainly the most efficient way to realize identification of data. ALGOL 68 and PL/I differ with respect to the balance between economy and security in programming.

PROGRAM SCHEMES, PROGRAMS AND LOGIC

by

D. C. Cooper

Introduction

Questions of termination, correctness and equivalence of programs and program schemes have been represented as questions of validity and satisfiability of formulae of the first order predicate calculus (Floyd 1967a, Manna 1968a,b, 1969a,b) and of the second order predicate calculus (Cooper 1969).

This note sets up a standard notation in which the theorems of Cooper (1969) and Manna (1968b) are expressed. All theorems are proved from the basic "Floyd Theorem" (Floyd 1967a) and a specified set of obvious properties of program schemes. We will not be formal about what we mean by a program, or a program scheme, or about the precise rules for deriving the formulae of the predicate calculus from the program. The main point is to state very precisely the relations between the formulae and properties of the program and to show how the various results are interrelated and all simply derivable from one basic result.

Program Schemes - a very brief summary

We only consider programs with assignment statements and conditional jumps. In a program scheme we leave the functions and predicates uninterpreted, i.e., they are just names. Thus Figure 1, if we do not explicitly state what functions f and g are or what predicates t and u are, is a program scheme. It can be made into a program by giving an <u>interpretation</u>, i.e., specifying some universe and some particular functions and predicates over that universe to be f, g, t, and u. Two program schemes are <u>equivalent</u> if all interpretations produce equivalent programs. A

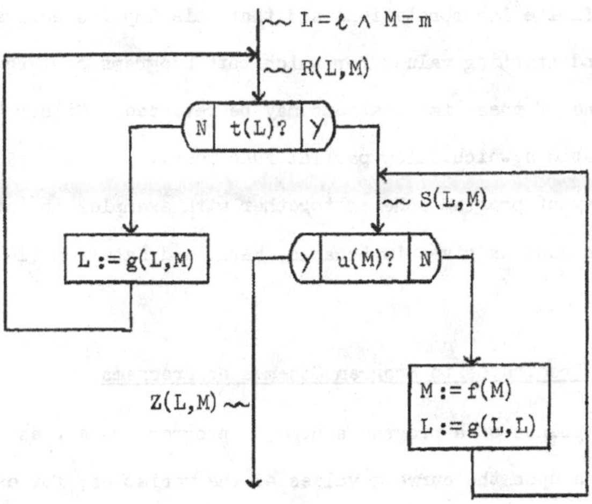

FIGURE 1

program scheme is <u>free</u> if, given any sequence through the program scheme, there is an interpretation and a set of starting values which cause that sequence to be the control sequence of the program. A program scheme is <u>progressive</u> if in all possible computation sequences through the scheme the location assigned to in one statement is a retrieval location for the next assignment statement, if any.

Basic decidability results about program schemes are that both equivalence and freedom are undecidable properties and that equivalence is decidable if schemes are restricted to one of the following classes:

Loop-free.

Halting under all interpretations and starting values.

One register.

Non-intersecting loops and monadic functions.

Progressive.

It is conjectured that equivalence of free schemes is decidable.

We may restrict the interpretations to those over finite domains, or to those in which the functions and predicates are recursive. It is clear that equivalence over all functions implies equivalence over recursive functions, that this implies equivalence over finite interpretations and that this implies equivalence over those interpretations and starting values for which both programs converge (so-called weak equivalence). None of these implications may be reversed. Milner (1970) has considered interpretations which allow partial functions.

A brief survey of program schemes together with examples is Paterson (1968). More detailed treatment is given in Luckham, Park, and Paterson (1970) and in Paterson (1967).

The Attachment of Predicates to Program Schemes or Programs

With various points of a program scheme or program, we may associate a predicate whose value depends upon the current values of the variables; for example, in Figure 1 predicates R and S are attached as indicated and Z is attached to the end point. We may regard these either as uninterpreted, or as specified predicates. It is intended that attaching a particular predicate S(L,M) at the indicated point shall imply that if, with some interpretation and starting value, control reaches that point, then S(L,M) must be true of the values of registers L and M at that point.

Motivated by this consideration, a condition can be formed for each path from one of these associated predicates to another, i.e.,

$$\{\text{predicate at start} \wedge \text{condition for path to be traversed}\} \supset \text{predicate at end}$$

where account must be taken of the effect of assignment statements. Location letters (e.g., L and M in Figure 1) will be free variables.

If predicates have been associated with points of the program such that all loops contain at least one predicate, then there is only a finite number of such conditions. Consider the conjunction of all these (include also paths starting at the beginning of the program with associated predicate stating the register values are all equal to some initial values) and consider this formula universally quantified over the register names.

For Figure 1 this gives:

$$
\begin{aligned}
(\forall L)(\forall M)[L = 1 \wedge M = m \supset R(L,M) & \quad \wedge \\
R(L,M) \wedge \neg t(L) \supset R(g(L,M),M) & \quad \wedge \\
R(L,M) \wedge t(L) \supset S(L,M) & \quad \wedge \\
S(L,M) \wedge \neg u(M) \supset S(g(L,L),f(M)) & \quad \wedge \\
S(L,M) \wedge u(M) \supset Z(L,M) & \quad] .
\end{aligned}
$$

We call this the Floyd condition associated with the program scheme and write it

$$FL_{PS}(I,x,P,Z)$$

where

PS is a program scheme;

x is a set of initial value names ($\{\ell,m\}$ in Figure 1);

I is a set of function and predicate names ($\{f,g,t,u\}$ in Figure 1);

P is a set of predicate names associated with points of the program so that there is at least one in every cycle ($\{R,S\}$ in Figure 1);

Z is a predicate name associated with the end point of the program.

The predicates of P (and also Z) have as many arguments as there are registers. It should be noted that a Floyd condition is a formula of first order logic.

An <u>interpretation</u> is a universe and an assignment of functions and predicates over that universe to the members of I . An <u>input</u> is an assignment of individuals of the universe to members of x .

This idea is reported and further developed independently by Floyd (1967 a) and Naur (1966). Techniques based on this idea have been used to prove properties of programs and to define the semantics of statements in programming languages. The idea has also been extended to cover other features of programming languages such as recursion and non-determinism.

Some Basic Notions and Their Properties

We shall define certain notions depending on PS , I , x , P , and Z as given above:

(a) $FL_{PS}(I,x,P,Z)$ can be defined for any program scheme as above, we omit formal details.

It is the "C" of Cooper (1969), and, in the notation of that paper, PS __does__ Z is $(\exists P)FL_{PS}(I,x,P,Z)$.

[Here and throughout if P is the set $\{P_1,P_2,\ldots,P_n\}$, then $(\exists P)$ stands for $(\exists P_1)(\exists P_2)\ldots(\exists P_n)$; also if x is $\{\ell,m\}$, I is $\{f,g,t,u\}$, then $(\forall x)$, $(\forall I)$ stand for $(\forall \ell)(\forall m)$ and $(\forall f)(\forall g)(\forall t)(\forall u)$.]

(b) $End_{PS}(I,x,Z)$

This is true if, with input x and interpretation I , the program scheme PS terminates and the predicate Z is true of the final values. It is false if the program scheme does not terminate, or if it terminates and Z is not true of the final values.

(c) $*_{PS}(I,x)$

This is true if, with input x and interpretation I , the program scheme PS terminates, false otherwise.

(d) $PS_1 \equiv PS_2$

This is the usual strong equivalence of program schemes. We may define it by

$$(\forall I)(\forall x)(\forall Z)[End_{PS_1}(I,x,Z) \equiv End_{PS_2}(I,x,Z)] .$$

[The $\equiv$ of this formula is the equivalence of logic.]

We may consider programs, rather than program schemes, by simply omitting all references to I . All theorems and results remain valid.

The following properties are all obvious:

$$*_{PS}(I,x) \equiv End_{PS}(I,x,true) \qquad (1)$$

$$End_{PS}(I,x,Z) \supset *_{PS}(I,x) \qquad (2)$$

$$\neg End_{PS}(I,x,false) \qquad (3)$$

$$End_{PS}(I,x,\neg Z) \equiv \neg End_{PS}(I,x,Z) \wedge *_{PS}(I,x) \qquad (4)$$

$$FL_{PS}(I,x,P,false) \equiv (\forall Z)FL_{PS}(I,x,P,Z) . \qquad (5)$$

__The Basic "Floyd" Theorem__

By the way $FL_{PS}(I,x,P,Z)$ has been defined, it is clear that if I , x , P , and Z are specified so as to make FL true and if, with that given I and x , PS terminates, then Z must be true of the final results. Conversely, if, with some I and x, the program terminates with Z true, or the program loops, then specific predicates may be defined to make FL true (just take for P predicates expressing precisely those and only those values attained at the particular program point). These remarks may be expanded into a proof of the following theorem:

$$(\exists P)FL_{PS}(I,x,P,Z) \equiv End_{PS}(I,x,Z) \vee \neg *_{PS}(I,x) \ . \qquad F1$$

Other versions of this theorem are

$$End_{PS}(I,x,Z) \equiv (\exists P)FL_{PS}(I,x,P,Z) \wedge *_{PS}(I,x) \qquad F2$$

and

$$*_{PS}(I,x) \supset [End_{PS}(I,x,Z) \equiv (\exists P)FL_{PS}(I,x,P,Z)] \ . \qquad F3$$

By equation (2), F2 and F3 are equivalent. They may be deduced from F1, again by using equation (2). However, it does not appear that F1 can be deduced from F2 or F3 and the above properties.

These theorems are the basis of the verifying compiler (see Floyd (1967 b) and King (1969)). In order to prove that a program's results satisfy some condition, specific predicates are attached to points of the program, and the corresponding Floyd predicate is proved true. A programmer should know these predicates; they represent relations between the current values and presumably were in his mind when he designed the program; he only has to formalize them. If this has been done, then $(\exists P)FL_{PS}(I,x,P,Z)$ has been proved true by exhibiting the P. The proof will be a first order proof, although $(\exists P)FL_{PS}(I,x,P,Z)$ is itself a formula of second order logic. (However, it should be remembered that the proof will depend upon properties of the specific functions used in the program, and these could themselves be expressed in second order logic.) If we have an independent proof that the program terminates, then we will have proved that the results satisfy the predicate Z.

Further Theorems

We shall now state and prove the main results from Cooper (1969) and Manna (1968). All these may be directly proved from F1 and equations (1) to (4).

(A) $\neg *_{PS}(I,x) \equiv (\exists P) FL_{PS}(I,x,P,false)$.

<u>Proof</u>. Put false for Z in F1 and use (3).

This is Theorem 2 of Cooper (1969); the idea of placing the predicate "false" on the exit of a program to investigate looping questions is due to Manna (see Manna (1968 a)). It relates the proof of looping or non-looping to a theorem of second order logic. In the form $*_{PS}(I,x) \equiv (\forall P) \neg FL_{PS}(I,x,P,false)$, it relates the proving of convergence to a theorem of first order logic, i.e., to proving that $\neg FL_{PS}(I,x,P,false)$ is valid.

(B) $PS_1 \equiv PS_2$ iff $(\forall I)(\forall x)(\forall Z)\{(\exists P)FL_{PS_1}(I,x,P,Z) \equiv (\exists P)FL_{PS_2}(I,x,P,Z)\}$.

<u>Proof</u>. By F1 the right hand side is

$$(\forall I)(\forall x)(\forall Z)\{End_{PS_1}(I,x,Z) \vee \neg *_{PS_1}(I,x) \equiv End_{PS_2}(I,x,Z) \vee \neg *_{PS_2}(I,x)\} , \quad B1$$

and by definition (d) the left hand side is

$$(\forall I)(\forall x)(\forall Z)\{End_{PS_1}(I,x,Z) \equiv End_{PS_2}(I,x,Z)\} . \quad B2$$

Now if we define B1' and B2' by removing the first two quantifiers from B1 and B2, respectively (so that B1 is $(\forall I)(\forall x)B1'$ and B2 is $(\forall I)(\forall x)(B2')$), then to show $B1 \equiv B2$, it is sufficient to show the stronger result $B1' \equiv B2'$. Consider four cases:

(i) $*_{PS_1}(I,x) \wedge *_{PS_2}(I,x)$

Clearly $B1' \equiv B2'$.

(ii) $\neg *_{PS_1}(I,x) \wedge *_{PS_2}(I,x)$

B1' is $(\forall Z)End_{PS_2}(I,x,Z)$, which is false by (3). B2', using $\neg *_{PS_1}(I,x)$ and (2), is $(\forall Z) \neg End_{PS_2}(I,x,Z)$, which is false by (1) and hypothesis. Therefore $B1' \equiv B2'$.

(iii) $*_{PS_1}(I,x) \wedge \neg *_{PS_2}(I,x)$

Proof similar to (ii).

(iv) $\neg *_{PS_1}(I,x) \wedge \neg *_{PS_2}(I,x)$

B1' is true. By (2), B2' is true. Therefore $B1' \equiv B2'$.

This is Theorem 4 of Cooper (1969) and shows that proving equivalence of program schemes can be done by proving a theorem of second order logic. The corresponding result and proof for equivalence of programs rather than program schemes is obtained by simply dropping the quantifier $(\forall I)$ and the argument I throughout.

(C) $(\forall x)[\phi(x) \supset End_{PS}(I,x,Z)]$ iff $(\forall x)(\forall P)[\phi(x) \supset \neg FL_{PS}(I,x,P,\neg Z)]$

<u>Proof</u>. By F1 and (4), the right hand side is

$$(\forall x)[\phi(x) \supset End_{PS}(I,x,Z) \wedge *_{PS}(I,x)]$$

which, using (2), is the left hand side.

This is Theorem 1 of Manna (1968 b). It shows that (given ϕ and Z) the problem of proving that, for all inputs satisfying ϕ , a program always terminates and Z is true of the answers is equivalent to proving a theorem of first order logic. This assumes, of course, that the necessary properties of the functions and predicates occurring in the interpretation I and ϕ and Z can all be expressed in first order logic.

(D) $(\forall x)[\phi(x) \supset *_{PS}(I,x)]$ iff $(\forall x)(\forall P)[\phi(x) \supset \neg FL_{PS}(I,x,P,\mathrm{false})]$

<u>Proof</u>. Put "true" for Z in (C) and use (1).

This is Manna's corollary to his Theorem 1 and shows that the problem of proving that, for all inputs satisfying ϕ , a program terminates is equivalent to proving a theorem of first order logic.

(E) $(\forall x)[\phi(x) \supset *_{PS_1}(I,x) \wedge *_{PS_2}(I,x) \wedge (\forall Z)\{End_{PS_1}(I,x,Z) \equiv End_{PS_2}(I,x,Z)\}]$ iff
$(\forall x)(\forall P)(\forall Q)(\forall Z)[\phi(x) \supset \neg FL_{PS_1}(I,x,P,Z) \vee \neg FL_{PS_2}(I,x,Q,\neg Z)]$

<u>Proof</u>. By F1 and (4), the right hand side is

$$(\forall x)(\forall Z)[\phi(x) \supset End_{PS_1}(I,x,\neg Z) \vee End_{PS_2}(I,x,Z)] \ .$$

In this last formula, use the rule of logic

$$(\forall Z)F(Z) \equiv (\forall Z)[F(Z) \wedge F(\neg Z)] \ ,$$

then use (4) to eliminate $\neg Z$, and then simplify to give

$$(\forall x)(\forall Z)[\phi(x) \supset \{End_{PS_1}(I,x,Z) \wedge End_{PS_2}(I,x,Z)\} \vee \{\neg End_{PS_1}(I,x,Z) \wedge \neg End_{PS_2}(I,x,Z) \wedge *_{PS_1}(I,x) \wedge *_{PS_2}(I,x)\}] \ .$$

By (2), this is equivalent to the left hand side of (E).

This is Theorem 2 of Manna (1968 b) and shows that the problem of proving that, for all inputs satisfying ϕ, both programs PS_1 and PS_2 converge and give the same answers is equivalent to proving a theorem of first order logic.

Acknowledgments

I wish to acknowledge the helpful suggestions of R. Milner in the preparation of this note. The material was first presented at "Journées d'Etudes sur l'Analyse Syntaxique" held at Fontainebleau, France, on the 10th and 11th of March, 1969.

ALGEBRAIC THEORIES AND PROGRAM SCHEMES

by

Calvin C. Elgot

1. Introduction

The objective of this paper is to point out that the "semantics" of cycle-free program schemes (of the kind intuitively described in Section 2) may be explicated by existing algebraic notions. More explicitly, equivalence classes of schemes correspond to morphisms in a free algebraic theory (cf., Lawvere (1963), and Eilenberg and Wright (1967)), and an interpretation of the collection of morphisms is a coproduct-preserving functor into an appropriate target category.

While some attempt has been made to be reasonably self-contained, the interested reader may wish to consult Elgot (1970), and Eilenberg and Wright (1967).

2. Program Schemes--Intuitive

The intuitive notion "program scheme" we have in mind is close in spirit to that of Ianov (1958), as generalized by Rutledge (1964). The "corresponding" precise notion will be called "machine scheme", described in section 4.

Roughly speaking, a program scheme consists of an indexed family of "instruction schemes", together with a specification of the *a priori* possible sequencing of instructions (i.e., interpreted instruction schemes). The indices are sometimes called "instruction labels" and sometimes "states". In addition, provision must be made for how the sequencing of instructions comes to a halt. In this context, an instruction may be understood as a function $X \to X \times [n]$, $n \geq 1$, where $[n] = \{1,2,\ldots,n\}$ and X is a set. The set X may be regarded as the set of possible states of "external memory" or, some may prefer to say, "storage". If external memory is in state $x \in X$ at some instant of time, if $f: X \to X \times [n]$ is executed at this instant of time, and if $f(x) = (x', i)$, $i \in [n]$, then after execution of f , external memory is in state x' , and the next instruction to be executed is determined by the number i . Thus execution of f simultaneously performs a test on X (with n possible <u>outcomes</u>) and an operation on X .

To illustrate the notion of "program", let $f_i: X \to X \times [2]$, $1 \leq i \leq 3$, be functions. The following "table" describes a program P .

→ 1. f_1;2,3
2. f_2;3,4
3. f_2;5,5
4. f_3;3,5
5. exit

Assuming instruction #1 is the "beginning" instruction, and given an "input" $x_0 \in X$, execution of the program procceds as follows: f_1 is executed; if the outcome is 1, the next instruction to be executed is #2, the first of the two indices following the semicolon on the top line; if the outcome is 2, the next instruction to be executed is #3. Suppose $f_1(x_0) = (x_1, 1)$, $f_2(x_1) = (x_2, 2)$, $f_3(x_2) = (x_3, 1)$; then the sequence of executions is: #1, #2, #3, #4, #5. When instruction #5 is reached, the execution sequence terminates. In this particular case, which is "cycle-free", execution sequences are all finite. If $f_2(x_3) = (x_4, i)$, $i = 1,2$, then the "output" of the execution is x_4 . The "external behavior" of P (cf., Elgot (1970) is a function $|P|: X \to X$ with $|P|(x_0) = x_4$.

The information in the table above may also be presented as follows.

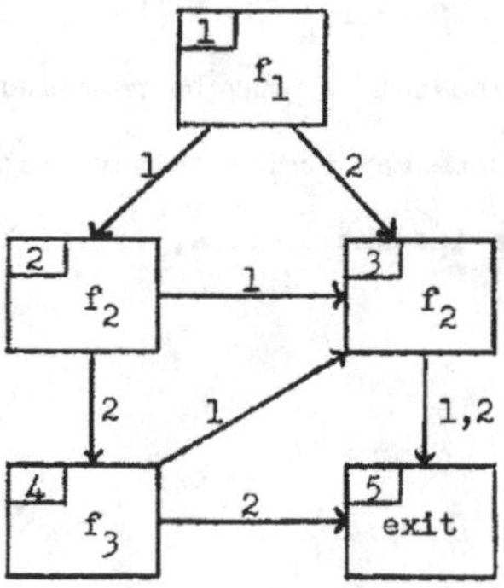

In this section, we will not bother to distinguish a program from the "isomorphism class" of programs which it determines.

The composite QR (cf., Elgot (1970)) of the following two programs

Q	R
→ 1. f_1;2,3	$\overset{1}{\to}$ 2. f_2;2,3
2. exit 1	$\overset{2}{\to}$ 3. f_2;3,4
3. exit 2	4. f_3;3,5
	5. exit

is the program P. The external behavior of Q, which has two exits, is $|Q| = f_1 : X \to X \times [2]$, while the external behavior of R which has two beginning states is $|R| : X \times [2] \to X$. If $|R|(x,i) = y$, $i = 1,2$, then execution of program R operating on input x, starting with the ith beginning state, yields output y. Moreover, $|P| = |Q||R| = f_1|R|$.

As a further illustration, let T be the following program:

$\overset{1}{\to}$ 2. exit 2

$\overset{2}{\to}$ 3. exit 1.

Then QT is the following:

→ 1. f_1;2,3

2. exit 2

3. exit 1,

while TR is the following:

$\overset{2}{\to}$ 2. f_2;2,3

$\overset{1}{\to}$ 3. f_2;3,4

4. f_3;3,5

5. exit.

It should also be noted that $|T| : X \times [2] \to X \times [2]$ satisfies $|T|(x,1) = (x,2)$, $|T|(x,2) = (x,1)$ for every $x \in X$. Of course, $|QT| = |Q||T|$ and $|TR| = |T||R|$.

3. Algebraic Theories

Categories (cf., MacLane and Birkhoff (1967) and Eilenberg and Wright (1967)) arise naturally in our considerations. A <u>category</u> $\mathcal{C}$ consists of a class Obj $\mathcal{C}$

whose elements $A, B, \ldots$ are called objects, a family of sets $\mathcal{A}(A,B)$ whose elements are called morphisms from A to B, and an operation called composition, required to be associative, which determines for each $\varphi \in \mathcal{A}(A,B)$ and $\psi \in \mathcal{A}(B,C)$ a unique element $\varphi\psi \in \mathcal{A}(A,C)$; in addition, it is assumed, for each B, there exists a (necessarily unique) element $1_B \in \mathcal{A}(B,B)$ such that $\varphi 1_B = \varphi$, $1_B \psi = \psi$. We write B in place of 1_B and $\varphi\colon A \to B$ or $A \xrightarrow{\varphi} B$ in place of $\varphi \in \mathcal{A}(A,B)$.

Let X be a set. We form a category $\mathcal{A} = (X)$ be taking as objects X^n, $n = 0,1,2,\ldots$ and taking as $\mathcal{A}(X^p, X^n)$ the set of all functions with domain X^p and image a subset of X^n.

For each $i \in [n] = \{1,2,\ldots,n\}$, the morphism $\pi_i\colon X^n \to X$ defined by $\pi_i(x_1, x_2, \ldots, x_n) = x_i$ is distinguished and the following basic property noted:

(3.1) for each family $\varphi_i\colon X^p \to X$, $i \in [n]$, of morphisms in (X), there is a unique morphism $\varphi\colon X^p \to X^n$ such that $\varphi_i = \varphi\pi_i$, i.e.,

$$\varphi_i\colon X^p \xrightarrow{\varphi} X^n \xrightarrow{\pi_i} X$$

for each $i \in [n]$.

Let $f\colon [n] \to [p]$, $g\colon [p] \to [q]$ be functions. Define $f\#\colon X^p \to X^n$ by $f\#(x_1, x_2, \ldots, x_p) = (x_{f(1)}, x_{f(2)}, \ldots, x_{f(n)})$. The reader may readily verify

(3.2) $(fg)\# = g\# f\#$;

(3.3) if X has at least two elements, the function $f \mapsto f\#$ preserves distinctness.

Now let $\mathcal{S}_F$ be the category whose objects are $[n]$, $n \geq 0$, and where morphisms $[n] \to [p]$ are the functions with domain $[n]$ and values in $[p]$. Let $\mathcal{S}_F^{op}$ be the category obtained from $\mathcal{S}_F$ by "reversing arrows". Properties (3.2) and (3.3) may then be reformulated:

(3.4) if X has at least two elements, the function $\mathcal{S}_F^{op} \to (X)$ which takes f into $f\#$ is an embedding.

For $i \in [n]$, we write $i\colon [1] \to [n]$ for the function with value i.

The above remarks are intended to motivate the following notion. A (non-degenerate) algebraic theory T is a category whose objects are $[n]$, $n \geq 0$, which satisfies

(3.4') $\mathcal{S}_F$ is a subcategory of T ;

(3.1') (a) if $\varphi,\psi\colon [n]\to[p]$ are in T, and if $i\varphi = i\psi$ for each $i\in[n]$, then $\varphi=\psi$;

(b) if $\varphi_i\colon [1]\to[p]$ is in T, $i\in[n]$, then there exists $\varphi\colon [n]\to[p]$ such that $i\varphi=\varphi_i$ for each $i\in[n]$.

Let $\Omega=\{\Omega_n\}$, $n\geq 0$, be a family of pairwise disjoint sets which are also disjoint from the set $\{x_1,x_2,x_3,\ldots\}$ of "variables". A revealing "concrete" example of an algebraic theory may be obtained by treating elements of Ω_n as "function symbols". A <u>term</u> is an element of the smallest class U of words such that

(3.5) $x_i\in U$, $i\geq 1$;

(3.6) if $u_1,\ldots,u_n\in U$ and $\omega\in\Omega_n$, then $\omega(u_1,\ldots,u_n)$ is in U.

The "words" indicated above employ parentheses and commas; as is well known, they may be suppressed without introducing ambiguity. We define a category T be taking as objects $[n]$, $n\geq 0$; taking $\varphi\colon[n]\to[p]$ in T to be $(u_1,\ldots,u_n)$, where u_i is a term in at most the variables $x_1,\ldots,x_p$; and defining $\varphi\psi$, where $\psi\colon[p]\to[q]$, to be the result of simultaneously substituting ψ_j for x_j in φ for each $j\in[p]$, where ψ_j is the jth term comprising ψ. By identifying $f\colon[n]\to[p]$ with $(x_{f(1)},x_{f(2)},\ldots,x_{f(n)})$ and identifying $\omega\in\Omega_n$ with the term $\omega(x_1,\ldots,x_n)$, we obtain an algebraic theory $T=\mathcal{S}_F[\Omega]$ in which $\Omega_n\subset T([1],[n])$ for all $n\geq 0$; briefly, $\Omega\subset T$.

Let T and T' be algebraic theories. By a <u>theory-morphism</u> or <u>morphism between theories</u>, we mean a function which takes $\varphi\colon[n]\to[p]$ in T into $\varphi'\colon[n]\to[p]$ in T' in such a way that

(3.7) if $j\colon[1]\to[p]$, $j\in[p]$, then $j'=j$;

(3.8) $(\varphi\psi)'=\varphi'\psi'$ where $\psi\colon[p]\to[q]$.

Clearly, a theory-morphism keeps $\mathcal{S}_F$ elementwise fixed.

It is not difficult to show

<u>Proposition 3.1</u>. Every family of functions

$$\Omega_n\to T([1],[n])$$

admits a unique extension to a theory-morphism

$$\mathcal{S}_F[\Omega] \to T' .$$

The theory $\mathcal{S}_F[\Omega]$ is freely generated by Ω.

Let X be a set. We construct a category $[X]_F$ with objects $[n]$, $n \geq 0$, as follows. The set $[X]_F([n],[p])$ of morphisms is the set of all functions $f: X \times [n] \to X \times [p]$. If $f: [n] \to [p]$ is in $\mathcal{S}_F$, we denote by $X \times f: X \times [n] \to X \times [p]$, the function which takes (x,i) into $(x,f(i))$. Note $X \times fg = (X \times f)(X \times g)$. If $X \neq \emptyset$, $\mathcal{S}_F$ may be embedded in $[X]_F$. In this way, $[X]_F$ becomes an algebraic theory.

By replacing functions by partial function or relations, we obtain algebraic theories $[X]_P$, $[X]$ respectively. Note that $[X]_F \subset [X]_P \subset [X]$. It may also be noted that $[X]_F$ fails to have morphisms $[1] \to [0]$, while $[X]_P$ and $[X]$ have exactly one.

Let $\mathcal{S}_P$ be the algebraic theory whose set $\mathcal{S}_P([n],[p])$ of morphisms is the set of all partial functions from $[n]$ into $[p]$. It is easy to see that $\mathcal{S}_P = \mathcal{S}_F[\Omega]$, where Ω_0 consists of a single element and $\Omega_i = \emptyset$ for $i \neq 0$. We also note that every theory-morphism $\mathcal{S}_P \to T$ preserves distinctness.

4. Ω-Machine Schemes (Total)

Let $\Omega = (\Omega_1, \Omega_2, \ldots)$ be a sequence of pairwise disjoint sets. A total Ω-machine scheme M from n to p, $n \geq 0$, $p \geq 0$, consists of

(1) a set S, whose elements are called (internal) states;

(2) a function $b: [n] \to S$; $b(i)$ is the ith beginning state;

(3) an injective function $e: [p] \to S$; $e(j)$ is the jth exit;

(4) a function τ which associates with each non-exit $s \in S$, a pair whose first member is $\omega \in \Omega_m$ and whose second member is a finite sequence $(s_1, s_2, \ldots, s_m)$, $s_j \in S$, $1 \leq j \leq m$, where m is some positive integer.

The scheme M is finite if S is finite.

Thus, for example, if $\omega_j \in \Omega_2$, $1 \leq j \leq 3$, the following table describes an Ω-machine scheme:

→ 1. $\omega_1;2,3$

2. $\omega_2;3,4$

3. $\omega_2;5,5$

4. $\omega_3;3,5$

5. exit .

Here $n=1=p$, $S=[5]$. The assignment $\omega_j \mapsto f_j$ determines the program P described earlier.

If $\tau(s)=(\omega,(s_1,s_2,\ldots,s_m))$, we say s is _ν-related_ to s_j for each j , $1 \leq j \leq m$, and write $(s,s_j) \in \nu$. Let

$$\nu^+ = \nu \cup \nu^2 \cup \nu^3 \cup \ldots$$

be the transitive closure of ν . A machine scheme M is _cycle-free_ if its _next-state relation_ ν is areflexive, i.e., $(s,s) \notin \nu^+$ for every $s \in S$. The machine scheme described above is cycle-free.

Let M' be another total Ω-machine scheme from n to p consisting of S' , ν' , e' , τ' . An _isomorphism_ $\theta: M \to M'$ is a bijection $\theta: S \to S'$ subject to the conditions

(a) $b': [n] \xrightarrow{b} S \xrightarrow{\theta} S'$;

(b) $e': [p] \xrightarrow{e} S \xrightarrow{\theta} S'$;

(c) for each non-exit $s \in S$, if $s' = \theta(s)$ and if

$$\tau(s) = (\omega,(s_1,\ldots,s_m))$$

then

$$\tau'(s') = (\omega,(s_1',\ldots,s_m')) .$$

Let $\bar{M}$ be the isomorphism class of machine schemes n to p determined by the machine scheme M . If N is a machine scheme from p to q , we define the composite $\bar{M}\bar{N} = \overline{M'N'}$; where $M' \in \bar{M}$, $N' \in \bar{N}$, the sets of states of M' and N' are disjoint and M'N' is defined as in Elgot (1970). If M" is isomorphic to M', and N" is isomorphic to N' , and the sets of states of M" and N" are disjoint, then M'N' is isomorphic to M"N" . Thus $\bar{M}\bar{N}$ is well defined. The operation of composition is associative.

At this point, we recall that the identity function on $[p]$ is also denoted by $[p]$.

For each function $f: [n] \to [p]$, we define a machine scheme Q_f whose set S_f of states is $[p]$ and such that $e_f = [p]$, $b_f = f$. Then

$$\bar{M}\bar{Q}_{[p]} = \bar{M}$$

and

$$\bar{Q}_{[p]}\bar{N} = \bar{N} .$$

We define the category $C_t[\Omega]$ of Ω-machine schemes as follows. The objects of $C_t[\Omega]$ are $[n]$, $n \geq 0$. A morphism $\varphi: [n] \to [p]$ of $C_t[\Omega]$ is an isomorphism class of finite machine schemes from n to p. Composition of these morphisms has been defined above.

(Actually $C_t[\Omega]$ is not a proper construct in the sense of Gödel-Bernays-von Neumann set theory. The difficulty may be removed by the ad hoc requirement that a "state" be a non-negative integer.)

Restriction of the morphisms of $C_t[\Omega]$ to isomorphism classes of cycle-free machines leads to the subcategory $C_{ct}[\Omega]$ of $C_t[\Omega]$ of <u>cycle-free Ω-machine schemes</u>.

Let $f: [n] \to [p]$ and $g: [p] \to [q]$ be functions. It is easy to see that $\bar{Q}_{fg} = \bar{Q}_f \bar{Q}_g$. Moreover, Q_f is cycle-free. It is convenient to embed $\mathcal{S}_F$ in $C_{ct}[\Omega]$ via the injective functor $f \mapsto Q_f$. Thus $\mathcal{S}_F \subset C_{ct}[\Omega]$.

Each $\omega \in \Omega_p$ determines a morphism $\omega: [1] \to [p]$ in $C_{ct}[\Omega]$ as follows. This morphism is the isomorphism class of the atomic machine

→ 0. ω;1,2,...,p

1. exit 1

2. exit 2

⋮

p. exit p

determined by ω. Clearly distinct elements of Ω_p go into distinct morphisms. We will identify $\omega \in \Omega_p$ with the <u>atomic</u> morphism $\omega: [1] \to [p]$ in $C_{ct}[\Omega]$ which it determines.

A given machine scheme M from n to p may possess non-exit states s which are inaccessible (from any beginning state), i.e., $(b(i),s) \notin \nu^+$ for all $i \in [n]$, where ν is the next-state relation of M. We turn our attention to this phenomenon.

Call a morphism $\gamma: [n] \to [p]$ in $C_t[\Omega]$ <u>f-like</u>, where $f: [n] \to [p]$ is a function, if $\gamma = \bar{M}$ and $f(i) = j$ implies $b_M(i) = e_M(j)$. Every non-exit state of M is inaccessible since an exit of a machine is ν_M-related to no state.

If M is a machine from n to p, its <u>accessible part</u> is the submachine M^{acc} from n to p whose set of states consist of the accessible states of M together with M's exits. The beginning and exit functions of M^{acc} and M agree. For convenience, assume $[s]$ is the set of states of M^{acc}; thus $[s] \subset S_M$.

By the inaccessible part M^{ina} of M, we mean the f-like machine from n to s whose set of states is S_M; whose beginning function is $b: [n] \xrightarrow{f} [s] \xrightarrow{I} S_M$, where f is the beginning function of M^{acc} and I is inclusion; whose exit function is I (so that the exits of M^{ina} are the states of M^{acc}); and whose τ-function is inherited from M.

Let M' be the machine from s to p with set $[s]$ of states whose τ- and exit-functions agree with M^{acc}, but whose beginning function is the identity $[s]$. Let $\psi = \bar{M}'$; then $f\psi = \overline{M^{acc}}$.

<u>Proposition 4.1</u>. Every morphism $\varphi: [n] \to [p]$ in $C_t[\Omega]$ admits a decomposition

$$\varphi: [n] \xrightarrow{\gamma} [s] \xrightarrow{\psi} [p]$$

where $\gamma = \overline{M^{ina}}$ and ψ is defined above.

From the intuitive discussion of section 2, it is clear that every "interpretation" of an f-like morphism, $f: [n] \to [p]$, over a set X should be the function $X \times f: X \times [n] \to X \times [p]$. This consideration motivates the following definition:

A <u>(machine) interpretation</u>, $F: C_{ct}[\Omega] \to T$, where T is an algebraic theory, assigns to each morphism, $\varphi: [n] \to [p]$, a morphism, $F(\varphi): [n] \to [p]$, in such a way that

(a) $F(\varphi\psi) = F(\varphi)F(\psi)$ where $\psi: [p] \to [q]$;

(b) if $\gamma: [n] \to [p]$ is f-like, then $F(\gamma) = f$.

The following is an immediate consequence of Proposition 4.1.

Corollary 4.2. If $\varphi: [n] \to [p]$ is in $C_{ct}[\Omega]$, if F is an interpretation, and if $\psi: [n] \to [p]$ is the accessible part of φ, then $F(\varphi) = F(\psi)$.

Our main assertion concerning $C_{ct}[\Omega]$ states its free nature. Let T be an algebraic theory.

Theorem 4.3. Each family of functions

$$\Omega_n \to T([1],[n]) , \quad n > 0 ,$$

has a unique extension to an interpretation

$$F: C_{ct}[\Omega] \to T .$$

The uniqueness part of the argument essentially depends upon the following points: Let $\varphi: [n] \to [p]$ be in $C_{ct}[\Omega]$. Then $F(i\varphi) = F(i)F(\varphi) = iF(\varphi)$. Since $F(\varphi)$ is in an algebraic theory, it is determined by the n values $F(i\varphi)$, $i \in [n]$. If $n = 1$ and if φ is not f-like for some f, then $\varphi: [1] \to [p]$ admits a unique factorization

$$[1] \overset{\omega}{\to} [m] \overset{\psi}{\to} [p]$$

where $\omega \in \Omega_m$. Cf., also Theorem 15.1 of Elgot (1970).

5. Congruences and Quotients

Let $\mathcal{A}$ and $\mathcal{B}$ be categories with the same class of objects. A functor $F: \mathcal{A} \to \mathcal{B}$ which is the identity on the class of objects will be called strict. The functors of section 4 are strict. A theory-morphism may be described as a strict functor which keeps $\mathcal{S}_F$ pointwise fixed. Strict functors behave much like homomorphisms between general algebras.

By the kernel $\varkappa(F)$ of F, we shall mean the family of relations $K = \{K_{A,B}\}$, where A,B are objects, defined by

$$(\varphi,\varphi') \in K_{A,B} \Leftrightarrow F(\varphi) = F(\varphi') ,$$

where $\varphi,\varphi': A \to B$ are morphisms in $\mathcal{A}$. The kernel of F is compatible with composition in the sense:

(5.1) $(\varphi,\varphi') \in K_{A,B} \wedge (\psi,\psi') \in K_{B,C} \Rightarrow (\varphi\psi,\varphi'\psi') \in K_{A,C}$.

Moreover, where $\mathcal{A}(A,B)$ is the set of morphisms in $\mathcal{A}$ from A to B , clearly:

(5.2) each component $K_{A,B}$ of K is an equivalence relation on $\mathcal{A}(A,B)$.

Call a family $K = \{K_{A,B}\}$, $K_{A,B} \subset \mathcal{A}(A,B) \times \mathcal{A}(A,B)$, a <u>congruence</u> on $\mathcal{A}$ if it satisfies (5.1) and (5.2). In analogy with the general algebraic situation, we may construct a quotient category $\mathcal{A}/K$. There is then a natural strict functor

$$N: \mathcal{A} \to \mathcal{A}/K$$

with the property

(5.3) for every strict functor $F: \mathcal{A} \to \mathcal{B}$ with the property $\varphi,\varphi' \in K_{A,B} \Rightarrow F(\varphi) = F(\varphi')$, there is a unique strict functor $G: \mathcal{A}/K \to \mathcal{B}$ such that

$$F: \mathcal{A} \overset{N}{\to} \mathcal{A}/K \overset{G}{\to} \mathcal{B} ,$$

i.e., $F = NG$.

Now let $\mathcal{A} = C_t[\Omega]$, and let K be the smallest congruence on $\mathcal{A}$ containing the family of relations $R = \{R_{[n],[p]}\}$, where

$$R_{[n],[p]} = \{(\gamma,f) \mid \gamma,f: [n] \to [p] , f \in \mathcal{S}_F , \gamma \text{ is f-like}\} ,$$

K is the componentwise intersection of all congruences which componentwise contain R .

<u>Proposition 5.1</u>. Let $\varphi,\psi: [n] \to [p]$ be in $C_t[\Omega]$. We have

$$(\varphi,\psi) \in K_{[n],[p]} \text{ iff } \varphi^{acc} = \psi^{acc} .$$

The proof of the porposition depends upon Proposition 4.1 and the following considerations:

(a) Let $M = M_1 M_2$ where M_1 is a machine scheme from n to 2 and M_2 is a machine scheme from 2 to p . Suppose exit 1 of M_1 is accessible, but exit 2 of M_1 is not. Then the set of states of M^{acc} is the union of the set of non-exit states of M_1 which are accessible from the beginning states of M_1 , the set of states of M_2 which are accessible from the first beginning state of M_2 and the set of exits of M_2 .

(b) If M_2^{acc} is isomorphic to M_3^{acc}, then $(1M_2)^{acc}$ is isomorphic to $(1M_3)^{acc}$, where 1 is the machine from 1 to 2 consisting of two exits and whose beginning state is exit 1.

We call the quotient $\mathcal{M}_t[\Omega] = C_t[\Omega]/K$, the <u>category of accessible Ω-machine schemes</u>. We denote by $\mathcal{M}_{ct}[\Omega] = C_{ct}[\Omega]/K$ the subcategory of $\mathcal{M}_t[\Omega]$ consisting of morphisms which have cycle-free representatives, i.e., the image of the composite functor IN :

$$C_{ct}[\Omega] \xrightarrow{I} C_t[\Omega] \xrightarrow{N} \mathcal{M}_t[\Omega] ,$$

where I is the inclusion functor.

<u>Corollary 5.2</u>. $\mathcal{S}_F$ is a subcategory of $\mathcal{M}_{ct}[\Omega]$.

From (5.3) and Theorem 4.3, we obtain

<u>Theorem 5.3</u>. Each family of functions

$$\Omega_n \to T([1],[n]) , \quad n > 0 ,$$

has a unique extension to a strict functor

$$\mathcal{M}_{ct}[\Omega] \to T$$

which keeps $\mathcal{S}_F$ pointwise fixed.

The category $\mathcal{A} = \mathcal{M}_t[\Omega]$ or $\mathcal{A} = \mathcal{M}_{ct}[\Omega]$ comes close to being an algebraic theory. Not only does it contain $\mathcal{S}_F$ as a subcategory, but it has the property:

(5.4) for every family of morphisms $\varphi_i: [1] \to [p]$, $i \in [n]$, in $\mathcal{A}$, there is a morphism $\varphi: [n] \to [p]$ in $\mathcal{A}$ such that $i\varphi = \varphi_i$ for every $i \in [n]$.

Note, though, if $G: \mathcal{A} \to T$ is a strict functor which keeps $\mathcal{S}_F$ elementwise fixed, and if the hypothesis of (5.5) is satisfied, then

$$iG(\varphi) = G(i)G(\varphi) = G(i\varphi) = G(i\psi) = G(i)F(\psi) = iG(\psi) .$$

Since (5.5) is satisfied in T, and since $iG(\varphi) = iG(\psi)$ for every $i \in [n]$, we conclude that $G(\varphi) = G(\psi)$. This motivates the introduction of the congruence L.

Let L be the smallest congruence on $\mathcal{M}_t[\Omega]$ which satisfies for φ,ψ: $[n] \to [p]$: if $i\varphi = i\psi$ for each $i \in [n]$ then $(\varphi,\psi) \in L_{[n],[p]}$. It is clear that $\mathcal{M}_t[\Omega]/L$ is an algebraic theory--the theory of Ω-machine schemes. We fix attention on the subtheory $\mathcal{M}_{ct}[\Omega]/L$ of cycle-free machine schemes.

Theorem 5.4. The theory of cycle-free total Ω-machine schemes is (theory-isomorphic to) $\mathcal{S}_F[\Omega]$, the algebraic theory freely generated by Ω.

Proof. From (5.3) and Theorem 5.3, we have: each family of functions

$$\Omega_n \to T([1],[n]) \ , \quad n>0 \ ,$$

has a unique extension to a strict functor

$$\mathcal{M}_{ct}[\Omega]/L \to T$$

which keeps $\mathcal{S}_F$ pointwise fixed. But this property characterizes $\mathcal{S}_F[\Omega]$.

We have seen that every interpretation G: $\mathcal{C}_{ct}[\Omega] \to T$ admits a factorization

$$G\colon \mathcal{C}_{ct}[\Omega] \xrightarrow{N_1} \mathcal{M}_{ct}[\Omega] \xrightarrow{N_2} \mathcal{S}_F[\Omega] \xrightarrow{H} T \ ,$$

where H is a theory-morphism. It will be noted later (section 8) that there is a theory-morphism J: $\mathcal{S}_F[\Omega] \to [X]_P$, for an appropriate set X, which is injective, i.e., preserves distinctness of morphisms. It follows, therefore, $\bigcap_G \mathcal{K}(G)$ as G runs through all interpretations G: $\mathcal{C}_{ct}[\Omega] \to [X]_P$ is exactly $\mathcal{K}(N_1N_2)$.

6. 0-Object Theories

By a 0-object theory, we shall mean an algebraic theory T such that the set $T([1],[0])$ consists of a single morphism 0: $[1] \to [0]$. It follows that each set $T([n],[0])$ consists of a single morphism 0: $[n] \to [0]$. In any theory T, $T([0],[p])$ consists of a single morphism 0: $[0] \to [p]$. Thus $[0]$ is a 0-object in the usual categorical sense. For each n, p there is a unique morphism 0: $[n] \to [p]$ in a 0-object theory which factors through $[0]$, i.e., 0: $[n] \to [0] \to [p]$.

The theories $[X]$ and $[X]_P$, but not $[X]_F$, are 0-object theories. We wish to describe and show the existence of free 0-object theories, for they will be related to machine schemes.

Let $\Omega = \{\Omega_n\}$, $n > 0$, and let $\Omega' = \{\Omega_n\}$, $n \geq 0$. The following assertion is evident.

(6.1) Each family of functions

$$\Omega_n \to T([1],[n]) \ , \quad n > 0 \ ,$$

where T is a 0-object theory, has a unique extension to a theory-morphism

$$\mathcal{S}_F[\Omega'] \to T \ .$$

If K is the kernel of the theory morphism and $\varphi, \varphi' \colon [n] \to [p]$, factor through $[0]$, then $(\varphi, \varphi') \in K_{[n],[p]}$.

Let $Z = \{Z_{[n],[p]}\}$ be the family of relations defined by: $(\varphi, \varphi') \in Z_{[n],[p]}$ iff φ and φ' both factor through $[0]$. Then Z is a theory-congruence (cf., Eilenberg and Wright (1967)), and $\mathcal{S}_F[\Omega']/Z$ is an algebraic theory. If $\Omega_0 \neq \phi$, then this quotient is a 0-object theory $\mathcal{S}_P[\Omega]$ (which clearly does not depend upon the cardinality of Ω_0). From (6.1), we infer

(6.2) each family of functions

$$\Omega_n \to T([1],[n]) \ , \quad n > 0 \ ,$$

where T is a 0-object theory, has a unique extension to a theory-morphism

$$\mathcal{S}_P[\Omega] \to T \ .$$

The theory $\mathcal{S}_P[\Omega]$ is <u>freely generated</u> as a 0-object theory by Ω. If $\varphi \colon [n] \to [p]$ is in the subtheory $\mathcal{S}_F[\Omega]$ of $\mathcal{S}_F[\Omega']$, it does not factor through $[0]$. Thus $\mathcal{S}_F[\Omega]$ is a subtheory of $\mathcal{S}_P[\Omega]$.

We shall require the following assertions:

<u>Theorem 6.1</u>. Let T be a 0-object theory, and let $\Omega \subset T$. Suppose Ω generates T as a 0-object theory, and suppose bijective maps $\Gamma_n \to \Omega_n$, $n > 0$, are given. If T also satisfies (6.3) below, then the unique extension of these maps to a theory-morphism $\mathcal{S}_P[\Gamma] \to T$ is bijective.

(6.3) If $\varphi \in T([1],[p])$, $\varphi \neq 0$, admits factorizations

$$\varphi: [1] \xrightarrow{\omega} [m] \xrightarrow{\psi} [p]$$

$$\varphi: [1] \xrightarrow{\omega'} [n] \xrightarrow{\psi'} [p] ,$$

where $\omega \in \Omega_m$, $\omega' \in \Omega_n$, then $m = n$, $\omega = \omega'$, $\psi = \psi'$.

Theorem 6.2. It is possible to assign to each morphism $\psi: [n] \to [p]$ in $\mathcal{S}_p[\Omega]$ a non-negative integer $\deg \psi$ satisfying:

(6.4) $\psi \in \mathcal{S}_p \Leftrightarrow \deg \psi = 0$;

(6.5) $\deg \psi = \Sigma_i \deg i \psi$;

(6.6) if $\psi \neq 0$ and $\omega \in \Omega_n$, then $\deg \omega \psi = 1 + \deg \psi$.

A variant of Theorem 6.2:

Theorem 6.2'. It is possible to assign to each morphism $\psi: [n] \to [p]$ in $\mathcal{S}_p[\Omega]$ a non-negative integer $ht(\psi)$ satisfying (6.4), (6.6), and

(6.5') $ht\, \psi = \sup_i ht(i\psi)$.

7. The Embedding Theorem

Let $T = \mathcal{S}_p[\Omega]$. For each $j \in [p]$, define $e_j: [p] \to [1]$ by the requirements

(7.1) $je_j = [1]: [1] \to [1]$, $je_k = 0: [1] \to [1]$ if $j \neq k$.

Let $\Sigma = \{\omega e_j \mid \omega \in \Omega_p , j \in [p]\}$. Now $T([1],[1])$ is a monoid under composition; let M be the smallest submonoid of $T([1],[1])$ containing Σ .

Proposition 7.1. The monoid M is freely generated by Σ .

Proof. Let $x \in M$, and suppose $x \neq 0$. Then, by (7.1), (6.4), (6.5), we have $\deg(e_j x) = \deg x$ so that, by (6.4), $e_j x \neq 0$ and $\deg(\omega e_j x) = 1 + \deg x$ by (6.6) and again by (6.4) $\omega e_j x \notin \mathcal{S}_p$. It follows $0 \notin M$. If $\omega e_j x = \omega' e_k y \neq 0$, we infer by (6.3) that $\omega = \omega'$ and $e_j x = e_k y \neq 0$. Thus $je_k y = je_j x = x \neq 0$. By (7.1), $j = k$ so that $x = y$. It follows that Σ freely generates M .

Let $\hat{M}$ be the monoid whose elements X are subsets of M and whose multiplication is inherited from M . Specifically, if $X, Y \subset M$, then $XY = \{xy \mid x \in X , y \in Y\}$. Let $X + Y \subset M$ be the union of X and Y . We form a category $[\hat{M}]$ whose objects are $[n]$, $n \geq 0$, and whose morphisms from $[n]$ to $[p]$ are $n \times p$ matrices whose entries are elements of $\hat{M}$. Composition of morphisms in $[\hat{M}]$ is taken as ordinary matrix multiplication. The mapping which takes $f: [n] \to [p]$ in $\mathcal{S}_F$ into the $n \times p$ matrix (f_{ij}) defined by

$$f_{ij} = \{1\} \text{ , if } f(i) = j$$
$$f_{ij} = \emptyset \text{ , otherwise ,}$$

embeds $\mathcal{S}_F$ in $[\hat{M}]$. Given row matrices $\varphi_i: [1] \to [p]$, $i \in [n]$, the $n \times p$ matrix φ whose ith row is φ_i is described by the requirement $i\varphi = \varphi_i$ for each $i \in [n]$. Thus $[\hat{M}]$ is an algebraic theory.

Theorem 7.2. The unique theory morphism

$$G: \mathcal{S}_P[\Omega] \to [\hat{M}]$$

which takes $\omega \in \Omega_p$ into the $1 \times p$ matrix

$$[\omega e_1, \omega e_2, \ldots, \omega e_p]$$

is injective.

Proof. The image T of G is a 0-object algebraic theory which is theory-generated by $G(\Omega)$, together with $0: [1] \to [0]$. Let $\varphi \in T([1],[p])$, $\varphi \neq 0$, and suppose

$$\varphi: [1] \xrightarrow{G(\omega)} [m] \xrightarrow{\psi_{ij}} [p]$$

$$\varphi: [1] \xrightarrow{G(\omega')} [n] \xrightarrow{\psi'_{kj}} [p] \text{ .}$$

Since $\varphi \neq 0$, for some j, $\Sigma_i \omega e_i \psi_{ij} = \Sigma_k \omega' e_k \psi'_{kj} \neq \emptyset$. Thus a word in this set has as first "letter" $\omega e_i = \omega' e_k$ for some i,k. It follows $m = n$, $\omega = \omega'$, and $i = k$.

It is now easy to deduce that the matrices (ψ_{ij}) and $\psi'_{kj})$ are equal. By Theorem 6.1, T is isomorphic to $\mathcal{S}_p[\Omega]$.

We note in passing that if $F: \mathcal{S}_P[\Omega] \to [X]_P$ is a theory-morphism, then the sequence

$$[F(\omega e_1), F(\omega e_2), \ldots, F(\omega e_p)]$$

consists of partial functions whose domains are pairwise disjoint.

8. Ω-Machine Schemes

Let $\Omega = \{\Omega_n\}$, $n > 0$ and $\Omega' = \{\Omega_n\}$, $n \geq 0$. There is no difficulty in defining as in section 4 Ω'-machine schemes, even though Ω_0 may be non-empty. They determine a category $C_t[\Omega']$ as in section 4. The kernel of any strict functor of $C_t[\Omega']$ into a 0-object theory T contains the congruence Z which identifies any two morphisms $[n] \to [p]$ which factor through $[0]$. The quotient category $C[\Omega] = C_t[\Omega']/Z$ depends only upon Ω, since $C_t[\Omega']$ possesses morphisms $[1] \to [0]$. The quotient $C_{ct}[\Omega']/Z$ depends not only upon Ω, but also upon whether or not $\Omega_0 = \emptyset$. If $\Omega_0 = \emptyset$, then $C_{ct}[\Omega']$ has no morphisms from [1] to [0].

In the case $\Omega_0 = \emptyset$, the quotient $C_{ct}[\Omega']/Z$ was previously called $C_{ct}[\Omega]$. In case $\Omega_0 \neq \emptyset$, we call the quotient $C_c[\Omega]$. As before, we define $\mathcal{M}_c[\Omega] = C_c[\Omega]/K$ and, in analogy with Theorem 5.4, we have:

Theorem 8.1. The theory $\mathcal{M}_c[\Omega]/L$ of cycle-free Ω-machine schemes is (theory isomorphic) to $\mathcal{S}_P[\Omega]$, the 0-object theory freely generated by Ω.

We now fill the gap left at the end of section 5. Let W be the underlying set of the monoid M of the previous section.

Theorem 8.2. The unique theory-morphism $\mathcal{S}_P[\Omega] \to [W]_P$, which extends the family of functions

$$\Omega_p \to [W]_P([1],[p])$$

defined by "$\omega \in \Omega_p$ goes into the partial function $\omega e_j x \mapsto (x,j)$, $j \in [p]$", is injective.

The reader may wish to verify this assertion. See Proposition 12.2 of Elgot (1970).

Corollary 8.3. There is an injective theory-morphism $\mathcal{S}_F[\Omega] \to [W]_P$.

9. Concluding Remarks

Let $\varphi: [n] \to [p]$ in $C_c[\Omega]$. The morphism φ, together with an assignment $\Omega \to T$ where $T = [X]$ "corresponds" to "cycle-free program". According to the intuitive discussion of section 2, the "external behavior", defined via a notion of "computation", satisfies: (a) the external behavior of a composite of programs is the composite of their external behaviors; (b) the external behavior of an f-like program is f . According to theorem 4.3, an interpretation F is determined by an assignment $\Omega \to T$; by virtue of (a) and (b), the external behavior of the program determined by φ and this assignment is $F(\varphi)$. Thus "external behavior" may be characterized without the intervention of "computation".

Machines with cycles may be treated as in Elgot (1970) by making use of a suitable subtheory of $[\hat{M}]$.

STRUCTURE AND MEANING OF ELEMENTARY PROGRAMS

by

Erwin Engeler*

We know that every program for the register machine of Shepherdson-Sturgis (1963) computes a partial recursive function. The usual textbook proofs go through a process of arithmetization to secure this fact, but for all its other merits this process obscures rather than enlightens the interconnection between the program and the definition of the corresponding μ-recursive function.

How should we go about finding a better method, one that could be taught to undergraduates as an effective capability for "reading off" the definition of a function from any given program? It was realized early by several people that the proper abstract framework to attack this and similar problems was that of considering programs over quite arbitrary relational structures, without regard to the elements or the relations and functions being given "constructively" in any sense. This attitude is taken in various theories of program schemes such as Yanov (1958), McCarthy (1963 a), Floyd (1967 a), and others. The contribution of the present author flowed out of two observations.

First, we should realize the complexity of connectivity of possible programs, say, as visualized by the graph of its flow-chart. Our first goal should be to bring structure into this amorphous mass of more and more complicated programs. We show in the normal form theorem 1.1 below that all programs can be generated (up to strong equivalence) by two simple processes of program composition and modification.

Second, we do need a formal language (other than the one in which we write the programs) for expressing various relations and attributes. For example, we would

*Research supported in part by National Science Foundation grant number GP-8829.

like to express the following relations in a formal language which admits their symbolic handling:

$$\Phi_\pi(x)$$

[true iff the program π terminates on input x] ,

$$\Psi_\pi(x,y)$$

[true iff the program π terminates on input x with output y] ,

$$\Phi^i_\pi(x,y)$$

[true iff y is the present content of registers whenever computation passes through the ith link of the program π and the input was x].
But we should beware of artificial restraints that we impose upon ourselves by making too simpleminded a choice for formalisms. The most natural requirement to make of an appropriate language is that its logical connectives reflect the processes of program composition and modification: as the programs are built up, so is the corresponding formula, and vice versa. The language that results is infinitary--but it is infinitary only in the inessential, constructive sense that makes the power series $\sum_{k=0}^{\infty} \frac{x^k}{k!}$ an infinite formula.[1]

For some purposes the logical languages approach is not needed at all. In particular, the problem of finding the recursive definition corresponding to a program is solved very directly using the normal form theorem. Another type of problem which can be attacked directly is that of proving termination, correctness, and equivalence of programs. We shall give a complete proof theory solely in terms of manipulations of programs.

1. Normal Form Theorem

Let $\underline{A} = \langle A; R_1, R_2, \ldots; f_1, \ldots; c_1, \ldots \rangle$ be a relational structure, i.e., let A be a non-empty set. R_i are n_i-ary relations on A , $i = 1, \ldots, m$; f_j are m_j-ary

[1]The first exposition of the formal language approach is in Engeler (1967). This approach has been used in Engeler (1968) to give a survey of the basic notions and results of abstract computer science, and in Engeler (1969, 1970) to treat questions of approximation and accuracy.

operations on A , $j=1,\ldots,n$; c_k are distinguished elements of A , $k=1,\ldots,k$. We assume that the relations R_i and operations f_j are constructive in the sense that the basic <u>instructions</u> below are executable whatever assignment of elements of A to the variables $x_1,x_2,\ldots$ is given.

r: <u>if</u> $R_i(x_{k_1},\ldots,x_{k_{n_i}})$ <u>then go to</u> p <u>else to</u> q ;

r: <u>do</u> $x_i := f_j(x_{k_1},\ldots,x_{k_{m_j}})$ <u>then go to</u> p ;

r: <u>do</u> $x_i := c_k$ <u>then go to</u> p ;

r: <u>do</u> $x_i := x_j$ <u>then go to</u> p ;

r: <u>go to</u> p .

The r,p,q are numerals, used as labels for the instructions and for <u>go to</u> directions. <u>Programs</u> are finite sets of instructions, no two of which have the same label. In each program one instruction is singled out as the <u>entrance</u>, that is the first instruction to be followed. If p occurs in the context <u>go to</u> p in a program without being the label of an instruction, then p is called an <u>exit</u>.

{All our subsequent work carries over, with trivial modifications, to <u>non-deterministic</u> programs. These arise if we allow instructions of the form

r: <u>if</u>...<u>then go to</u> p_1 <u>or</u> p_2 <u>or</u>...<u>or</u> p_m <u>else go to</u> q_1 <u>or</u> q_2 <u>or</u>...<u>or</u> q_n ;
r: <u>do</u>...<u>then go to</u> p_1 <u>or</u> p_2 <u>or</u>...<u>or</u> p_k .}

We also present programs in the form of directed graphs in the familiar manner. This allows us to indicate composition and other manipulation of program structures in an easily mnemonic manner.

The <u>meaning</u> of a program π is obtained by finding a mathematical characterization of the relation

$$\Psi_\pi(x,y) ,$$

where Ψ_π is true of x and y iff the program π terminates with output y if it is given input x . (x and y are in general finite sequences of elements of the relational structure considered; they represent the assignment of such elements to the variables occurring in the program at the beginning and end of the execution of the program.)

For some programs, such as

1: go to 1   (program ν)

the meaning is obvious: the program never halts, and the relation is empty.

Two programs, π_1 and π_2, are called equivalent (for a given relational structure $\underline{A}$) if

$$\Psi_{\pi_1}(x,y) \text{ is the same relation as } \Psi_{\pi_2}(x,y) ;$$

they are strongly equivalent if they are equivalent for all relational structures on which the programs are defined.

After these preliminary definitions, we now come to the central notion of this paper: normal forms for programs. Intuitively speaking, a program is in normal form if its flow diagram has the form of a tree in which some leaves are bent back to earlier nodes of the branch on which they sit. For example:

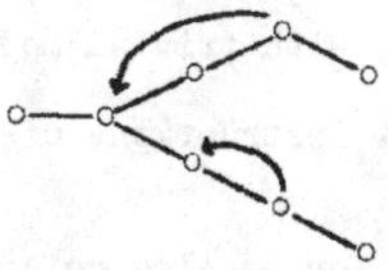

More formally, we define the class of programs in normal form recursively as the smallest class of programs such that

(i) each program consisting of only one instruction is in normal form;

(ii) if

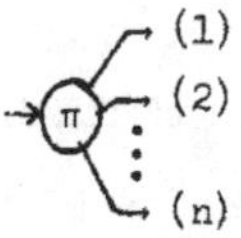

is in normal form and has n exits, and if $\pi_1,\ldots,\pi_n$ are also in normal form, then the composition

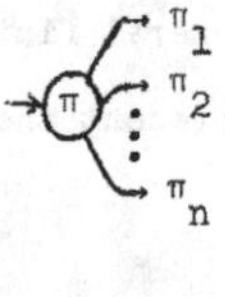

is a program in normal form;

(iii) if

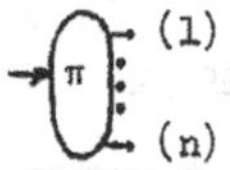

is in normal form, then so is

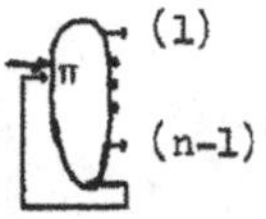

the result of leading one of the exits back to the entrance of the program.

Theorem 1.1. For every program π there exists a program π' in normal form such that π is strongly equivalent to π'.

The proof of this theorem exhibits an effective procedure to obtain π' from π; it proceeds by induction on the number of instructions in π. Assume that for each program of length $< n$ we can find an equivalent one in normal form which has the same number of exits and such that the two programs remain equivalent if corresponding exits are led to ν. Now let π be a program with n instructions, and suppose that the added instruction is of the form

k: do ψ then go to p .

Let

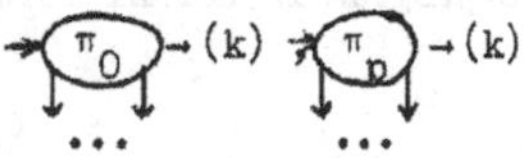

be the programs that are obtained from π by dropping the instruction with label k and choosing as entrance, respectively, the same entrance as π and the instruction p. Clearly the composite program

is equivalent to π. It remains to use the induction assumption on π_0, π_p, replacing these programs by programs π_0', π_p' in normal form. We then obtain a program π' in normal form which is equivalent to π. In case the added instruction is of the form

k: <u>if</u> φ <u>then</u> <u>go</u> <u>to</u> p <u>else</u> <u>go</u> <u>to</u> q ,

we argue similarly. (To be quite precise, we have to distinguish the various <u>go</u> <u>to</u> instructions in π_0 which lead to k and keep them separate, thus π_0 is really of the form

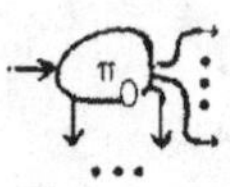

and the composition should look like this:

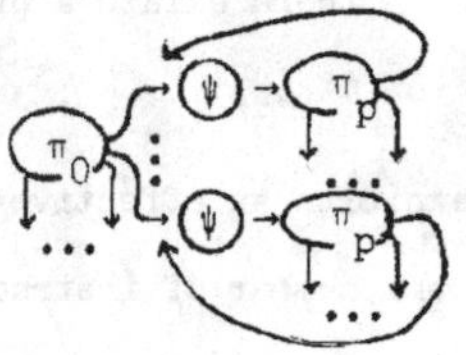

.)

By a <u>simple</u> <u>program</u>, usually denoted by σ, we understant a program of the form

$$\rightarrow \pi_1 \rightarrow \ldots \rightarrow \pi_n \rightarrow$$

where each π_i is a program consisting of one instruction only and having only one exit.

<u>Corollary 1.2</u>. Each non-simple program in normal form is equivalent to a program in one of the following forms:

(a) 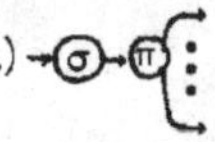 where π has at least two exits;

(b)  .

2. <u>Definitions from Programs</u>

A very straightforward application of the normal form theorem 1.1 gives us a new proof of the fact that all functions computed on the register machine of Shepherdson-Sturgis (1963) are partial recursive.

Thus let π be a given program for the register machine. We may assume that π is in normal form and proceed by induction to the structure of π as follows.

<u>Induction assumption</u>. There exists a partial recursive function f_π and a partial recursive function g_i for each exit i of π such that

(i) $f_\pi(x)$ is defined iff π halts on x ; then $f_\pi(x) = \pi(x)$;

(ii) for each i

$$g_i(x) = \begin{cases} 0 & \text{if exit } i \text{ is taken} \\ 1 & \text{if some other exit is taken} \\ \text{undefined otherwise} . \end{cases}$$

The induction assumption is easily checked for the case that π consists of one instruction only.

In the case of composite programs, consider first the composition of 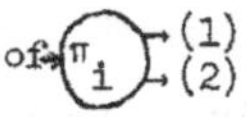 and 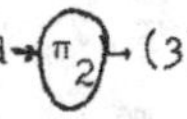 , resulting in 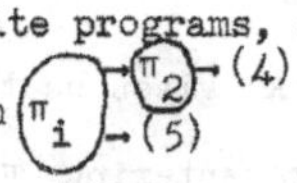 . Let f_1 , f_2 be the functions computed by the first two programs, and let g_1 , g_2 , g_3 be the "characteristic functions" associated to the respective exits of these programs. Then f , g_4 , g_5 are found as follows.

Let

$$g_4(x_1,\dots,x_n) = \begin{cases} 0 \text{ if } g_1(x_1,\dots,x_n) = 0 \text{ and } g_3(f_1(x_1,\dots,x_n)) = 0 \\ 1 \text{ if } g_1(x_1,\dots,x_n) = 1 \\ \text{undefined otherwise} , \end{cases}$$

$$g_5(x_1,\dots,x_n) = \begin{cases} 0 \text{ if } g_1(x_1,\dots,x_n) = 0 \\ 1 \text{ if } g_1(x_1,\dots,x_n) = 0 \text{ and } g_3(f_1(x_1,\dots,x_n)) = 0 \\ \text{undefined otherwise} . \end{cases}$$

These functions are easily composed from g_1 , g_2 , g_3 , f_1 , f_2 , and f_3 by means of well-known (primitive) recursive functions. Let

$$d(x) = \begin{cases} 1 & \text{if } x = 0 \\ 0 & \text{otherwise} . \end{cases}$$

Then, for example,

$$f(x) = d(g_4(x)) \cdot f_2(f_1(x))) + d(g_5(x)) \cdot f_1(x) .$$

Next, let us consider the case of looping a program 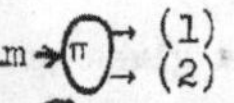 resulting in  (3) . Let f_0 be the function associated to 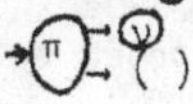 and f_1 the function associated to 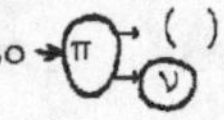 . Let e be the zero function, and let h be defined by

$$h(y,x_1,\ldots,x_n) = f_0^{(y)}(x_1,\ldots,x_n) ,$$

the y-fold iteration of f_0 . Finally, let k be defined by

$$k(x_1,\ldots,x_n) = (\mu y)[g_1(h(y,x_1,\ldots,x_n)) = 0] .$$

Now let us put

$$f(x_1,\ldots,x_n) = f_1(h(k(x_1,\ldots,x_n),x_1,\ldots,x_n)) ;$$

$$g(x_1,\ldots,x_n) = e(f(x_1,\ldots,x_n)) .$$

Namely, for input $(x_1,\ldots,x_n)$, the function k gives us the smallest number y of times that the loop has to be run through before entering π for the last time, since under these circumstances $h(y,x_1,\ldots,x_n)$ is the n-tuple of values before this last input, and $g_1(h(y,x_1,\ldots,x_n),x_1,\ldots,x_n)$ is zero for the first time. {The general cases are inessential variants of the cases considered here.}

<u>Remark</u>. The above proof is the specialization to register machines of a proof for the generalization of the theorem in question to arbitrary algorithmic bases in Engeler (1968), Chapter 4.

<u>Example</u>. The technique used in the proof above allows us to assign "meanings" to programs for register machines in a mechanical fashion. For the following program, this meaning could easily be guessed, but, as has been often remarked, this guesswork is far from trivial for programs that have any reasonable complexity at all.

The program π

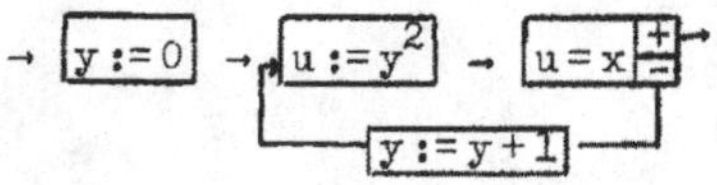

is in normal form. The composition sequence of π ends with the programs π_1 , π_2 , π_3 below. We apply the ideas of the above proof to each of these programs in turn as follows.

π_1 :

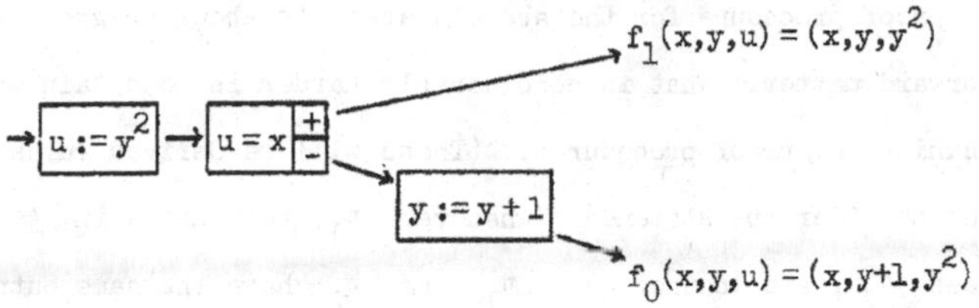

Let $g(x,y,u) = \{ {0 \text{ if } x = y^2 \atop 1 \text{ otherwise}}$. This function is used as characteristic function for the two exits of π_1 .

π_2 :

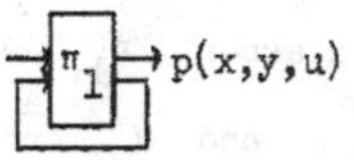

Let $h(z;x,y,u) = f_0^{(z)}(x,y,u) = (x, y+z, (y+z-1)^2)$ and

$$\begin{aligned} k(x,y,u) &= (\mu z)[g(h(z;x,y,u)) = 0] \\ &= (\mu z)[g(x, y+z, (y+z-1)^2) = 0] \\ &= (\mu z)[x = (y+z)^2] \ . \end{aligned}$$

Then

$$\begin{aligned} p(x,y,z) &= f_1(h(k(x,y,u);x,y,u)) \\ &= f_1(h((\mu z)[x=(y+z)^2] \ ; x,y,u)) \\ &= f_1(x, y+(\mu z)[x=(y+z)^2] \ , (y+(\mu z)[x=(y+z)^2])^2) \\ &= (x, y+(\mu z)[x=(y+z)^2] \ , (y+(\mu z)[x=(y+z)^2])^2) \ . \end{aligned}$$

π_3 :

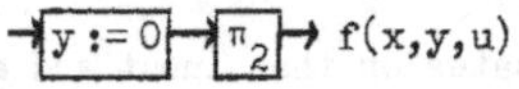

$$f(x,y,u) = P(x,0,u) = (x, (\mu z)[x=z^2] \ , ((\mu z)[x=z^2])^2) \ .$$

Hence $y := (\mu z)[x = z^2]$ is the function computed by π $(=\pi_3)$.

3. A Complete Proof System

There is a great variety of statements that we may wish to prove about a program; we mention the termination problem, the problem of equivalence of programs, and the work that went into finding proof procedures for "certifying" programs by giving formal proofs of correctness. Our goal here is to illustrate that obtaining a complete proof procedure for the above statements about programs is a rather straightforward matter. What is considerably harder is to obtain practically workable, mechanizable, proof procedures. (These will be derived rules of our system.)

Let us consider the statement "whenever π_1 terminates in $\underline{A}$ for some input, then so does π_2, and in that case π_1 and π_2 have the same output". We abbreviate this statement by writing

$$\pi_1 \underset{\underline{A}}{\Vdash} \pi_2 .$$

For very simple programs π_1 and π_2, we often can assert $\pi_1 \underset{\underline{A}}{\Vdash} \pi_2$ offhand by inspection. The problem becomes interesting when π_1 and π_2 are more complex. We may assume, without loss of generality, that all programs to be considered are in normal form. Then we may hope that statement $\pi_1 \underset{\underline{A}}{\Vdash} \pi_2$ <u>derives</u> from statements between the <u>components</u> of π_1 and π_2 in a manner analogous to the proof procedures of mathematical logic. That this is indeed so is the content of theorem 3.1 below.

To formulate the appropriate axioms and rules of proof, we need the following conventions:

Let M and N be countable (finitely or denumerably infinite) sets of programs. We write

$$M \underset{\underline{A}}{\Vdash} N$$

if, for every input on which all programs in M terminate with the same output, there is a program in N which also terminates on that input and also has that same output. ν denotes, as before, the non-terminating program 1: <u>go</u> <u>to</u> 1 . σ denotes a simple program.

Finally, if

is a program containing a loop as indicated and if k is any positive integer, then

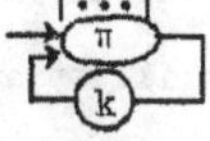

denotes the following program

k times

The format of our proof system is that of Gentzen's natural deduction: There are formulated a number of axioms or axiom schemes of the form $M \Vdash_{\underline{A}} N$ and some rules of proof of the form

$$\frac{M_1 \Vdash_{\underline{A}} N_1 \, , \, M_2 \Vdash_{\underline{A}} N_2 \, , \ldots}{M \Vdash_{\underline{A}} N}$$

which allow to conclude the statement $M \Vdash_{A} N$ if the statements $M_i \Vdash_{\underline{A}} N_i$, $i = 1,2,\ldots$, have already been deduced. A <u>proof</u> is a tree-like structure; at the leaves are axioms, the nodes connect statements that are related by a rule of proof, and all branches are of finite length. The statement at the root of a proof-tree is said to be provable. In this fashion we parallel the semantical notion $\Vdash\!\!=_{\underline{A}}$ with the proof-theoretic, manipulatory notion $\Vdash_{A}$. Our goal is to obtain:

<u>Theorem 3.1</u>. For any countable sets M, N of programs, we have $M \models\!\!\models_{\underline{A}} N$ if and only if $M \Vdash_{\underline{A}}$ is provable.

In our search for the appropriate axioms and rules of proof for this theorem to hold, we observe the pattern that underlies the formulation of rules in other Gentzen-type proof systems: To each syntactical formation rule (for building up expressions that go into M and N) there correspond two rules of proof. The first rule indicates under what hypotheses

$$M_1 \Vdash_{\underline{A}} N_1 \, , \ldots \, .$$

The conclusion

$$M \Vdash_{\underline{A}} N$$

can be deduced when the formation rule in question is the rule last applied to get a given expression in M; the second rule deals with the case that this expression is in N . In logical proof systems of this sort, the M_i and N_i consist of expressions out of which the expressions in M and N are composed (often by the very syntactical formation rule that gave rise to the rule of proof); in any case, the expressions in M_i and N_i are at most as complex as the expressions in M and N , and some are less so. Thus, the axioms consist only of expressions of minimal complexity. This, roughly, is the pattern after which the proof system below is fashioned.

We recall the observation that every non-simple program in normal form may be assumed to have one of the following two forms (corollary 1.2):

(a) 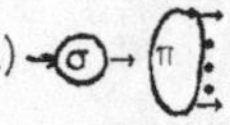 where π has at least two exits, or

(b) 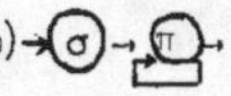 .

Thus it is natural to take as "expressions of minimal complexity" the simple programs

A. <u>Axioms</u>. Whenever M , N are countable sets of simple programs such that $M \VDash_{\underline{A}} N$, then $M \Vdash_{\underline{A}} N$ is an axiom.

B. <u>Rules of Proof</u>.

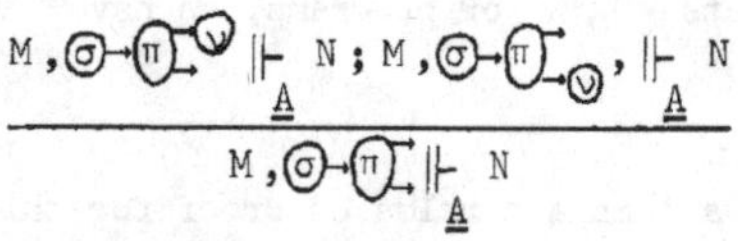

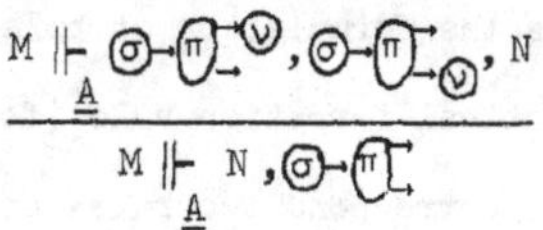

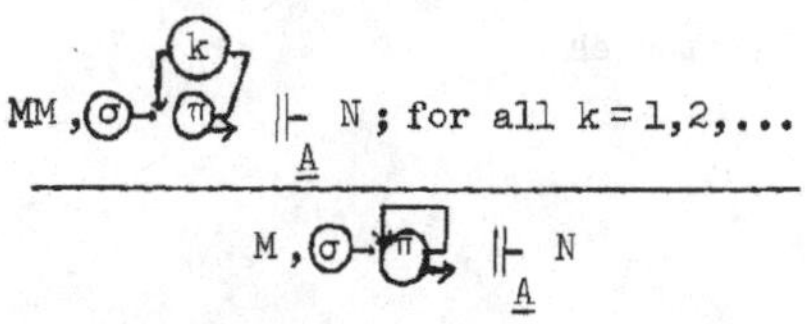

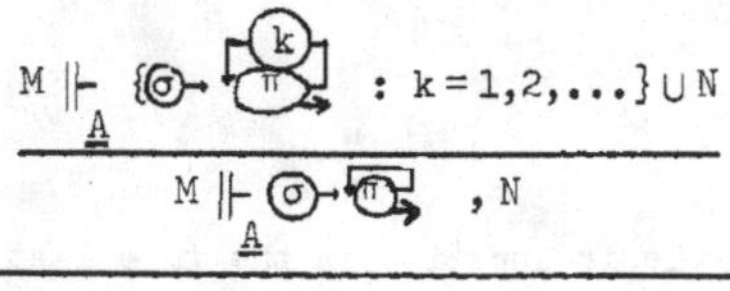

TABLE 1

σ , and as syntactical formation rules the ones indicated in (a), (b) above. This suggests the proof system of Table 1 above. The proof of theorem 3.1 is easily adapted from proofs of the similar theorem in the translation to infinitary logic, for example in Engeler (1968).

PROCEDURES AND PARAMETERS: AN AXIOMATIC APPROACH

by

C. A. R. Hoare

Introduction

It has been suggested, Hoare (1969), that an axiomatic approach to formal language definition might simultaneously contribute to the clarity and reliability of programs expressed in the language, and to the efficiency of their translation and execution on an electronic digital computer. This paper gives an example of the application of the axiomatic method to the definition of procedure and parameter passing features of a high-level programming language. It reveals that ease of demonstrating program correctness and high efficiency of implementation may be achieved simultaneously, provided that the programmer is willing to observe a certain familiar and natural discipline in his use of parameters.

CONCEPTS AND NOTATIONS

The notations used in this paper are mainly those of symbolic logic and particularly natural deduction. They are supplemented by conventions introduced in Hoare (1969). The more important of them are summarized below.

1. $P\{Q\}R$ --where P and R are propositional formulae of logic and Q is a part of a program. Explanation: If P is true of the program variables before executing the first statement of the program Q, and if the program Q terminates, then R will be true of the program variables after execution of Q is complete.

2. S^x_e --where S is an expression or formula, x is a variable, and e is an expression. Explanation: The result of replacing all free occurrences of x in S by e. If e is not free for x in S, a preliminary systematic change of bound variables of S is assumed to be made.

3. $\frac{A, B}{C}$ --where A, B, and C are propositional formulae. Explanation: A rule of inference which states that if A and B have been proved, then C may be deduced.

4. $\frac{A, B \vdash C}{D}$ --where A, B, C, and D are propositional formulae. Explanation: A rule of inference which permits deduction of D if A and C are proved; however it also permits B to be assumed as a hypothesis in the proof of C. The deduction of C from B is known as a subsidiary deduction.

It is assumed that, with the exception of program material (usually enclosed in braces), all letters stand for formulae of some suitably chosen logical system. The formulae of this system are presumed to include:

(a) all expressions of the programming language;

(b) the familiar notations of predicate calculus (truth-functions, quantifiers, etc.).

The properties of the basic operands, operators and built-in functions of the language are assumed to be specified by some suitably chosen axiom set; and the proof procedures are assumed to be those of the first-order predicate calculus.

As a simple example of the use of these notations, let Q stand for the single assignment statement $k := (m+n) \div 2$. We wish to prove that after execution of this statement, k will take a value between m and n, whenever $m \leq n$; or more formally that the desired result R of execution of Q is $m \leq k \leq n$. In Hoare (1969) there was introduced the axiom schema

$$R^x_e \; \{x := e\} \; R \; . \qquad \text{Axiom of Assignment}$$

This indicates that for an assignment $x := e$, if R is the desired result of the assignment, a sufficient precondition for this result to obtain is that R^x_e is true before execution. R^x_e is derived by replacing all occurrences of the target variable x in R by the assigned expression e. In the present case, the target variable is k, and the assigned expression is $(m+n) \div 2$. Thus we obtain, as an instance of the axiom schema,

$$m \leq (m+n) \div 2 \leq n \; \{k := (m+n) \div 2\} \; m \leq k \leq n \; .$$

It is an obvious theorem of mathematics that $m \leq (m+n) \div 2 \leq n$ can be inferred from the truth of $m \leq n$. This may be written using the notation explained above

$$m \leq n \vdash m \leq (m+n) \div 2 \leq n .$$

Another obvious rule mentioned in Hoare (1969) states that if S{Q}R has been proved, and also that the truth of S may be inferred from the truth of P, then P{Q}R is a valid inference, or more formally

$$\frac{S\{Q\}R,\ P \vdash S}{P\{Q\}R} \quad . \qquad \text{Rule of Consequence}$$

In applying this rule to our example, we take $m \leq n$ for P and $m \leq (m+n) \div 2 \leq n$ for S, and obtain

$$m \leq n \ \{k := m+n \div 2\}\ m \leq k \leq n ,$$

which is what was required to be proved.

A full list of rules of inference, together with associated conventions, is given in the Appendix. They are not supposed to give a "complete" proof procedure for correctness of programs, but they will probably be found adequate for the proof of most practical algorithms.

PROCEDURES

Before embarking on a treatment of parameters, it is convenient first to consider the simpler case of a procedure without parameters. Suppose p has been declared as a parameterless procedure, with body Q. We introduce the notation

p <u>proc</u> Q

to represent this declaration. A call of this procedure will take the form

<u>call</u> p .

It is generally accepted that the effect of each call of a procedure is to execute the body Q of the procedure in the place of the call. Thus if we wish to prove that a certain consequence R will follow from a call of p (under some precondition

P), all we have to do is to prove that this consequence will result from execution of Q (under the same precondition). This reasoning leads to an inference rule of the form

$$\frac{p\ \underline{\text{proc}}\ Q\ ,\ P\{Q\}R}{P\ \{\underline{\text{call}}\ p\}\ R} \quad . \qquad \text{Rule of Invocation}$$

PARAMETERS

In dealing with parametrization, we shall treat in detail the case where all variables of the procedure body Q (other than locally declared variables) are formal parameters of the procedure. Furthermore, we shall make a clear notational distinction between those formal parameters which are subject to assignment in the body Q , and those which do not appear to the left of an assignment in Q nor in any procedure called by it.

These decisions merely simplify the discussion; they do not involve any loss of generality, since any program can fairly readily be transformed to one which observes the conventions.

Let $\underline{x}$ be a list of all non-local variables of Q which are subject to change by Q . Let $\underline{v}$ be a list of all other non-local variables of Q . We extend the notation for procedure declarations, thus

$$p(\underline{x}) : (\underline{v})\ \underline{\text{proc}}\ Q\ .$$

This asserts that p is the name of a procedure with body Q and with formal parameters $\underline{x}$, $\underline{v}$.

The notation for a procedure call is similarly extended:

$$\underline{\text{call}}\ p(\underline{a}) : (\underline{e})\ .$$

This is a call of procedure p with actual parameters $\underline{a}$, $\underline{e}$ corresponding to the formals $\underline{x}$, $\underline{v}$, where $\underline{a}$ is a list of variable names, and $\underline{e}$ is a list of expressions; and $\underline{a}$ is the same length as $\underline{x}$, and $\underline{e}$ is the same length as $\underline{v}$.

As before, we assume that P{Q}R has been proved of the body Q . Consider the call

$$\underline{\text{call}}\ p(\underline{x}) : (\underline{v})$$

in which the names of the formal parameters have been "fed back" as actual parameters, thus effectively turning the procedure back into a parameterless one. It is fairly obvious that this call has the same effect as the execution of the procedure body itself; thus we obtain the rule

$$\frac{p(\underline{x}) : (\underline{v})\ \underline{\text{proc}}\ Q\ ,\ P\{Q\}R}{P\ \{\underline{\text{call}}\ p(\underline{x}) : (\underline{v})\}\ R} \quad . \qquad \text{Rule of Invocation}$$

Of course, this particular choice of actual parameters is most unlikely to occur in practice; nevertheless, it will appear later that the rule of invocation is a useful basis for further advance.

Consider next the more general call

$$\underline{\text{call}}\ p(\underline{a}) : (\underline{e}) \ .$$

This call is intended to perform upon the actual parameters $\underline{a}$ and $\underline{e}$ exactly the same operations as the body Q would perform upon the formal parameters $\underline{x}$ and $\underline{v}$. Thus it would be expected that $R\,\frac{\underline{x}\,,\,\underline{v}}{\underline{a}\,,\,\underline{e}}$ would be true after execution of the call, provided that the corresponding precondition $P\,\frac{\underline{x}\,,\,\underline{v}}{\underline{a}\,,\,\underline{e}}$ is true before the call. This reasoning leads to the rule

$$\frac{P\ \{\underline{\text{call}}\ p(\underline{x}) : (\underline{v})\}\ R}{P\,\frac{\underline{x}\,,\,\underline{v}}{\underline{a}\,,\,\underline{e}}\ \{\underline{\text{call}}\ p(\underline{a}) : (\underline{e})\}\ R\,\frac{\underline{x}\,,\,\underline{v}}{\underline{a}\,,\,\underline{e}}} \quad . \qquad \text{Proposed Rule of Substitution}$$

Unfortunately, this rule is not universally valid. If the actual parameter list $\underline{a}\,,\,\underline{e}$ contains the same variable more than once, the proof of the body of the subroutine is no longer valid as a proof of the correctness of the call. This may be shown by a trivial counter example, contained in Table 1. In order to prevent such contradictions, it is necessary to formulate the conditions that all variables in $\underline{a}$ are distinct, and none of them is contained in $\underline{e}$. We shall henceforth insist that every procedure call satisfy these readily tested conditions, and thus reestablish the validity of the rule of substitution. In a programming language standard, we shall see later that there are other reasons for leaving undefined the effect of a procedure call which fails to satisfy the conditions.

Counterexample

Assume:	$p(x) : (v)$ <u>proc</u> $x := v+1$		(1)
	$v+1=v+1$ $\{x := v+1\}$ $x=v+1$	[Assignment]	(2)
	<u>true</u> $\vdash v+1=v+1$	[Logical theorem]	(3)
From 2, 3:	<u>true</u> $\{x := v+1\}$ $x=v+1$	[Consequence]	(4)
From 1, 4:	<u>true</u> {<u>call</u> $p(x) : (v)$} $x=v+1$	[Invocation]	(5)
From 5:	<u>true</u> {<u>call</u> $p(a) : (a)$} $a=a+1$	[Substitution]	(6)

Since the conclusion is an obvious contradiction, we must prohibit calls of the form <u>call</u> $p(a) : (a)$.

TABLE 1

As an example of the successful use of the rule of substitution, assume that a declaration has been made,

$$\text{random } (k) : (m, n) \ \underline{\text{proc}} \ Q \ ,$$

where Q is a procedure body of which it has been proved

$$m \leq n \ \{Q\} \ m \leq k \leq n \ .$$

The rule of invocation permits deduction of

$$m \leq n \ \{\underline{\text{call}} \text{ random } (k) : (m, n)\} \ m \leq k \leq n \ ,$$

and applying the rule of substitution to a particular call, it is possible to obtain

$$1 \leq q+1 \ \{\underline{\text{call}} \text{ random } (r) : (1, q+1)\} \ 1 \leq r \leq q+1 \ .$$

In some cases it is necessary to use a slightly more powerful rule of substitution. Suppose P and/or R contain some variables <u>k</u> which do not occur in <u>x</u> or <u>v</u> , but which happen to occur in <u>a</u> or <u>e</u> . In such a case it is necessary first to substitute some entirely fresh variables, <u>k'</u> for <u>k</u> in P and R , before applying the rule given above. This is justified by a more powerful version of the rule

$$\frac{P\ \{\underline{\text{call}}\ p(\underline{x}) : (\underline{v})\}\ R}{P^{\underline{k}\,,\,\underline{x}\,,\,\underline{v}}_{\underline{k}'\,,\,\underline{a}\,,\,\underline{e}}\ \{\underline{\text{call}}\ p(a) : (e)\}\ R^{\underline{k}\,,\,\underline{x}\,,\,\underline{v}}_{\underline{k}'\,,\,\underline{a}\,,\,\underline{e}}\ .} \qquad \text{Rule of Substitution}$$

DECLARATION

In most procedures it is highly desirable to declare that certain of the variables are local to the procedure, or to some part of it, and that they are to be regarded as distinct from any other variables of the same name which may exist in the program. Since local variables do not have to be included in parameter lists, a considerable simplification in the structure of the program and its proof may be achieved. We will introduce the notation for declarations,

begin new x ; Q **end** ,

where x stands for the declared variable identifier, and Q is the program statement (scope) within which the variable x is used; or, in ALGOL terms, the block to which it is local.

The effect of a declaration is merely to introduce a new working variable, and its introduction is not intended to have any effect on any of the other variables of the program; it cannot therefore affect the truth of any provable assertion about these variables.

Thus in order to prove

P {**begin new** x ; Q **end**} R ,

all that is in principle necessary is to prove the same property of the body of the block, namely,

P{Q}R .

However, this rule is not strictly valid if the variable x happens to occur in either of the assertions P or R . In this case, the validity of the rule can be reestablished by first replacing every occurrence of x in Q by some entirely fresh variable y , which occurs neither in P , Q , nor R . It is a general property of declarations that such a systematic substitution can have no effect on the meaning of the program. Thus the rule of declaration takes the form

$$\frac{P\ \{Q^x_y\}\ R}{P\ \{\underline{new}\ x\,;\ Q\}\ R}\ , \qquad \text{Rule of Declaration}$$

where y is not free in P or R , nor does it occur in Q (unless y is the same variable as x).

In practice it is convenient to declare more than one variable at a time, so that the rule of declaration needs to be strengthened to apply to lists $\underline{x}$ and $\underline{y}$ rather than single variables.

RECURSION

The rules of inference given above are not sufficient for the proof of the properties of recursive procedures. The reason is that the body Q of a recursive procedure contains a call of itself, and there is no way of establishing what are the properties of this recursive call. Consequently, it is impossible to prove any properties of the body Q . This means that it is impossible to use even the simple rule of invocation,

$$\frac{p\ \underline{proc}\ Q\ ,\ P\{Q\}R}{P\ \{\underline{call}\ p\}\ R}\quad ,$$

since the proof of the second premise P{Q}R remains forever beyond our grasp.

The solution to the infinite regress is simple and dramatic: to permit the use of the desired conclusion as a hypothesis in the proof of the body itself. Thus we are permitted to prove that the procedure body possesses a property, on the assumption that every recursive call possesses that property, and then to assert categorically that every call, recursive or otherwise, has that property. This assumption of what we want to prove before embarking on the proof explains well the aura of magic which attends a programmer's first introduction to recursive programming.

In formal terms, the rule of invocation for a recursive procedure is

$$\frac{p(\underline{x}) : (\underline{v})\ \underline{proc}\ Q,\quad P\{\underline{call}\ p(\underline{x}) : (\underline{v})\}\ R \vdash P\ \{Q\}R}{P\ \{\underline{call}\ p(\underline{x}) : (\underline{v})\}\ R} \qquad \text{Rule of Recursive Invocation}$$

Unfortunately, this relatively simple rule is not adequate for the proof of the properties of recursive procedures. The reason is that it gives no grounds for supposing that the local variables of the procedure (other than those occurring in the left hand parameter list) will remain unchanged during a recursive call. What is required is a rather more powerful rule which permits the assumed properties of a recursive call to be adapted to the particular circumstances of that call. The formulation of a rule of adaptation is designed in such a way as to permit a mechanically derived answer to the question, "If S is the desired result of executing a procedure call, call p(a) : (e) , and P {call p(a) : (e)} R is already given, what is the weakest precondition W such that W {call p(a) : (e)} S is universally valid?"

It turns out that this precondition is

$$\exists \underline{k}(P \wedge \forall \underline{a}\ (R \supset S)) \text{ ,}$$

where $\underline{k}$ is a list of all variables free in P , R but not in $\underline{a}$, $\underline{e}$, or S . This fact may be formalized

$$\frac{P\ \{\underline{\text{call}}\ p(\underline{a}) : (\underline{e})\}\ R}{\exists \underline{k}(P \wedge \forall \underline{a}\ (R \supset S))\ \{\text{call } p(a) : (e)\}\ S \text{ .}} \qquad \text{Rule of Adaptation}$$

In the case where $\underline{k}$ is empty, it is understood that the $\exists$ will be ommitted.

The rule of adaptation is also extremely valuable when applied to non-recursive procedures, since it permits a single proof of the properties of the body of a procedure to be used again and again for every call of the procedure. In the absence of recursion, the rule of adaptation may be justified as a derived inference rule, since it can be shown that every theorem proved with its aid could also have been proved without it. However, successful use of the rule of adaptation to simplify proofs still depends on observance of the conditions of the disjointness of actual parameters.

Example

As an example of the application of the rules given above, we will take the trivial, but familiar, problem of the computation of the factorial $\underline{r}$ of a non-negative integer a . The procedure is declared:

```
fact (r) : (a) proc
        if a = 0 then r: = 1
              else begin new w ;
                       call fact (w) : (a - 1) ;
                       r: = axw
              end .
```

It is required to prove that

$$a \geq 0 \ \{\underline{\text{call}} \ \text{fact} \ (r) : (a)\} \ r = a! \quad \dots \ (I) \ .$$

This is achieved by proving

$$a \geq 0 \ \{Q\} \ r = a!$$

where Q stands for the body of the procedure, on the hypothesis that (I) already holds for the internal recursive call. The proof is given in Table 2; it contains a number of lemmas which can be readily proved as theorems in the arithmetic of integers. A list of inference rules used is contained in the Appendix.

Proof of Factorial Program

1.	$a \times w = a! \ \{r := a \times w\} \ r = a!$	D0
2.	$a - 1 \geq 0 \wedge \forall w (w = (a-1)! \supset a \times w = a!) \{\underline{\text{call}} \ \text{fact} \ (w) : (a-1)\} a \times w = a!$	D6, D7, Hypothesis
3.	$a > 0 \vdash a - 1 \geq 0 \wedge \forall w \ (w = (a-1)! \supset a \times w = a!)$	Lemma 1
4.	$a > 0 \ \{\underline{\text{begin}} \ \underline{\text{new}} \ w ; \underline{\text{call}} \ \text{fact} \ (w) : (a-1) ; r := a \times w \ \underline{\text{end}}\} \ r = a!$	D1, D2, D8 (1, 2, 3)
5.	$1 = a! \ \{r := 1\} \ r = a!$	D0
6.	$\underline{\text{if}} \ \ a = 0 \ \ \underline{\text{then}} \ \ 1 = a! \ \ \underline{\text{else}} \ \ a > 0 \ \{Q\} \ r = a!$	D4 (5, 4)
7.	$a \geq 0 \vdash \underline{\text{if}} \ \ a = 0 \ \ \underline{\text{then}} \ \ 1 = a! \ \ \underline{\text{else}} \ \ a > 0$	Lemma 2
8.	$a \geq 0 \ \{Q\} \ r = a!$	D1 (7, 6)
9.	$a \geq 0 \ \{\underline{\text{call}} \ \text{fact} \ (r) : (a)\} \ r = a!$	D5 (8)

TABLE 2

IMPLEMENTATION

It has been suggested by Floyd (1967 a) that a specification of proof techniques for a language might serve well as a formal definition of the semantics of that language, for it lays down, as an essential condition on any computer implementation, that every provable property of any program expressed in the language shall, in practice, obtain when the program is executed by the implementation. It is, therefore, interesting to inquire what implementation methods for parameter passing will be valid in a language whose definition includes the inference rules described in the previous sections. It appears that there is a wide choice of valid implementation methods, covering all the standard methods which have been used in implementations of widely familiar languages.

This means that each implementor of the language defined by these rules can make his selection of method in order to maximize the efficiency of his implementation, taking into account not only the characteristics of his particular machine, but also varying his choice in accordance with the properties of each particular procedure and each particular type of parameter; and he is free to choose the degree of automatic optimization which he will bring to bear, without introducing any risk that an optimized program will have different properties from an unoptimized one.

The remainder of this section surveys the various familiar parameter mechanisms, and assesses the circumstances under which each of them gives the highest efficiency. The reader may verify that they all satisfy the requirements imposed by the proof rules stated above, bearing in mind that every procedure call,

$$\underline{\text{call}}\ p(\underline{a}):(e)\ ,$$

conforms to the condition that all the parameters $\underline{a}$ are distinct from each other, and none of them appears in any of the expressions $\underline{e}$. The observance of this restriction is the necessary condition of the validity of most of the commonly used parameter passing methods.

1. Compile-time macro-substitution. Macro-substitution is an operation which replaces all calls of a procedure by a copy of the body of the procedure, after this copy has itself been modified by replacing each occurrence of a formal parameter

within it by a copy of the corresponding actual parameter. The normal process of translation to machine code takes place only after these replacements are complete. This technique has been comonly used for Assembly Languages and for Business Oriented Languages.

Macro-substitution will be found to be a satisfactory technique in any of the following circumstances:

(1) The procedure body is so short that the code resulting from macro-substitution is not appreciably longer than the parameter planting and calling sequence would have been.

(2) The procedure is called only once from within the program which incorporates it.

(3) The procedure is called several times, but on each occasion some of all of the parameters are identical. Substitution can be applied only to those parameters which are identical, leaving the remaining parameters to be treated by some run-time mechanism.

(4) In a highly optimizing compiler, macro-substitution will ensure not only that no time is wasted on parameter passing, but also that each call can be fully optimized in the knowledge of the identity and properties of its actual parameters.

The technique is not applicable to recursive procedures.

2. Run-time code construction. An alternative to substitution in the source code at compile time is the execution of a logically identical operation on the object code at run time. This involves planting the addresses of the actual parameters within the machine code of the procedure body on each occasion that the procedure is called. The technique may be favored whenever all the following conditions are satisfied:

(1) The computer has an instruction format capable of direct addressing of the whole store.

(2) The actual parameter is an array (or other large structure) which would be expensive to copy.

(3) The called procedure is not recursive.

(4) The called procedure contains at least one iteration.

This technique was used in FORTRAN implementations on the IBM 704/709 series of computers.

3. Indirect addressing. Before jumping to the procedure body, the calling program places the addresses of the actual parameters in machine registers or in the local workspace of the called procedure. Whenever the procedure refers to one of its parameters, it uses the corresponding address as an indirect pointer (modifier).

This technique is suitable in the following circumstances:

(1) The computer has an instruction format with good address-modification facilities.

(2) The actual parameter is an array (or other large structure) which would be expensive to copy.

If a single parameter mechanism is to be used in all circumstances, this is undoubtedly the correct one. However, on fast computers with slave stores, operand pre-fetch queues, or paging methods of storage control, it could cause some unexpected inefficiencies.

This technique is used in PL/1 and in many implementations of FORTRAN on the IBM 360 series of computers.

4. Value and result. Before jumping to the subroutine, the calling program copies the current values of the actual parameters into machine registers or local workspace of the called subroutine. After exit from the subroutine, the calling program copies back all values that might have been changed (i.e., those to the left of the colon in the actual parameter list).

This technique is to be preferred when either of the following conditions hold:

(1) The size of the parameter is sufficiently small that copying is cheap, accomplished in one or two instructions.

(2) The actual parameter is "packed" in such a way that it cannot readily be accessed by indirect addressing.

This technique is available in ALGOL W, and is used in several current implementations of FORTRAN.

5. Call by name (as in ALGOL 60). The calling program passes to the procedure the addresses of portions of code corresponding to each parameter. When the procedure wishes to access or change the value of a parameter, it jumps to the corresponding portion of code.

Since the restrictions on the actual parameters prohibit the use of Jensen's device, there is no reason why this technique should ever be used. It is difficult to envisage circumstances in which it could be more efficient than the other techniques listed.

CONCLUSION

It has been shown that it is possible by axiomatic methods to define an important programming language feature in such a way as to facilitate the demonstration of the correctness of programs and at the same time to permit flexibility and high efficiency of implementation. The combination of these two advantages can be achieved only if the programmer is willing to observe certain disciplines in his use of the feature, namely that all actual parameters which may be changed by a procedure must be distinct from each other, and must not be contained in any of the other parameters. It is believed that this discipline will not be felt onerous by programmers who are interested in the efficient and reliable solution of practical problems in a machine-independent fashion.

It is interesting to note that the discipline imposed is combination of the disciplines required by the ISO standard FORTRAN and by the IFIP recommended subset of ALGOL 60. The former insists on the distinctness of all parameters changed by a procedure, and the latter insists that each of them be an unsubscripted identifier.

Acknowledgement

The basic approach adopted in this paper was stimulated by an investigation reported in Foley (1969).

Appendix

P, P_1, P_2, R, S stand for propositional formulae.

Q, Q_1, Q_2 stand for program statements.

x, y stand for variable names (y not free in P or R).

e stands for an expression.

B stands for a Boolean expression.

p stands for a procedure name.

$\underline{x}$ stands for a list of non-local variables of Q which are subject to change in Q.

$\underline{v}$ stands for a list of other non-local variables of Q.

$\underline{a}$ stands for a list of distinct variables.

$\underline{e}$ stands for a list of expressions, not containing any of the variables $\underline{a}$.

$\underline{k}$ stands for a list of variables not free in $\underline{x}, \underline{v}$.

$\underline{k}'$ stands for a list of variables not free in $\underline{a}, \underline{e}, S$.

D0 $\quad R^x_e \; \{x := e\} \; R$ — Assignment

D1 $\quad \dfrac{P\,\{Q\}\,S,\; S \vdash R}{P\,\{Q\}\,R} \qquad \dfrac{P \vdash S,\; S\,\{Q\}\,R}{P\,\{Q\}\,R}$ — Consequence

D2 $\quad \dfrac{P\,\{Q_1\}\,S,\; S\,\{Q_2\}\,R}{P\,\{Q_1 ; Q_2\}\,R}$ — Composition

D3 $\quad \dfrac{P\,\{Q\}\,S,\; S \vdash \underline{\text{if}}\ B\ \underline{\text{then}}\ P\ \underline{\text{else}}\ R}{P\,\{\underline{\text{while}}\ B\ \underline{\text{do}}\ Q\}\,R}$ — Iteration

D4 $\quad \dfrac{P_1\,\{Q_1\}\,R,\; P_2\,\{Q_2\}\,R}{\underline{\text{if}}\ B\ \underline{\text{then}}\ P_1\ \underline{\text{else}}\ P_2\ \{\underline{\text{if}}\ B\ \underline{\text{then}}\ Q_1\ \underline{\text{else}}\ Q_2\}\,R}$ — Alternation

D5 $\quad \dfrac{p(\underline{x}) : (\underline{v})\ \underline{\text{proc}}\ Q,\; P\,\{\underline{\text{call}}\ p(\underline{x}) : (\underline{v})\}\,R \vdash P\,\{Q\}\,R}{P\,\{\underline{\text{call}}\ p(\underline{x}) : (\underline{v})\}\,R}$ — Recursion

D6 $\quad \dfrac{P\,\{\underline{\text{call}}\ p(\underline{x}) : (\underline{v})\}\,R}{P^{\underline{k},\,\underline{x},\,\underline{v}}_{\underline{k}',\,\underline{a},\,\underline{e}}\,\{\underline{\text{call}}\ p(\underline{a}) : (\underline{e})\}\,R^{\underline{k},\,\underline{x},\,\underline{v}}_{\underline{k}',\,\underline{a},\,\underline{e}}}$ — Substitution

D7 $\quad \dfrac{P\,\{\underline{\text{call}}\ p(\underline{a}) : (\underline{e})\}\,R}{\exists\, k'(P \wedge \forall \underline{a}(R \supset S))\,\{\underline{\text{call}}\ p(\underline{a}) : (\underline{e})\}\,S}$ — Adaptation

D8 $\quad \dfrac{P\,\{Q^x_y\}\,R}{P\,\{\underline{\text{new}}\ x ; Q\}\,R}$ (where y is not in Q unless y and x are the same) — Declaration

SEMANTICS OF ALGOL-LIKE STATEMENTS*

by

Shigeru Igarashi

1. INTRODUCTION

This paper is intended to describe an axiomatic approach to the semantics of Algol-like statements, which is mainly based on the axiomatic treatments of the equivalence of Algol-like statements by Igarashi (1964).

In section 2, the class of Algol-like statements of our concern is defined syntactically, in order to clarify the scope of the present paper, which class is essentially generated by simple variables of a type, go to statements, labels, assignment statements with a set of functions, if-then-else with a set of predicates, semicolons for concatenation, and parentheses to compose compound statements.

Besides McCarthy's operator, namely $(\rightarrow ,)$ for if-then-else, some notations different from usual ones will be introduced for the sake of conciseness, which will possibly help us to apply our mathematical intuition, though the writer has no intention of proposing such a notation for a general use. It must be noted that we use only different symbols and do not change the syntax. (Otherwise, it might become uncertain that we are working on algorithmic languages.)

We use a generalized inductive definition in order to define the class of our concern, which, although a little unnatural, constitutes a basis for defining and proving some things related to that class, by the help of the apparent induction principle induced by it.

In section 3, the interpretation (that might be seen to be already a kind of semantics) of the statements belonging to the above mentioned class is given, which

*The research reported here was supported in part by the Advanced Research Project Agency of the Office of the Department of Defense (SD-183).

is done using induction on the class, and the result has a somewhat analytical appearance. Actually, we shall define the interpretation as a function on the class into a certain set of partial functions, and, presumably, one can prove everything about these Algol-like statements using this function.

Some results included in the work by Manna and McCarthy (1970) will be taken into consideration, when we define the interpretation of conditional statements.

In section 4, categories of a kind whose objects are Algol-like statements, the interpretation being fixed, will be introduced in order to clarify the meaning of the relations which have been used in equivalence theories of Algol-like statements by Yanov (1958), Igarashi (1964), de Bakker (1968), etc., (McCarthy (1963 a) discussed the equivalence of conditional forms, which was also related to Algol-like statements, because the latter contain conditional statements), together with the correspondence between these relations and the notion of correctness introduced by Floyd (1967a) and refined by Manna (1968a, 1969a) which is also related to the discussions by Hoare (1969). (Also, cf. Kaplan (1968a).)

The relations $\underset{X}{\cong}$ and $\cong$ defined by Igarashi (1964) become special cases of isomorphisms in one of these categories. (On the one hand, these categories, whose objects are defined in sections 2 and 3, are intended to serve as a model of the formal system described in the later sections, though we shall not enter into this point. On the other hand, they can possibly be regarded as a basis for further algebraic theories concerning programs, as a branch of mathematical theory of computation.)

In section 5, a formal system representing the relations $\underset{X}{\cong}$ and $\cong$ will be presented, which is a revision of the main formal system (L.4) in the paper by Igarashi (1964), of which the latter will sometimes be called 'the previous system'. Besides minor refinements, it is so extended that partial functions and partial predicates may be allowed in statements and that the ability of the formalism may be considerably improved, although it is incomplete (which is inevitable). Expecially, inference rule 9 is new, for which McCarthy's notion of homomorphisms of programs (unpublished) and Floyd's above mentioned work are taken into consideration, as well as the obvious relationship between program schemata, firstly treated by Yanov (1958), and finite automata discussed by Igarashi (1963) and Rutledge (1964). This rule is, however, still a result of compromise between capability and simplicity.

Axioms related to go to statements have been entirely reformed.

In section 6, a number of elementary metatheorems concerning the formal system of section 5 are proved. These metatheorems show that any theorem in the previous system becomes a theorem also in the present system. Therefore, each of the completeness theorems for the previous system remains valid, though we shall not enter into this point.

It must be noted that the incompleteness of the formal system does not imply that this formalism gives only an inadequate description of semantics, for describing or defining the meaning of a program can be regarded as a rather special case of equivalence. In fact, for any Algol-like statement A (in the sense of section 2) in which variable symbols $x_1,\ldots,x_n$ occur, and for any variable-free arithmetic expressions (constants in effect) $c_1,\ldots,c_n,d_1,\ldots,d_n$, the following holds:

Let $\tilde{c}_1,\ldots,\tilde{c}_n,\tilde{d}_1,\ldots,\tilde{d}_n$ be the values corresponding to $c_1,\ldots,c_n,d_1,\ldots,d_n$, respectively. Then A stops and gives the final values $\tilde{d}_1,\ldots,\tilde{d}_n$ to $x_1,\ldots,x_n$, respectively, provided that the initial values of $x_1,\ldots,x_n$ are $\tilde{c}_1,\ldots,\tilde{c}_n$, respectively, <u>if and only if</u> the formula

$$x_1 := c_1; \ldots ; x_n := c_n ; A \cong x_1 := d_1 ; \ldots ; x_n := d_n$$

is provable in the formal system of our concern. (See theorem 55 by Igarashi (1964).)

Thus the formalism has an ability no less than the explicit definition of the interpretation given in section 3. (Namely, $J[A](\tilde{c}_1,\ldots,\tilde{c}_n,\iota) = (\tilde{d}_1,\ldots,\tilde{d}_n,\iota)$ if and only if the above formula is provable.)

In section 7, we shall define a special transformation of the class of Algol-like statements of our concern. On the one hand, this transformation can be regarded as a representation of a conceptual compiler. On the other hand, it demonstrates how the meaning of each statement can be defined in terms of certain primitive actions on a conceptual machine. (Therefore, this transformation itself might be regarded as a 'constructive' definition of semantics.)

In section 8, we shall formally prove the validity of the above transformation, (which mathematically means that each program is transformed into a program equivalent to it), in the system presented in section 5. On the one hand, this can be regarded as a kind of proof of compiler correctness (at least most of the essential features of the proof of compiler correctness being included), which has been done firstly by McCarthy and Painter (1967) for arithmetic expressions using induction on expressions. On the other hand, this can be regarded as a sufficient proof of the validity of the particular description of semantics in section 7 which is based on primitive actions. (Also, cf. Painter (1967).)

Notation and Terminology. We shall use the following notations and terminology.

1. Sets. Symbol $\emptyset$ denotes the null set. $S+S'$ denotes set $S \cup S'$ whenever $S \cap S' = \emptyset$. $\mathfrak{n} = \{0,1,2,\ldots,\}$. $\mathfrak{n}^+ = \{1,2,\ldots,\}$. $[0] = \emptyset$. If $n \geq 1$, then $[n] = \{1,2,\ldots,n\}$.

2. Functions. We shall use the word function to mean a possibly partial function.

   (f1) Expression $f\colon S \to S'$ reads as follows.

   (i) $f(a)$ may or may not be defined, for each $a \in S$.

   (ii) If $f(a)$ is defined, then $f(a) \in S'$.

   (iii) If $a \notin S$, then $f(a)$ is undefined.

   (f2) $\mathrm{Dom}\, f = \{a \mid f(a) \text{ is defined}\}$.

   (f3) Let $f\colon S \to S'$ and $S_0 \subseteq S$, then $f|S_0$ means the function g defined as follows: $g\colon S_0 \to S'$. $\mathrm{Dom}\, g = \mathrm{Dom}\, f \cap S_0$. $g(a) = f(a)$ for each $a \in \mathrm{Dom}\, g$.

   (f4) We note that $f|\mathrm{Dom}\, f$ is a total function for any f.

   (f5) $f = g$ means that f and g are defined on the same set and that $f|\mathrm{Dom}\, f = g|\mathrm{Dom}\, g$, while the latter equality means the equivalence of the total functions in the usual sense.

   (f6) If $f\colon S \to S'$ and $g\colon S' \to S''$, then $g \circ f$ or gf means the function h defined as follows: $h\colon S \to S''$. $\mathrm{Dom}\, h = \mathrm{Dom}\, f \cap \{a \mid f(a) \in \mathrm{Dom}\, g\}$. $h(a) = g(f(a))$ for each $a \in \mathrm{Dom}\, g$.

   (f7) If $f\colon S \to S$, then f^n denotes the function $f \circ \ldots \circ f$ (n times). $\lim_{n\to\infty} f^n$ means the function g defined as follows: $g\colon S \to S$. $a \in \mathrm{Dom}\, g$ if and only if there exists $M_a \in \mathfrak{n}$ such that $f^{M_a}(a) = f^{M_a+1}(a)$, so that $f^m(a) = f^n(a)$ for any $m \geq M_a$ and $n \geq M_a$. $g(a) = f^{M_a}(a)$ for each $a \in \mathrm{Dom}\, g$.

(f8) If $f: S \to S'$ and $g: S \to S'$, then $f+g$ means the function h defined as follows: $h: S \to S'$. $\mathrm{Dom}\, h = (\mathrm{Dom}\, f - \mathrm{Dom}\, g) \cup (\mathrm{Dom}\, g - \mathrm{Dom}\, f) \cup \{a \mid a \in \mathrm{Dom}\, f \cap \mathrm{Dom}\, g \text{ and } f(a) = g(a)\}$. $h(a) = \begin{cases} f(a), & a \in \mathrm{Dom}\, f \\ g(a), & a \in \mathrm{Dom}\, g - \mathrm{Dom}\, f \end{cases}$.

3. <u>Predicates</u>. We shall use the word predicate to mean a possibly partial predicate. We shall write $p(a) = T$, $p(a) = F$, and $p(a) = U$ to mean $p(a)$ is true, false, and undefined, respectively. For each predicate p, ∇p denotes the total predicate defined by

$$(\nabla p)(a) = \begin{cases} T, & p(a) = U \\ F, & \text{otherwise}. \end{cases}$$

Similarly, for each function f, ∇f denotes the total predicate defined by

$$(\nabla f)(a) = \begin{cases} T, & f(a) \text{ is undefined} \\ F, & \text{otherwise}. \end{cases}$$

(Here p and f are assumed to be unary and defined on a certain fixed set, for simplicity's sake.) Thus $(\nabla f)(a)$ means $\neg {*f(a)}$ used by Manna and McCarthy (1970), while we shall use $*$ for various purposes in the present paper.

4. <u>Truth Tables</u>. Since we are going to treat partial predicates, we have to define the meaning of logical connectives $\neg, \vee, \wedge, \supset$, and $\equiv$, for three-valued logic, for which we shall use the truth tables by Łukasiewicz (1941) denoted by Γ_{ℓ}, and that by McCarthy (1963 b) denoted by Γ_{m}. Γ_{ℓ} for the value U is as follows:

$(\neg U) = U$. $(U \wedge T) = (T \wedge U) = U$. $(U \wedge F)^* = (F \wedge U) = F$. $(U \wedge U) = U$. $(U \vee T)^* = (T \vee U) = T$. $(U \vee F) = (F \vee U) = U$. $(U \vee U) = U$. $(U \supset T)^* = T$. $(U \supset F) = U$. $(U \supset U) = U$. $(T \supset U) = U$. $(F \supset U) = T$. $(U \equiv T) = (T \equiv U) = F$. $(U \equiv F) = (F \equiv U) = F$. $(U \equiv U) = T$.

In Γ_{m} the asterisked members, the remaining members being the same, become as follows:

$(U \wedge F) = U$. $(U \vee T) = U$. $(U \supset T) = U$. ($F \wedge U = F$ and $T \vee U = T$.)

In order to indicate the truth tables considered, logical connectives will be suffixed by Γ_{ℓ} or Γ_{m}. Thus, for instance, $\wedge_{\Gamma_{m}}(U,F) = U$.

5. Structures. By a structure $\mathfrak{R}$, we shall mean a collection of functions and predicates defined on a set, which is called the underlying set of $\mathfrak{R}$ and denoted by $|\mathfrak{R}|$, together with that set. In the present paper, these functions and predicates are possibly partial. We shall consider two structures (or two similarity classes strictly) $\mathfrak{R}$ and $\mathfrak{A}$ in the text.

2. FORMATION OF ALGOL-LIKE STATEMENTS

Alphabet. Let $\mathcal{L}$, $\mathcal{V}$, $\mathcal{F}$, and $\mathfrak{R}$ be four disjoint sets whose elements are called label symbols, variable symbols, function symbols, and predicate symbols, respectively. The set $\mathcal{F}$ is the union of disjoint sets $\mathcal{F}^{(0)}, \mathcal{F}^{(1)}, \ldots,$ and the elements of $\mathcal{F}^{(n)}$ are called n-ary function symbols. Similarly, ρ is the union of disjoint sets $\rho^{(0)}, \rho^{(1)}, \ldots,$ and the elements of $\rho^{(n)}$ are called n-ary predicate symbols. The alphabet of Algol-like statements consists of all the elements of $\mathcal{L}$, $\mathcal{V}$, $\mathcal{F}$, and ρ, together with the following special symbols: Λ $^{-1}$ $:=$; ($\rightarrow$,) . In some cases described below, the logical symbols, $\neg \wedge \vee \forall \exists$, will be also contained.

Algol-like Statements. Algol-like statements, or statements, are defined together with a function denoted by $(\ )^-$, which sends each statement onto a finite subset of $\mathcal{L}$, by generalized inductive definition as follows:

Atomic Statements.

(a1) Λ is an atomic statement. $(\Lambda)^- = \phi$.

(a2) For each $\sigma \in \mathcal{L}$, σ and σ^{-1} are both atomic statements. $(\sigma)^- = \phi$. $(\sigma^{-1})^- = \{\sigma\}$.

(a3) For each $x \in \mathcal{V}$ and each $y \in \mathcal{V}$, $x := y$ is an atomic statement. $(x := y)^- = \phi$.

Statements. An atomic statement is a statement. Any other word on the above alphabet is a statement if and only if it is defined to be a statement by a repeated use of the following rules:

(b1) If A and B are two statements such that $(A)^- \cap (B)^- = \phi$, then A;B is a statement. $(A;B)^- = (A)^- + (B)^-$.

(b2) If $x := f_1, \ldots, x := f_n$ are n statements and $\pi^{(n)} \in \mathcal{F}^{(n)}$, then $x := \pi^{(n)} f_1 \ldots f_n$ is a statement. $(x := \pi^{(n)} f_1 \ldots f_n)^- = \phi$.

(b3) If $x:=f_1,\ldots,x:=f_n$, A , and B are $n+2$ statements such that $(A)^{-}\cap(B)^{-}=\phi$ and $\rho^{(n)}\in P^{(n)}$, then $(\rho^{(n)}f_1\ldots f_n\to A,B)$ is a statement. $((\rho^{(n)}f_1\ldots f_n\to A,B))^{-}=(A)^{-}+(B)^{-}$.

A statement which is defined to be so only by the above rules will be calles a <u>basic statement</u>.

(c1) If $(p\to A,B)$ is a statement, then $(\neg\, p\to A,B)$ is a statement. $((\neg\, p\to A,B))^{-}=((p\to A,B))^{-}$.

(c2) If $(p\to A,B)$ and $(q\to A,B)$ are two statements, then $(p\wedge q\to A,B)$ and $(p\vee q\to A,B)$ are both statements. The values of $(\ )^{-}$ are both identical with $((p\to A,B))^{-}$.

(c3) If $(p\to A,B)$ is a statement such that $x\in\mathcal{V}$ occurs in p and neither $\forall x$ nor $\exists x$ occurs in p , then $(\forall xp\to A,B)$ and $(\exists xp\to A,B)$ are both statements. The values of $(\ )^{-}$ are both identical with $((p\to A,B))^{-}$.

Parentheses and commas will be used also auxiliarly to avoid syntactic ambiguity and to improve readability. Expecially $\pi^{(n)}f_1\ldots f_n$ and $\rho^{(n)}f_1\ldots f_n$ are written as $\pi^{(n)}(f_1,\ldots,f_n)$ and $\rho^{(n)}(f_1,\ldots,f_n)$, respectively. Semicolons will be abbreviated if there is no possibility of ambiguity.

<u>Representation by ALGOL 60</u>. The statements in the above sense are intended to mean the statements in the sense of ALGOL 60 (Naur et al., 1960, 1963) as follows:

Λ corresponds to a dummy statement (empty).

σ corresponds to <u>go</u> <u>to</u> σ .

σ^{-1} corresponds to σ: (dummy statement labelled by σ) .

$(p\to A,B)$ corresponds to <u>if</u> p <u>then</u> A <u>else</u> B .

:= ; $\neg$ $\wedge$ and $\vee$ mean the same as in ALGOL 60.

The parentheses used to avoid ambiguity either correspond to <u>begin</u> and <u>end</u> delimiting compound statements or mean the same as in ALGOL 60.

$(A)^{-}$ denotes the set of labels standing in A .

Thus each statement can be regarded as a statement in the sense of ALGOL 60 in so far as neither $\forall$ nor $\exists$ occurs in that. Thus we shall call σ,σ^{-1}, f such that $x:=f$ is a statement, and p such that $(p\to A,B)$ is a statement, respectively, a go-to, a labelling, an arithmetic expression, and a Boolean expression.

Notations. Statements are denoted by A,B,C,... . Arithmetic expressions and Boolean expressions are denoted by f,g,h,... and p,q,r,... , respectively. Label symbols and variable symbols are denoted by $\sigma,\tau,\upsilon,\ldots$ and x,y,z,... , respectively. We shall use a number of functions and predicates defined on the statements which describe elementary syntactic properties. The function $(\ )^-$, being a typical example, was already defined in the above. All other functions and predicates listed below can be effectively defined in a similar manner.

a. Sets of Labels. By an occurrence of $\sigma \in \mathcal{L}$ in a statement A , we mean only such an occurrence as is different from the occurrences in the statements of the form σ^{-1} occurring in A .

$A^+ = \{\sigma | \sigma \text{ occurs in } A\}$.

$A^- = \{\sigma | \sigma^{-1} \text{ occurs in } A\}$.

$A^{\pm} = A^+ \cup A^-$.

$A^{++} = A^+ - A^-$.

$A^{-+} = \{\sigma | \sigma \in A^+ \cap A^- \text{ and } \sigma^{-1} \text{ occurs textually earlier than an occurrence of } \sigma \text{ in } A\}$.

Thus A^+ means the set of labels which are used for the purpose of designating the destinations of the go to statements occurring in A . If $A^{++} \neq \phi$, then the control may leave A by executing a go to statement whose destination is not within A . Such a go to statement will be called an exit of A . If $A^{-+} = \phi$, there are no loops in A .

b. Sets of Variables.

$V[A] = \{x | x \text{ occurs in } A\}$,

$V[f] = \{x | x \text{ occurs in } f\}$,

and

$V[p] = \{x | x \text{ occurs in } p\}$.

$L[A] = \{x | \text{ a statement of the form } x := f \text{ occurs in } A\}$.

R[A] is defined by induction as follows:

For each atomic statement such that $V[A] = \phi$, $R[A] = \phi$.

$R[x := f] = V[f]$.

$R[A;B] = R[A] \cup R[B]$.

$R[(p \to A,B)] = V[p] \cup R[A] \cup R[B]$.

Thus L[A] means the set of variables whose values may be changed by the execution of A , while R[A] means the set of variables whose values may affect the course of action and the results of executing A .

c. Substitution. Let $B_1,\dots,B_n$ and A be n+1 statements such that B_i occurs in A m_i times, $(m_i \geq 0)$, where the occurrences may be overlapped by each other unless they are not the same. Let $\hat{B}_i^j$, $j \in [m_i]$, denote the jth occurrence of B_i , where the order is defined by the position of the occurrence of the first symbol. Let $C_1,\dots,C_n$ be n statements. Then, by

$$A_{B_1,\dots,B_n}[C_1,\dots,C_n]$$

or (omitted commas)

$$A_{B_1\dots B_n}[C_1,\dots,C_n]$$

is meant an arbitrary statement that is obtained from A by substituting C_i for $\hat{B}_i^{h(i,1)},\dots,\hat{B}_i^{h(i,\ell_i)}$ for each $i \in [n]$, with the following restrictions:

$0 \leq \ell_i \leq m_i$.

$1 \leq h(i,1) < \dots < h(i,\ell_i) \leq m_i$.

The occurrence $\hat{B}_i^{h(i,j)}$ and $\hat{B}_{i'}^{h(i',j')}$ do not overlap each other, for any distinct pairs (i,j) and (i',j') .

The result of the substitution is a statement.

By

$$A_{B_1\dots B_n}[C_1,\dots,C_n]^o$$

is meant the unique statement that is obtained in the case that $\ell_i = m_i$, for every $i \in [n]$ in the above, which does not always exist because of the restriction concerning overlapping and the requirement that the result should be a statement.

We shall use the same notation also for arithmetic expressions and Boolean expressions.

d. Copies. Let $\sigma_1,\dots,\sigma_n$ be arbitrary distinct elements of $A^{\pm} - A^{++}$, and let $\tau_1,\dots,\tau_n$ be distinct and $\tau_i \notin A^{\pm}$ for any $i \in [n]$. Then

$$A_{\sigma_1 \ldots \sigma_n \sigma_1^{-1} \ldots \sigma_n^{-1}}[\tau_1, \ldots, \tau_n, \tau_1^{-1}, \ldots, \tau_n^{-1}]^{o}$$

is called a copy of A . If A_1 is a copy of A , and A_2 is a copy of A_1 , then A_2 is also called a copy of A . Copies of A are denoted by A', A", A"', ...

e. Go-to and Labelling.

A <u>begins with a labelling</u>, if A is of the form $\sigma^{-1}B$.

A <u>ends with a go-to</u>, if either A is of the form $B\sigma$ or A is of the form $(p \to B, C)$ and B and C both end with go-tos.

An occurrence of statement B in A is <u>preceded by a go-to</u> (equivalently, B is preceded by a go-to in A), if A is of the form $C_{\Lambda}[\sigma B]$.

3. INTERPRETATION OF ALGOL-LIKE STATEMENTS

By an interpretation of statements, we shall mean $(U, \varkappa, \mathfrak{R}, \Gamma^{o}, J)$ defined as follows.

Let U be a subset of $\mathcal{V}$, $\mathcal{A}_U$ the set of statements $\{A | V[A] \subseteq U\}$, and $\varkappa$ a bijection (i.e., 1-1 and onto function) such that

$$\varkappa: U \to I ,$$

where I is either [s] for an s , or $\mathfrak{N}^{+}$ in accordance with the cardinality of U . Let $\mathcal{L}^{\iota}$ denote $\mathcal{L} + \{\iota\}$, where ι is a new fixed symbol.

Let $\mathfrak{R}$ be a structure that satisfies the following conditions.

(1) $|\mathfrak{R}| \neq \phi$.

(2) For each $\pi^{(n)} \in \mathcal{F}^{(n)}$, an n-ary partial function denoted by $\pi_{\mathfrak{R}}^{(n)}$ is defined. I.e.,

$$\pi_{\mathfrak{R}}^{(n)}: |\mathfrak{R}|^{n} \to |\mathfrak{R}| .$$

(3) For each $\rho^{(n)} \in \mathcal{R}^{(n)}$, an n-ary partial relation denoted by $\rho_{\mathfrak{R}}^{(n)}$ is defined. I.e.,

$$\rho_{\mathfrak{R}}^{(n)}: |\mathfrak{R}|^{n} \to \{T, F\} .$$

The elements of $|\mathfrak{R}|$ will be denoted by $a_1, b_1, c_1, a_2, b_2, c_2, \ldots$. Thus by $\mathfrak{R}$ will be meant the total functions by which $\pi^{(n)} \mapsto \pi_{\mathfrak{R}}^{(n)}$ and $\rho^{(n)} \mapsto \rho_{\mathfrak{R}}^{(n)}$ as well as the structure itself, strictly.

Let Γ^{o} be a set of truth tables for logical connectives. Let $|\mathfrak{S}|$ denote $|\mathfrak{R}|^{s} \times \mathscr{L}^{\iota}$, i.e.,

$$|\mathfrak{R}| \times \underbrace{\ldots}_{s \text{ times}} \times |\mathfrak{R}| \times (\mathscr{L} + \{\iota\}) ,$$

if U is finite. The elements of $|\mathfrak{S}|$ will be denoted by a,b,c,... . For each $a \in |\mathfrak{S}|$ such that

$$a = (a_1, \ldots, a_s, \sigma)$$

and each $u \in U$, a_u denotes $a_{\varkappa(u)}$ and a_{χ} denotes σ . We write $(a)_u$ instead of a_u frequently for the readability's sake. If U is infinite, the infinite dimensional direct product $|\mathfrak{R}|^{I}$ will be used instead of $|\mathfrak{R}|^{s}$, namely, s is considered to be infinite.

The total function J defined below sends each statement $A \in \mathfrak{A}_U$ onto a partial function, $J[A]$, from $|\mathfrak{S}|$ into $|\mathfrak{S}|$. $J[A]$ will be written as $A_{\mathfrak{S}}$, thus

$$A_{\mathfrak{S}}: |\mathfrak{S}| \to |\mathfrak{S}| .$$

Two partial functions, one sending each arithmetic expression f such that $V[f] \subseteq U$ onto a partial function

$$f_{\mathfrak{S}}: |\mathfrak{S}| \to |\mathfrak{R}| ,$$

and the other sending each Boolean expression p such that $V[p] \subseteq U$ onto a partial predicate

$$p_{\mathfrak{S}}: |\mathfrak{S}| \to \{T, F\} ,$$

will be defined simultaneously for the readability's sake.

For a partial function

$$\varphi: |\mathfrak{S}| \to |\mathfrak{S}| ,$$

$\bar{\varphi}$ denotes the function defined by

$$\bar{\varphi}: |\mathfrak{D}| \to |\mathfrak{D}|$$

and

$$\bar{\varphi}(a) = \begin{cases} a & , a_\chi = \iota \\ \varphi(a) & , \text{otherwise} . \end{cases}$$

Definition of J .

The definition of $J[A]$, i.e., $A_\mathfrak{D}$, given in accordance with the last rule which should be used in order to define A to be a statement (section 2), which defines $J[A]$ for every $A \in \mathfrak{A}_U$ effectively, by the induction principle induced by the definition of statements, is as follows.

Atomic Statements.

(a1) $A = \Lambda$.

$$A_\mathfrak{D}(a) = a \quad \text{for any} \quad a \in |\mathfrak{D}| .$$

Hereafter the phrase like 'for any $a \in |\mathfrak{D}|$ ' will be omitted.

(a2) (i) $A = \sigma$.

$$(A_\mathfrak{D}(a))_u = \begin{cases} \sigma & , u = \chi \\ a_u & , u \in U \end{cases} , \text{ if } a_\chi = \iota ;$$

and

$$A_\mathfrak{D}(a) = a , \text{ otherwise} .$$

(ii) $A = \sigma^{-1}$.

$$(A_\mathfrak{D}(a))_u = \begin{cases} \iota & , u = \chi \\ a_u & , u \in U \end{cases} , \text{ if } a_\chi = \sigma ;$$

and

$$A_\mathfrak{D}(a) = a , \text{ otherwise} .$$

(a3) $A = x := y$.

$$y_\mathfrak{D}(a) = a_y .$$

$$(A_{\mathfrak{D}}(a))_u = \begin{cases} y_{\mathfrak{D}}(a) \ , \ u = x \\ a_u \ , \ u \in U - x \ \text{ or } \ u = \chi \end{cases} \ , \text{ if } \ a_\chi = \iota \ ;$$

and

$$A_{\mathfrak{D}}(a) = a \ , \text{ otherwise} \ .$$

Statements (non-atomic).

(b1) $A = B;C$.

$$A_{\mathfrak{D}}(a) = \lim_{n \to \infty} (\overline{C_{\mathfrak{D}} \circ B_{\mathfrak{D}}})^n ((C_{\mathfrak{D}} \circ B_{\mathfrak{D}})(a)) \ .$$

(b2) $A = x := \pi^{(n)} f_1 \ldots f_n$.

$$(\pi^{(n)} f_1 \ldots f_n)_{\mathfrak{D}}(a) = \pi^{(n)}_{\mathfrak{R}}(f_{1,\mathfrak{D}}(a), \ldots, f_{n,\mathfrak{D}}(a)) \ .$$

$$(A_{\mathfrak{D}}(a))_u = \begin{cases} (\pi^{(n)} f_1 \ldots f_n)_{\mathfrak{D}}(a) \ , \ u = x \\ a \ , \ u \in U - \{x\} \ \text{ or } \ u = \chi \end{cases} \ , \text{ if } \ a_\chi = \iota \ ;$$

and

$$A_{\mathfrak{D}}(a) = a \ , \text{ otherwise} \ .$$

(b3) $A = (\rho^{(n)} f_1 \ldots f_n \to B, C)$.

$$(\rho^{(n)} f_1 \ldots f_n)_{\mathfrak{D}}(a) = \rho^{(n)}_{\mathfrak{R}}(f_{1,\mathfrak{D}}(a), \ldots, f_{n,\mathfrak{D}}(a)) \ .$$

$$(1) \qquad A_{\mathfrak{D}}(a) = \begin{cases} \lim_{n \to \infty} (\bar{B}_{\mathfrak{D}} + \bar{C}_{\mathfrak{D}})^n \, B_{\mathfrak{D}}(a) \ , \ a_\chi = \iota \ \text{ and } \ (\rho^{(n)} f_1 \ldots f_n)_{\mathfrak{D}}(a) = T \ , \text{ or } \ a_\chi \in B^- \ ; \\ \lim_{n \to \infty} (\bar{B}_{\mathfrak{D}} + \bar{C}_{\mathfrak{D}})^n \, C_{\mathfrak{D}}(a) \ , \ a_\chi = \iota \ \text{ and } \ (\rho^{(n)} f_1 \ldots f_n)_{\mathfrak{D}}(a) = F \ , \text{ or } \ a_\chi \in C^- \ ; \\ a \ , \ a_\chi \notin B^- \cup C^- \cup \{\iota\} \ ; \\ \text{undefined} \ , \ \text{otherwise} \ . \end{cases}$$

(c1) $A = (\neg\, p \to B, C)$.

$$(\neg\, p)_{\mathfrak{D}}(a) = \neg_{\Gamma^{\circ}}(p_{\mathfrak{D}}(a)) \ .$$

(See section 1.)

$A_{\wp}$ if defined by the same rule as (1) of (b3) above, except that $(\rho^{(n)} f_1 \dots f_n)_{\wp}(a)$, occurring twice in it, should be replaced by $(\neg p)_{\wp}(a)$.

(c2) (i) $A = (p \wedge q \to B, C)$.

$$(p \wedge q)_{\wp}(a) = \wedge_{\Gamma^0}(p_{\wp}(a), q_{\wp}(a)).$$

$A_{\wp}$ is defined by the same rule as (1) of (b3) above, except that $(\rho^{(n)} f_1 \dots f_n)_{\wp}(a)$ should be replaced by $(p \wedge q)_{\wp}(a)$.

The case $A = (p \vee q \to B, C)$ as well as the case (c3) will be omitted, for it suffices to define $(p \vee q)_{\wp}$, $(\forall x p)_{\wp}$, and $(\exists x p)_{\wp}$ similarly and use (1) as the above.

Intuitive Meaning of J.

Practivally, $J[A]$, namely $A_{\wp}$, has the following meaning. We consider a computational process denoted by (A,a) as follows:

1. Suppose

$$a = (a_1, \dots, a_s, \sigma)$$

(s may be infinite). Assign the value $a_x = a_{\kappa(x)}$ to the variable x (identified with the variable symbol x) as the initial value for each $x \in U$.

2. Execute A from the point labelled by σ, while the leftmost point of A is chosen as the entry if $\sigma = \iota$, and if $\sigma \notin A^-$, then we consider A has no effect (i.e., identity transformation).

Then the following hold.

If the process (A,a) terminates at the exit whose destination is τ, giving the final value b_x to the variable x for each $x \in U$, then

$$(J[A](a))_x = b_x \quad \text{for each } x \in U$$

and

$$(J[A](a))_\chi = \tau,$$

and vice versa.

If (A,a) terminates at the normal exit, i.e., the rightmost point of A, then

$$(J[A](a))_\chi = \iota,$$

while the relationship concerning the values remains unchanged, and, if (A,a) does not terminate, then J[A] is undefined. The converse are also valid.

Choice of Γ^o.

As studied by Manna and McCarthy (1970), the choice of Γ^o is an important problem. We shall assume $\Gamma_{\mathcal{L}}$ as the foundation hereafter, unless we specify Γ^o. However, it must be noted that all the axioms of the formal system presented in section 5 are valid, whichever set of truth tables we may use. From the practical point of view, the processes of most implementations are related to $\Gamma_{\mathcal{M}}$ rather than to $\Gamma_{\mathcal{L}}$.

On the other hand, they make no difference in so far as all $f_{\mathfrak{R}}^{(n)}$ and $p_{\mathfrak{R}}^{(n)}$ are total and neither $\forall$ nor $\exists$ is involved, which is also the usual case when we consider actual ALGOL 60 programs which contain no recursive calls of procedures.

Remark.

Function J is an extension of J_1 for T_1-statements and J for T_2-statements (Igarashi (1964). For instance, J[A](a) defined above is identical with

$$(J_1[A](a_1,\ldots,a_s),\iota) .$$

The reader may notice that $|\mathfrak{R}|$ in the present paper corresponds to $\mathfrak{D}$ in that paper, while $\mathfrak{D}$ in this paper is used in a different meaning.

4. CATEGORY OF PROGRAMS

Programs in the General Sense

It seems to be convenient for us to consider more general programs as the background for the treatments of the properties of Algol-like statements. By a program, let us mean a partial function from an arbitrary set to another set together with its denotation. This definition does not exclude those partial functions which cannot be defined effectively. Instead, we shall describe it explicitly whenever the definability or constructiveness matters.

Programs will be denoted by A,B,C,... . For each A , J[A] denotes the partial function corresponding to A , and G[A] the graph of J[A] . Let D be an Algol-like statement such that $D \in \mathfrak{a}_U$ and $(U,\varkappa,\mathfrak{R},\Gamma^o,J)$ be an interpretation. Then the pair $(D, (U,\varkappa,\mathfrak{R},\Gamma^o,J))$ is a program, for a unique partial function J[D] , namely $D_{\mathfrak{D}}$, is determined by it. Therefore we shall assume the interpretation is fixed hereafter, so that each $D \in \mathfrak{A}_U$ represents a unique program. Thus we identify an Algol-like statement with the program represented by it, and the set of such programs will be denoted by $\mathfrak{A}$.

What we shall do firstly is almost the same as considering a subcategory of $\mathcal{E}ns$ (the category of sets) whose objects are graphs of partial functions. The only difference lies in that the denotations are distinguished in our treatments. For instance, we do not say A and B are identical nor $A = B$, even if $J[A] = J[B]$, while we may say A and B are isomorphic.

<u>Category $\mathcal{P}r$</u>

Each program will be called an <u>object</u> of category $\mathcal{P}r$. The class of all the objects, namely programs, is denoted by $\mathrm{Ob}\,\mathcal{P}r$. For each pair A and B belonging to $\mathrm{Ob}\,\mathcal{P}r$, $\mathrm{Hom}_{\mathcal{P}r}(A,B)$ denotes the set of triples of the form (A,ζ,B) such that

$$\zeta\colon G[A] \to G[B]$$

and that ζ is a <u>total</u> function. The elements of $\mathrm{Hom}_{\mathcal{P}r}(A,B)$ are called <u>morphisms</u> of $\mathcal{P}r$. If there is no possibility of confusion, the morphism (A,ζ,B) will be abbreviated by ζ . We frequently write $\zeta\colon A \to B$ or $A \xrightarrow{\zeta} B$ instead of $\zeta \in \mathrm{Hom}_{\mathcal{P}r}(A,B)$. If $A \xrightarrow{\xi} B \xrightarrow{\eta} C$, then $(A,\eta\xi,C) \in \mathrm{Hom}_{\mathcal{P}r}(A,C)$ is defined as the <u>composition</u> of morphisms (A,ξ,B) and (B,η,C) , where $\eta\xi$ in $(A,\eta\xi,C)$ denotes the composition of functions ξ and η in the usual sense. Let $\mathrm{id}_{G[A]}$ denote the identity function of $G[A]$ onto itself. The morphism $(A,\mathrm{id}_{G[A]},A)$ is called the identity morphism of A and is denoted by 1_A .

We shall see that $\mathcal{P}r$ satisfies the axioms of category as follows:

1. <u>Associativity of composition</u>. If

$$A \xrightarrow{\xi} B \xrightarrow{\eta} C \xrightarrow{\zeta} D ,$$

then $\zeta(\eta\xi) = (\zeta\eta)\xi$ as morphisms.

2. <u>Identity</u>. If $A \xrightarrow{\xi} B$, then $\xi = \xi 1_A$. If $C \xrightarrow{\eta} A$, then $\eta = 1_A\eta$.
3. If the pairs (A_1,B_1) and (A_2,B_2) are distinct, then

$$\mathrm{Hom}_{\mathcal{P}r}(A_1,B_1) \cap \mathrm{Hom}_{\mathcal{P}r}(A_2,B_2) = \phi \ .$$

Category $\mathcal{P}r^{\iota}$

Let $\mathcal{P}r^{\iota}$ denote the full subcategory of $\mathcal{P}r$ such that $\mathrm{Ob}\,\mathcal{P}r^{\iota}$ consists of only those programs A such that

$$\mathrm{Dom}(J[A]) \subseteq |\mathcal{D}^{\iota}| \ ,$$

where

$$|\mathcal{D}^{\iota}| = \{a \mid a \in |\mathcal{D}| \text{ and } a_{\chi} = \iota\} \ .$$

(See the below modification of J.) For each $A \in \mathrm{Ob}\,\mathcal{P}r^{\iota}$ and $B \in \mathrm{Ob}\,\mathcal{P}r^{\iota}$,

$$\mathrm{Hom}_{\mathcal{P}r^{\iota}}(A,B) = \mathrm{Hom}_{\mathcal{P}r}(A,B) \ ,$$

by definition (of full subcategory).

We consider a map

$$\mathrm{Ob}\,\mathcal{P}r \to \mathrm{Ob}\,\mathcal{P}r^{\iota}$$

which sends each $A \in \mathrm{Ob}\,\mathcal{P}r$ onto ${}_{\iota}A \in \mathrm{Ob}\,\mathcal{P}r^{\iota}$ such that

$$J[{}_{\iota}A] = J[A] \mid |\mathcal{D}^{\iota}| \ .$$

That is to say, we shall forget computational processes starting from any entry different from the normal one, namely the leftmost point, if A is an Algol-like program, modifying $J[A]$ into $J[A] \mid |\mathcal{D}^{\iota}|$.

Hereafter, we shall be concerned with $\mathcal{P}r^{\iota}$, so that $A,B,C,\ldots$ will be understood as ${}_{\iota}A, {}_{\iota}B, {}_{\iota}C,\ldots$ if the former do not belong to $\mathrm{Ob}\,\mathcal{P}r^{\iota}$. Apparently the morphism (A,ζ,B) is a <u>monomorphism</u>, <u>epimorphism</u>, or <u>isomorphism</u>, according as the function ζ is univalent (1-1), onto, or univalent and onto. We shall write $\zeta\colon A \tilde{\to} B$ or $A \overset{\zeta}{\tilde{\to}} B$ to express that $\zeta\colon A \to B$ is an isomorphism, and $A \simeq B$ to express that there is an isomorphism from A to B, namely A and B are isomorphic.

Value-Preserving Monomorphisms

We pay special attention to such a monomorphism ζ that has the following property:

Suppose $\zeta: A \to B$, and the function $\zeta: G[A] \to G[B]$ sends $(a,b) \in G[A]$ onto $(c,d) \in G[B]$ such that

$$a = c$$

and

$$b_u = d_u \quad \text{for each} \quad u \in X + \{x\} ,$$

for a subset X of U , for any $a \in |\mathfrak{D}^2|$.

In such a case, ζ (as a morphism and as a function) will be said to preserve the values of X , or to <u>preserve</u> X , and we shall frequently write ζ_X instead of ζ in order to indicate that ζ preserves X . Moreover, if the choice of ζ itself does not matter, we write $A \underset{X}{\to} B$ instead of $\zeta_X: A \to B$. Similarly, we shall frequently write $A \underset{X}{\cong} B$ or $A \underset{X}{\cong} B$ instead of $\zeta_X: A \cong B$, and $A \cong B$ instead of $\zeta_U: A \cong B$, that is $A \underset{U}{\cong} B$.

<u>Remarks</u>

(i) $A \underset{X}{\to} B \underset{Y}{\to} C$ implies $A \underset{X \cap Y}{\to} C$.

(ii) $\xi_X \zeta = \eta_Y$ implies that ζ preserves $X \cap Y$.

(iii) $\zeta \xi_X = \eta_Y$ implies that the function $\zeta | \mathrm{Im}\, \xi_X$ preserves $X \cap Y$.

(iv) $\eta_Y \xi_X = 1_A$ implies that ξ and η both preserve $X \cup Y$.

(v) In an arbitrary category $\mathcal{C}$, a morphism γ is an isomorphism if and only if there exists a morphism δ and $c,d \in \mathrm{Ob}\, \mathcal{C}$ such that

$$\delta\gamma = 1_c \quad \text{and} \quad \gamma\delta = 1_d .$$

Such a δ is unique and usually denoted by γ^{-1} .

<u>Proposition 1</u>. If $A \underset{X}{\to} B$ and $B \underset{Y}{\to} A$, then $A \underset{X \cup Y}{\cong} B$.

<u>Proof</u>. By definition of $\underset{X}{\to}$, there exists $\xi_X: A \to B$. Then

$$\xi_X(a, J[A](a)) = (a, J[B](a))$$

for any $a \in \mathrm{Dom}\, J[A]$, because the right side is the unique element of the form (a,b) belonging to $G[B]$. Similarly there exists $\eta_Y: B \to A$ such that

$$\eta_Y(a, J[B](a)) = (a, J[A](a))$$

for any $a \in \mathrm{Dom}\, J[B]$. Thus ξ_X is an isomorphism, for $\eta_Y \xi_X = 1_A$ and $\xi_X \eta_Y = 1_B$ (cf., remark (v)). Besides, ξ preserves $X \cup Y$, by remark (iv). Q.E.D.

Proposition 2. $A \underset{X}{\tilde{\to}} B$ if and only if $A \underset{X}{\to} B$ and $B \underset{X}{\to} A$.

Proof. Sufficiency: apparent from Proposition 1. Necessity: if $A \underset{X}{\tilde{\to}} B$, there exists $\zeta_X : A \to B$ and $\zeta^{-1} : B \to A$ such that $\zeta^{-1} \zeta_X = 1_A$ (cf., remark (v)). ζ^{-1} preserves X , by remark (iv). Q.E.D.

For each $A \in \mathfrak{A}$ and $B \in \mathfrak{A}$, these value-preserving monomorphisms or isomorphisms have the practical meanings listed below. The reader may recall that ${}_{\iota}A$ is understood whenever A denotes such a program that $\mathrm{Dom}\, J[A] \subseteq |\mathfrak{D}^{\iota}|$ is not satisfied.

1. Relation $\underset{X}{\to}$

The following relationships are equivalent with each other.

(a) $A \underset{X}{\to} B$.

(b) $\mathrm{Dom}\, J[A] \subseteq \mathrm{Dom}\, J[B]$, and for any $a \in \mathrm{Dom}\, J[A]$,

$$(J[A](a))_u = (J[B](a))_u$$

for each $u \in X + \{\chi\}$.

(c) For each $a \in |\mathfrak{D}^{\iota}|$, if the process (A,a) (see section 3) terminates with the result b , $b \in |\mathfrak{D}|$, then the process (B,a) terminates with the result c satisfying

$$b_x = c_x \quad \text{for each } x \in X ,$$

namely the values of variables coincide variable-wise, and

$$b_\chi = c_\chi ,$$

namely the destinations of the exits are identical.

2. Relation $\underset{\emptyset}{\to}$

The relationship $A \underset{\emptyset}{\to} B$ holds if and only if the following conditions are satisfied.

If (A,a) terminates, then (B,a) terminates for any $a \in |\mathfrak{D}^{\iota}|$. Besides, the destinations of the exits are identical.

3. Relation $\underset{X}{\cong}$

The following relationships are equivalent with each other.

(a) $A \underset{X}{\cong} B$.

(b) $A \underset{X}{\rightarrow} B$ and $B \underset{X}{\rightarrow} A$, or, by proposition 1, $A \underset{X}{\rightarrow} B$ and $B \underset{\emptyset}{\rightarrow} A$.

(c) $\mathrm{Dom}\, J[A] = \mathrm{Dom}\, J[B]$, and, for any $a \in \mathrm{Dom}\, J[A]$,

$$(J[A](a))_u = (J[B](a))_u$$

for each $u \in X + \{\chi\}$.

(d) The process (A,a) terminates if and only if (B,a) terminates, and the same conditions as 1(c) above are satisfied by the results of these processes.

4. Strong Equivalence and Ordering

The relationship $A \cong B$ holds if and only if A and B are <u>strongly equivalent</u> in the usual sense. The relationship $A \underset{U}{\rightarrow} B$ holds if and only if $J[A] \leq J[B]$ in the natural ordering of partial functions, namely $\varphi \leq \psi$ if and only if φ is a restriction of ψ . $A \cong B$ if and only if $A \underset{U}{\rightarrow} B$ and $B \underset{U}{\rightarrow} A$, which are still weaker than $J[A] = J[B]$ in the original sense of $J[A]$ and $J[B]$, being equivalent to $J[{}_\iota A] = J[{}_\iota B]$, i.e.,

$$A_{\mathfrak{D}}|\ |\mathfrak{D}^{\iota}| = B_{\mathfrak{D}}|\ |\mathfrak{D}^{\iota}| \ .$$

5. Correctness

Firstly, the concept of correctness of programs introduced by Floyd (1967a) and extended by Manna (1969a) will be explained in our notation so that the comparison becomes easier. Manna's definitions are as follows:

Program A is said to be <u>partially correct w.r.t.</u> predicates $p_{\mathfrak{D}}$ and $q_{\mathfrak{D}}$ if and only if

$$p_{\mathfrak{D}}(a) = T \text{ implies } q_{\mathfrak{D}}(J[A](a)) = T \text{ , for any } a \in \mathrm{Dom}\, J[A] \ . \quad (1)$$

Program A is said to be <u>correct w.r.t.</u> $p_{\mathfrak{D}}$ and $q_{\mathfrak{D}}$ if and only if

$$p_{\mathfrak{D}}(a) = T \text{ implies } a \in \mathrm{Dom}\, J[A] \ , \quad (2)$$

besides (1) above.

Let δ denote either σ or $\sigma^{-1}\sigma$ for an arbitrary σ such that $\sigma \notin A^{\pm}$. Then apparently (1) and (2) are equivalent to the following relationships in this order.

$$(p \to A, \delta) \underset{\emptyset}{\cong} (p \to A ; (q \to \Lambda, \delta), \delta) . \qquad (1')$$

$$(p \to A, \delta) \underset{\emptyset}{\cong} (p \to \Lambda, \delta) . \qquad (2')$$

6. Representations of $\underset{X}{\to}$ and $\underset{X}{\cong}$ by $\cong$

Since we shall consider a formal system which represents (although incompletely) the concept of equivalence, namely relations $\underset{X}{\cong}$ and $\cong$, we shall see that $\underset{X}{\to}$ and $\underset{X}{\cong}$ can be defined by $\cong$ here. We shall use, however, $\underset{X}{\cong}$ as well as $\cong$ in the formal system because of its practical applicability.

Let $\Sigma^{x_1 \ldots x_n}(f_1,\ldots,f_n)$ denote the statement

$$x_1 := f_1 ; \ldots ; x_n := f_n .$$

Relationship $A \underset{X}{\cong} B$ holds if and only if

$$A;\Sigma^{t_1 \ldots t_m}(c,\ldots,c) \cong B;\Sigma^{t_1 \ldots t_m}(c,\ldots,c) ,$$

for an arithmetic expression c such that $V[c] = \emptyset$ and $t_1,\ldots,t_m$ such that $\{t_1,\ldots,t_m\} = V[A] \cup V[B] - X$.

Relationship $A \underset{X}{\to} B$ holds if and only if

$$A \underset{X}{\cong} \Sigma^{v_1 \ldots v_n}(u_1,\ldots,u_n);(A\sigma_1^{-1}\ldots\sigma_k^{-1})';\Sigma^{u_1 \ldots u_n}(v_1,\ldots,v_n);B ,$$

where the following conditions are satisfied:

$$\{u_1,\ldots,u_n\} = V[A] \cap V[B] .$$

$$\{v_1,\ldots,v_n\} \cap (V[A] \cup V[B] \cup X) = \emptyset .$$

$$\{\sigma_1,\ldots,\sigma_k\} = A^{++} .$$

$(A\sigma_1^{-1}\ldots\sigma_k^{-1})'$ is a copy of $A\sigma_1^{-1}\ldots\sigma_k^{-1}$ (see section 2) such that $(A\sigma_1^{-1}\ldots\sigma_k^{-1})'^{\pm} \cap B^{\pm} = \emptyset$.

Inductive Limits

The concept of inductive limits is useful in $\mathcal{P}r$ and $\mathcal{P}r^{\ell}$. For instance, we can frequently use the following method in order to prove $A \underset{Z}{\rightarrow} B$.

We find two sequences of programs $(A_i)_{i \in n}$ and $(B_i)_{i \in n}$ with morphisms such as

$$\xi_X^{ij}: A_i \to A_j ,$$

$$\eta_Y^{ij}: B_i \to B_j ,$$

$$(A, \xi_X^{i}) = \underrightarrow{\lim} (B_i, \eta_X^{ij}) ,$$

and

$$\zeta_Z^{i}: A_i \to B_i ,$$

for each $i \in n$ and $j \in n$. This is a sufficient condition for a ζ such that $\zeta: A \to B$ and that ζ preserves X to exist. If p and q contradict each other, then $(p \to A, (q \to B, \Delta))$ is a <u>sum</u> of $(p \to A, \Delta)$ and $(q \to B, \Delta)$, in the sense of the terminology of category being a special case of inductive limit, where Δ is a statement of the form $\sigma^{-1}\sigma$ such that $\sigma \notin A^{\pm} \cup B^{\pm}$ and $A^{++} \cap B^{-} = A^{-} \cap B^{++} = \phi$. This fact may be considered as a justification of writing $p \cdot A + \bar{p} \cdot B$ instead of $(p \to A, B)$ conveniently used in the proof of the completeness of L.3 by Igarashi (1964).

5. FORMAL SYSTEM REPRESENTING THE EQUIVALENCE OF STATEMENTS

Well-formed Formulas

For two arbitrary Algol-like statements A and B belonging to $\mathcal{A}_U$ and an arbitrary subset X of U,

$$A \cong B$$

and

$$A \underset{X}{\cong} B$$

are well-formed formulas, or wffs. (Cf., Intended Interpretation below.)

Substitution Rules

In the following schemata of axioms and inference rules, arbitrary statements, variable symbols, label symbols, arithmetic expressions, Boolean expressions, and sets of variable symbols can be substituted in place of $A,B,C,\ldots$; $x,y,z,\ldots$; $\sigma,\sigma_1,\ldots,\ \tau,\tau_1,\ldots$; $f,g,\ldots$; $p,q,r,\ldots$; and $X,Y,Z,\ldots$; respectively, provided that the results of such substitutions constitute wffs, and that all the restrictions imposed on the schemata, immediately following each schema, are fulfilled.

An arbitrary copy of the statement that is substituted in place of C can be substituted in place of C' in axiom 12; any other occurrence of substitution operator indicated by brackets should be treated similarly; and an arbitrary statement of the form $\sigma^{-1}\sigma$ can be substituted in place of Δ, with the same proviso as the above.

A schema of wffs $\mathfrak{B}(i)$ in which i occurs as index of statements should be replaced by the line of the form

$$\mathfrak{B}(1)\ldots\mathfrak{B}(\nu)$$

before any other substitution, where $\nu\in\mathfrak{n}$ and ν should be substituted in place of n occurring in the restrictions.

The symbol 1 stands for a nullary predicate symbol such that $1_{\mathfrak{R}}=T$. Similarly $0_{\mathfrak{R}}=F$.

The formulas in the sense of predicate calculus that are obtained after the substitutions of the symbols $f,g,\ldots$, $p,q,\ldots$, and that constitute a part of restriction, except those expressions containing set-theoretic symbols, should be interpreted in one of the following ways:

(I) Let $\mathfrak{A}$ be a formula (in the sense of predicate calculus) that contains exactly n variables such as $x_1,\ldots,x_n$. Then we consider that the restriction expressed by $\mathfrak{A}$ is satisfied if and only if

$$(\forall x_1\ldots\forall x_n\mathfrak{A})_{\mathfrak{S}}=T .$$

(II) We presuppose an axiom system Γ (or theory) that is consistent (and semantically complete, preferably) and that contains all the symbols belonging to $\mathfrak{F}$ or $\mathfrak{P}$ and the two symbols $=$ and ∇. Then we consider the restriction expressed by $\mathfrak{A}$ as above is satisfied if and only if

$$\vdash_{\Gamma}\mathfrak{A} .$$

In the both methods, logical connectives occurring in the restrictions before substitutions should be read as the connectives of $\Gamma_{\not{\mathcal{L}}}$, and $\varphi = \psi$ interpreted as either both sides are defined and equal, or both sides are undefined.

Remark

Since semantically complete axiom systems do not always exist, we have to note (I).

Axioms and Theorems

Any wff that is a result of substitution into an axiom schema is an axiom. An axiom is a theorem. If

$$\frac{\mathfrak{F}_1 \cdots \mathfrak{F}_n}{\mathfrak{F}}$$

is a result of substitution into an inference rule schema, and $\mathfrak{F}_1, \ldots, \mathfrak{F}_n$ are theorems, then $\mathfrak{F}$ is also a theorem. All the theorems are defined to be so only by these rules. We shall frequently write

$$\vdash \mathfrak{F}$$

to mean that $\mathfrak{F}$ is a theorem.

Asterisks are used to emphasize a certain restriction, for the readability's sake, so that they are not parts of the formal system. Index like (Ia†), (IIIm†), etc., indicates that the same axiom or inference rule was used and indexed by Ia, IIm, etc., by Igarashi (1964), for the convenience of comparison.

Special Substitution

In the following schemata of axioms and inference rules, any occurrence of $\cong$ can be replaced by $\underset{U}{\cong}$, and vice versa.

Axiom 1. (a) $(AB)C \cong A(BC)$.

(b) $\sigma((AB)C) \cong \sigma(A(BC))$.

Axiom 2. (a) $A\Lambda \cong A$.

(b) $\Lambda A \cong A$. (Ie†)

Axiom 3. (a) $\sigma^{-1} \cong \Lambda$.

(b) $\sigma\sigma^{-1} \cong \Lambda$.

Axiom 4. $\sigma A \cong \sigma$.

$\sigma \notin A^{-}$.

Axiom 5. $A\Delta \cong \Delta$.

$A^{++} = \phi$.

Axiom 6. $x := x \cong \Lambda$. (Ia†)

Axiom 7. (a) $x := f; A; x := g \cong A_x[f]^*; x := g_x[f]^*$. (Ib†)

$A^{++} = \phi$.

$L[A] \cap (V[f] \cup \{x\}) = \phi$.

* $A_x[f]$ and $g_x[f]$ in the right side are restricted to be $A_x[f]^{\circ}$ and $g_x[f]^{\circ}$, respectively.

(b) $x := f; A; y := g \cong x := f; A_x[f]; y := g_x[f]$. (Ic†)

x and y are distinct.

$L[A] \cap (V[f] \cup \{x\}) = \phi$.

$x \notin V[f]$.

Axiom 8. $A \underset{X}{\cong} \Lambda$. (Id†)

$L[A] \cap X = \phi$.

$A^{++} = \phi$.

$A^{-+} = \phi$.

Every function or predicate symbol occurring in A represents a total function or predicate, by the interpretation.

Axiom 9. $(1 \rightarrow A, B) \cong A$. (IIIm†)

Axiom 10. $(p \rightarrow A, B) \cong (\neg\, p \rightarrow B, A)$. (IIIo†)

Axiom 11. (a) $(p \rightarrow (q \rightarrow A, B), C) \cong (p \wedge q \rightarrow A, (p \wedge \neg\, q \rightarrow B, C))$. (IIIp†)

(b) $(p \rightarrow (q \rightarrow A, B), C) \cong (p \rightarrow \Delta, C)$.

$p \supset \nabla q$.

Axiom 12. (a) $(p \rightarrow A, B)C \cong (p \rightarrow AC, BC')$. (IIIu†)

(b) $\sigma(p \rightarrow A, B)C \cong \sigma(p \rightarrow AC, BC')$.

$\sigma \notin C'^{-}$.

Axiom 13. $x := f; (p \to A, B) \cong (p_x[f]^* \to x := f; A, x := f; B)$.

*If $x \in V[f]$, then $p_x[f]$ is restricted to be $p_x[f]^o$.

Axiom 14. $(p \to A, B) \cong (p \to A_C[(p \to C, D)], B)$.

$$L[A] \cap V[p] = \emptyset .$$

Axiom 15. (a) $(p \to x := f, A) \cong (p \to x := g, A)$.

$$p \supset f = g .$$

(b) $(p \to x := f, A) \cong (p \to \Delta, A)$.

$$p \supset \nabla f .$$

Axiom 16. (a) $A \cong A_f[g]$.

$$f = g .$$

(b) $A \cong A_p[q]$.

$$p \equiv q .$$

Inference Rule 1.

$$\frac{A \underset{X}{\cong} B}{B \underset{X}{\cong} A} .$$

Inference Rule 2.

$$\frac{A \underset{X}{\cong} B \quad B \underset{X}{\cong} C}{A \underset{X}{\cong} C} .$$

Inference Rule 3.

$$\frac{A \underset{X}{\cong} B \quad A \underset{Y}{\cong} B}{A \underset{Z}{\cong} B} .$$

$$Z \subseteq X \cup Y .$$

Inference Rule 4.

$$\frac{(p \to A, C) \underset{X}{\cong} (p \to B, C) \quad (q \to A, D) \underset{X}{\cong} (q \to B, D)}{(r \to A, E) \underset{X}{\cong} (r \to B, E)} .$$

$$r \supset p \vee q .$$

Inference Rule 5. (a)

$$\frac{\sigma A \cong {}_{\tau}B}{C \cong C_{\sigma}[\tau]} .$$

A and B end with go-tos.

A and B occur in C .

(b)

$$\frac{\sigma A \cong B}{C \cong C_{\sigma}[B]} .$$

B ends with a go-to.

A occurs in C , or A is

$C_{A_1 \ldots A_n}[\Lambda, \ldots, \Lambda]$, where $A_1, \ldots, A_n$ are preceded by go-tos in C .

Inference Rule 6.

$$\frac{A \underset{X}{\cong} B}{AC \underset{X \cup L[C]}{\cong} BC} .$$

$R[C] \subseteq X$.

$C^{++} \cap A^{-} = C^{++} \cap B^{-} = \emptyset$.

Inference Rule 7.

$$\frac{A \underset{X}{\cong} B^{*} \quad \sigma_i A \underset{X}{\cong} \sigma_i B}{CA \underset{X}{\cong} CB} .$$

$C^{++} \cap A^{-} = C^{++} \cap B^{-} = \{\sigma_1, \ldots, \sigma_n\}$.

$A^{++} \cap C^{-} = B^{++} \cap C^{-} = \emptyset$.

*If C ends with a go-to or A and B both begin with labellings, then the upper left formula may be omitted, provided that $n \geq 1$.

Inference Rule 8.

$$\frac{A \cong B^{*} \quad \sigma_i A \cong \sigma_i B}{C \cong C_A[B]} . \qquad \text{(IVg†)}$$

$C_A[\Lambda]^{++} \cap A^{-} = C_A[\Lambda]^{++} \cap B^{-} = \{\sigma_1, \ldots, \sigma_n\}$.

*Same as above.

Inference Rule 9.

$$\frac{D^iA \cong A^iA \quad \tilde{D}^iB \cong B^iB \quad A^i_{\sigma_1 \ldots \sigma_n}[\tilde{\sigma}_1,\ldots,\tilde{\sigma}_n]^\circ \underset{X}{\cong} B^i \quad B^i \underset{X}{\cong} C^i}{D^kA \underset{X}{\cong} \tilde{D}^kB} \ .$$

$$k \in [n] \ .$$

1. The set $S = \{\sigma_1,\ldots,\sigma_n\}$ is a non-empty subset of A^- , and a total function

$$\zeta : S \to \mathcal{L}$$

sends each σ_i onto $\tilde{\sigma}_i$. ζ , together with S , satisfies the following conditions:

$$S \supseteq S'$$

and

$$\zeta(\sigma) = \sigma \quad \text{for each} \quad \sigma \in S \cap S'' ,$$

where

$$S' = \bigcup_i (A^i)^{++} \cap A^-$$

and

$$S'' = \bigcup_i (A^i)^{++} \cap A^{++} \ .$$

2. The following conditions are satisfied for each $i \in [n]$.
   (i) D^i is of the form $(p_i \to \sigma_i, \delta^i)$ and $\tilde{D}^i$ is of the form $(p_i \to \tilde{\sigma}_i, \delta^i)$, where δ^i is either τ_i or $\tau_i^{-1}\tau_i$ such that $\tau_i \notin A^{\pm} \cup B^{\pm}$.
   (ii) All the occurrences of σ_i in $A^1,\ldots,A^n$ are within the statements of the form $(p_i \to \sigma_i, \epsilon^i)$, or all the occurrences of $\tilde{\sigma}_i$ in $B^1,\ldots,B^n$ are within the statements of the form $(p_i \to \tilde{\sigma}_i, \epsilon^i)$ where ϵ^i subjects to the same restriction as δ^i above.
   (iii) $R[C^i] \subseteq X$.
3. If A does not begin with a labelling σ^{-1} such that $\sigma \in S$, then all of $A^1,\ldots,A^n$ must end with go-tos. If B does not begin with a labelling σ^{-1} such that $\sigma \in \zeta(S)$, then all of $B^1,\ldots,B^n$ must end with go-tos.

Intended Interpretation

A wff of the form

$$A \underset{X}{\cong} B$$

will be interpreted as the relationship $A \underset{X}{\cong} B$ in the sense of category Pr^{ι} (see section 4). Similarly, wff

$$A \cong B$$

will be interpreted as the relationship $A \cong B$ in Pr^{ι}.

Intuitively, it seems to be obvious that $\vdash A \underset{X}{\cong} B$ always implies that relationship $A \underset{X}{\cong} B$ in Pr^{ι} holds so that the above system is consistent. We shall not verify the consistency, however, in the present paper for which, presumably, the constructive definition of J will suffice. (See section 3.)

6. ELEMENTARY METATHEOREMS

Index such as (Th. 3†) shows the number of the same theorem for the formal systems treated by Igarashi (1964). The results of this section imply that every axiom of L.4 in that paper becomes a theorem in the present system, and that for every rule of inference of L.4 such as

$$\frac{\mathfrak{J}_1 \ldots \mathfrak{J}_n}{\mathfrak{J}},$$

the following holds:

$$\text{If } \vdash \mathfrak{J}_1, \ldots, \vdash \mathfrak{J}_n, \text{ then } \vdash \mathfrak{J}.$$

Therefore every theorem concerning completeness in that paper holds also for the present formal system.

Theorem 1. (Reflexivity) $\vdash A \underset{X}{\cong} A$. (Th. 3†)

Proof.

$\Lambda A \cong A$.	(Ax. 2a)	(1)
$A \cong \Lambda A$.	(Inf. 1, (1))	(2)
$A \cong A$.	(Inf. 2, (1), (2))	(3)
$A \underset{X}{\cong} A$.	(Inf. 3, (3))	

Q.E.D.

Thus $\underset{X}{\cong}$ satisfies the equivalence law formally, the symmetricity and the transitivity being Inf. 1 and Inf. 2.

Theorem 2. If

$$\vdash A_1 \underset{X_1}{\cong} A_2, \ldots, \vdash A_{n-1} \underset{X_{n-1}}{\cong} A_n ,$$

then

$$\vdash A_1 \underset{X}{\cong} A_n ,$$

for any X such that

$$X \subseteq \bigcap_{i \in [n]} X_i .$$

Proof. A repeated use of Inf. 3 and Inf. 2. Q.E.D.

Theorem 3. $\vdash (0 \to A, B) \cong B$. (Th. 25†, cf., McCarthy (1963 a))

Proof. $(0 \to A, B) \cong (\neg\ 1 \to A, B)$ (Axiom 16b)

$\cong (1 \to B, A)$ (Axiom 10)

$\cong B$. (Axiom 9) Q.E.D.

Theorem 4. If

$$\underset{\Gamma}{\vdash} p \vee \neg p ,$$

then

$$\vdash (p \to A, A) \cong A . \quad \text{(IIIn†, cf., McCarthy (1963 a))}$$

Proof. $(p \to (p \to A, A), \Lambda) \cong (p \wedge p \to A, (p \wedge \neg p \to A, \Lambda))$ (Axiom 11a)

$\cong (p \to A, (0 \to A, \Lambda))$ (Axiom 16b)

$\cong (p \to A, \Lambda)$. (Theorem 3, Inf. 8) (1)

Similarly

$$\vdash (\neg\ p \to (p \to A, A), \Lambda) \cong (\neg\ p \to A, \Lambda) . \quad (2)$$

Thus

$$(p \vee \neg p \to (p \to A, A), \Lambda) \cong (p \vee \neg p \to A, \Lambda) . \quad \text{((1),(2), Inf. 4)}$$

The premise of the theorem, Axiom 16b, and Axiom 10 give the conclusion. Q.E.D.

The premise of theorem 4, being the law of the excluded middle, holds if $p_{\mathcal{D}}$ is total and Γ is semantically complete. (See section 5, method (II).)

Theorem 5. $\vdash (p \to A, (q \to B, C)) \cong (p \to A, (\neg\ p \wedge q \to B, C))$, (IIIq†)

with the same premise as theorem 4.

Proof. $(p \to A,(q \to B,C)) \cong (\neg\, p \to (q \to B,C),A)$ (Axiom 10)

$\cong (\neg\, p \wedge q \to B,(\neg\, p \wedge \neg\, q \to C,A))$. (Axiom 11a) (1)

$(p \to A,(\neg\, p \wedge q \to B,C)) \cong (\neg\, p \to (\neg\, p \wedge q \to B,C),A)$ (Axiom 10)

$\cong (\neg\, p \wedge \neg\, p \wedge q \to B,(\neg\, p \wedge \neg(\neg\, p \wedge q) \to C,A))$ (Axiom 11a)

$\cong (\neg\, p \wedge q \to B,(\neg\, p \wedge \neg\, q \to C,A))$. (Axiom 16b) (2)

Statements (1) and (2) are identical. Q.E.D.

The above also implies that Axiom IIq of L.2 in the previous paper was dependent on others.

Theorem 6. If $\vdash_{\Gamma} f = g$, then

$$\vdash x := f \cong x := g \ . \qquad \text{(If†)}$$

Proof. A special case of Axiom 16a. Q.E.D.

Theorem 7. If $\vdash_{\Gamma} p \supset f = g$, then

$$\vdash (p \to x := f; A,B) \cong (p \to x := g; A,B) \ . \qquad \text{(IIIv†)}$$

Proof. $(p \to x := f, \Lambda) \cong (p \to x := g, \Lambda)$. (Axiom 15a)

Right multiplying both sides by A ,

$(p \to x := f; A,A') \cong (p \to x := g; A,A')$. (Axiom 12a)

By Inf. 4,

$(p \to x := f; A,B) \cong (p \to x := g\,; A,B)$. Q.E.D.

Theorem 8. If $\vdash_{\Gamma} p \equiv q$, $\vdash A \underset{X}{\cong} B$, and $\vdash C \underset{X}{\cong} D$, then

$$\vdash (p \to A,C) \underset{X}{\cong} (q \to B,D) \ . \qquad \text{(IIIs†)}$$

Proof. $(p \to A,C) \underset{X}{\cong} (p \to B,C)$ (Inf. 4, premise)

$\cong (\neg\, p \to C,B)$ (Axiom 10)

$\underset{X}{\cong} (\neg\, p \to D,B)$ (Inf. 4, premise)

$\cong (p \to B,D)$ (Axiom 10)

$\cong (q \to B,D)$. (Axiom 16b)

By theorem 2,

$$(p \rightarrow A, C) \underset{X}{\cong} (q \rightarrow B, D) \text{ .} \qquad \text{Q.E.D.}$$

<u>Theorem 9</u>. If $\vdash A \underset{X}{\cong} B$ and $\vdash A \underset{Y}{\cong} B$, then

$$\vdash A \underset{X \cup Y}{\cong} B \text{ .} \qquad (\text{Ih}\dagger)$$

<u>Proof</u>. A special case of Inf. 3. Q.E.D.

<u>Theorem 10</u>. If $\vdash A \underset{X \cup Y}{\cong} B$, then

$$\vdash A \underset{X}{\cong} B \text{ .} \qquad (\text{Ii}\dagger)$$

<u>Proof</u>. A special case of Inf. 3. Q.E.D.

<u>Theorem 11</u>. (Superfluous Labels) If $\sigma \notin A^+ \cup B^-$, then

$$\vdash AB \cong A\sigma^{-1}B \text{ .} \qquad (\text{IVa}\dagger)$$

<u>Proof</u>.

$\Lambda \cong \sigma^{-1}$.	(Axiom 3a)	(1)
$A \Lambda B \cong A\sigma^{-1}B$.	(Inf. 8, (1))	(2)
$A \cong A\Lambda$.	(Axiom 2a)	(3)
$\sigma A \cong (\sigma A)\Lambda$	(Axiom 2a)	
$\cong \sigma(A\Lambda)$.	(Axiom 1a)	(4)

By Inf. 8 with (3) and (4),

$$AB \cong A\Lambda B \text{ .} \qquad (5)$$

By (2) and (5),

$$AB \cong A\sigma^{-1}B \text{ .} \qquad \text{Q.E.D.}$$

<u>Theorem 12</u>. (Disconnected Statements) If $\sigma \notin B^-$ and $A^+ \cap B^- = \emptyset$, then

$$\vdash A\sigma B \cong A\sigma \text{ .} \qquad (\text{IVb}\dagger)$$

<u>Proof</u>. $\sigma B \cong \sigma$. (Axiom 4, premise) (1)

Also, the premise implies that $A^+ \cap (\sigma B)^- = A^+ \cap (\sigma)^- = \phi$, so that

$$A(\sigma B) \cong A\sigma \ . \qquad \text{(Inf. 8, (1))} \qquad \text{Q.E.D.}$$

<u>Theorem 13</u>. (Superfluous Go-Tos) $\vdash A \cong A_{\sigma^{-1}}[\sigma\sigma^{-1}]$.

<u>Proof</u>. For any τ ,

$$\tau\sigma^{-1} \cong \tau\sigma\sigma^{-1}$$

$$\cong \begin{cases} \Lambda , & \tau = \sigma \\ \tau , & \text{otherwise} , \end{cases} \qquad (1)$$

because the formula

$$\sigma\sigma \cong \sigma \qquad \text{(Theorem 12)}$$

and Inf. 8 give

$$(\sigma\sigma)\sigma^{-1} \cong \sigma\sigma^{-1} \ .$$

Inf. 8 with (1) gives the conclusion. Q.E.D.

<u>Theorem 14</u>. (Additional Exits) If $\vdash A\sigma \cong B\sigma$ for a σ such that $\sigma \notin A^{\pm} \cup B^{\pm}$, then

$$\vdash A \cong B \ . \qquad \text{(IVf†)}$$

<u>Proof</u>. Right multiplying both sides of the first formula by σ^{-1} , we obtain

$$A\sigma\sigma^{-1} \cong B\sigma\sigma^{-1} \ . \qquad \text{(Inf. 6)}$$

By the premise concerning σ and Inf. 8,

$$A \cong B \ . \qquad \text{Q.E.D.}$$

<u>Theorem 15</u>. (Copies) $\vdash A \cong A'$. (Th. 41†)

<u>Proof</u>. (i) The case that $A^{\pm} \cap (A')^- = \phi$ will be proved firstly. Suppose $A^{\pm} - A^{++} = \{\alpha_1, \ldots, \alpha_n\}$ and A' is

$$A_{\alpha_1 \ldots \alpha_n \alpha_1^{-1} \ldots \alpha_n^{-1}}[\beta_1, \ldots, \beta_n, \beta_1^{-1}, \ldots, \beta_n^{-1}]^{\circ} \ .$$

Let B be

$$A_{\alpha_1 \ldots \alpha_n \alpha_1^{-1} \ldots \alpha_n^{-1}}[\alpha_1^{-1}\beta_1^{-1}\gamma_1\gamma_1^{-1}, \ldots, \alpha_n^{-1}\beta_n^{-1}\gamma_n\gamma_n^{-1}] ,$$

where $\gamma_i \notin A^{\pm} \cup B^{\pm}$ for any $i \in [n]$. Then

$$\vdash A \cong B . \qquad \text{(Theorems 11 and 13)} \qquad (1)$$

But

$$\vdash B \cong B_{\alpha_i}[\beta_i] \qquad (2)$$

for each occurrence of α_i , because $\alpha_i^{-1}\beta_i^{-1}\gamma_i$ occurs in B , for which

$$\alpha_i(\alpha_i^{-1}\beta_i^{-1}\gamma_i) \cong \beta_i(\alpha_i^{-1}\beta_i^{-1}\gamma_i) \cong \gamma_i , \quad \text{(Axiom 3b, Theorems 11, 13)}$$

so that Inf. 5a gives (2). Since the number of occurrences of α_i in B is finite,

$$\vdash B \cong B_{\alpha_1 \ldots \alpha_n}[\beta_1, \ldots, \beta_n]^{\circ} \qquad (3)$$

by the repeated use of (2). But the right side of (3) is

$$A'_{\beta_1^{-1} \ldots \beta_n^{-1}}[\alpha_1^{-1}\beta_1^{-1}\gamma_1\gamma_1^{-1}, \ldots, \alpha_n^{-1}\beta_n^{-1}\gamma_n\gamma_n^{-1}]$$

by definition of A' , so that

$$\vdash (3) \cong A' , \qquad (4)$$

similarly to (1). Formulas (1), (3), and (4) give

$$A \cong A' .$$

(ii) The case that $A^{\pm} \cap (A')^{-} \neq \emptyset$ is reduced to (i) as follows: Consider another copy A" of A for which $A^{\pm} \cap (A'')^{-} = \emptyset$ and $(A')^{\pm} \cap (A'')^{-} = \emptyset$. Then $\vdash A \cong A''$ and $\vdash A' \cong A''$ according to (i), so that $\vdash A \cong A'$. Q.E.D.

Theorem 16. (Operating σ (1)) If B occurs in A and ends with a go-to, and $\sigma \in B^{-}$, then

$$\vdash \sigma A \cong (\sigma B)'A .$$

Proof. $\vdash \sigma B \cong (\sigma B)'$ by theorem 15. $(\sigma B)'$ ends with a go-to, so that

$$\sigma A \cong (\sigma A)_{\sigma}[(\sigma B)'] \qquad \text{(Inf. 5a)}$$
$$\cong (\sigma B)'A \text{ .} \qquad \text{Q.E.D.}$$

Theorem 17. (Operating σ (2)) If $\sigma A_{B_1 \ldots B_n}[\Lambda,\ldots,\Lambda]$ ends with a go-to and $B_1,\ldots,B_n$ are preceded by go-tos in A, then

$$\vdash \sigma A \cong (\sigma A_{B_1 \ldots B_n}[\Lambda,\ldots,\Lambda])'A \text{ ,}$$

for any σ such that $\sigma \notin \bigcup_{i\in[n]} B_i$.

Proof. Similar to the above, while we notice the latter alternative in the restrictions of Inf. 5b. Q.E.D.

Theorem 18. If

$$\vdash \sigma_i A \cong A^i A$$

for each $i \in [n]$ and

$$\vdash \sigma_i B \cong A^i B$$

for each $i \in [n]$, for a subset $S = \{\sigma_1,\ldots,\sigma_n\}$ of A^- such that $S \supseteq S'$, where

$$S' = \bigcup_{i\in[n]} (A^i)^{++} \cap A^-$$

and each A^i ends with a go-to, then

$$\vdash \sigma_i A \cong \sigma_i B$$

for any $i \in [n]$.

Proof. Let D^i be $(1 \to \sigma_i, \tau)$, where $\tau \notin A^{\pm} \cup B^{\pm}$, and $\hat{A}^i$ be $A^i_{\sigma_1 \ldots \sigma_n}[D^1,\ldots,D^n]^{\circ}$ for each $i \in [n]$. Then

$$D^i A \cong \sigma_i A \qquad \text{(Axiom 9, Inf. 6)}$$
$$\cong A^i A \qquad \text{(premise)}$$
$$\cong \hat{A}^i B \text{ .} \qquad \text{(repeat Inf. 8, Axiom 9)} \qquad (1)$$

Similarly

$$D^i B \cong \hat{A}^i B \ . \tag{2}$$

In order to use Inf. 9, σ_i is defined as $\tilde{\sigma}_i$, and $\hat{A}^i$ is substitutes in place of A^i , B^i , and C^i of that schema. The left two schemata of wffs become (1) and (2), and the right two

$$\hat{A}^i_{\sigma_1 \ldots \sigma_n}[\tilde{\sigma}_1, \ldots, \tilde{\sigma}_n] \underset{X}{\cong} \hat{A}^i \tag{3}$$

and

$$\hat{A}^i \underset{X}{\cong} \hat{A}^i \ . \tag{4}$$

But the left side of (3) is $\hat{A}^i$ itself, so that (3) as well as (4) holds because of the reflexivity (theorem 1). We examine the restrictions.

<u>Condition 1</u>. $\zeta(\sigma_i) = \sigma_i$ for each $i \in [n]$, so that the second condition, namely

$$\zeta(\sigma) = \sigma \quad \text{for each} \quad \sigma \in S \cap S'' \ ,$$

where

$$S'' = \bigcup_{i \in [n]} (A^i)^{++} \cap A^{++} \ ,$$

is satisfied, while the first condition is explicitly included in the premise of the theorem.

<u>Condition 2</u>. (i), (ii) apparent. (iii) We define U as X .

<u>Condition 3</u>. Apparent. Thus, by Inf. 9,

$$\vdash D^i A \cong D^i B$$

for any $i \in [n]$. By the derivations for (1) and (2),

$$\sigma_i A \cong D^i A \cong D^i B \cong \sigma_i B \ . \qquad \text{Q.E.D.}$$

<u>Theorem 19</u>. (Interchange of Copies) If B ends with a go-to and B and B' occur in A , then

$$\vdash A \cong A_{BB'}[B', B] \ . \tag{IVc†}$$

Proof. (i) The case that A begins with a labelling and that B and B' are preceded by go-tos in A is proved firstly. Let C be $A_{BB'}[\Lambda,\Lambda]$ and D be the right side of the conclusion of the theorem. Let τ be a label such that $\tau \notin A^{\pm}$.

$$\sigma A\tau \cong \begin{cases} (\sigma B)'A\tau\ , & \sigma \in B^{-} \qquad \text{(Theorem 16)} \\ (\sigma B')'A\tau\ , & \sigma \in (B')^{-} \qquad \text{(Theorem 16)} \\ (\sigma C)'\tau A\tau\ , & \sigma \in C^{-}\ . \qquad \text{(Theorem 17)} \end{cases} \tag{1}$$

Similarly

$$\sigma D\tau \cong \begin{cases} (\sigma B)'D\tau\ , & \sigma \in B^{-} \\ (\sigma B')'D\tau\ , & \sigma \in (B')^{-} \\ (\sigma C)'\tau A\tau\ , & \sigma \in C^{-}\ , \end{cases} \tag{2}$$

because $D_{B'B}[\Lambda,\Lambda]$ is also C. By theorem 18, (1), and (2),

$$\sigma A\tau \cong \sigma D\tau$$

for any $\sigma \in A^{-}$. Therefore,

$$\sigma A \cong \sigma D \qquad \text{(Theorem 14)}$$

for any $\sigma \in A^{-}$. Choosing σ_o such that σ_o^{-1} occurs at the leftmost of A,

$$A \cong D\ . \qquad \text{(Theorem 13)}$$

(ii) If A does not begin with a labelling, then we prove

$$\tau\tau^{-1}A \cong \tau\tau^{-1}A_{BB'}[B',B] \tag{3}$$

for a τ such that $\tau \notin A^{\pm}$, which is a special case of (i). Formula (3) and theorem 13 give the conclusion. If B or B' is not preceded by go-tos in A, then we insert $\alpha\alpha^{-1}$ and $\beta\beta^{-1}$ before B and B', where $\alpha \notin A^{\pm}$ and $\beta \notin A^{\pm}$. $\beta^{-1}B'$ being a copy of $\alpha^{-1}B$, (i) implies

$$A_{BB'}[\alpha\alpha^{-1}B,\beta\beta^{-1}B'] \cong A_{BB'}[\alpha\beta^{-1}B',\beta\alpha^{-1}B]\ . \tag{4}$$

Because of $\alpha(\alpha^{-1}B) \cong \beta(\beta^{-1}B')$ and Inf. 5a, used twice,

$$(4) \cong A_{BB'}[\beta\beta^{-1}B',\beta\alpha^{-1}B]$$
$$\cong A_{BB'}[\beta\beta^{-1}B',\alpha\alpha^{-1}B] \; . \qquad (5)$$

Deleting $\alpha\alpha^{-1}$ and $\beta\beta^{-1}$ from the left sides of (4) and (5) by theorem 11 and theorem 13, we get

$$A \cong A_{BB'}[B',B] \; . \qquad \text{Q.E.D.}$$

Theorem 21. (Go-To leading to usual statements) If $\sigma^{-1}B\tau^{-1}$ occurs in A , then

$$\vdash A \cong A_{\sigma}[B'\tau] \; . \qquad (\text{IVd}\dagger)$$

Proof. (i) The case that $\sigma \notin B^{++}$ is proved firstly. Let C be $A_{\tau^{-1}}[\tau\tau^{-1}]$. Then

$$A \cong C \; . \qquad \text{(Theorem 13)} \qquad (1)$$

$\sigma^{-1}B\tau$ occurs in C and

$$\sigma(\sigma^{-1}B\tau) \cong B\tau \qquad \text{(Theorem 11, Theorem 13)}$$
$$\cong B'\tau \; . \qquad \text{(Theorem 15)} \qquad (2)$$

By (2) and Inf. 6b,

$$C \cong C_{\sigma}[B'\tau]$$
$$\cong A_{\sigma}[B'\tau] \; . \qquad \text{(Theorem 13)} \qquad (3)$$

Formulas (1) and (3) give the conclusion.

(ii) The case that $\sigma \in B^{++}$ will be proved. Suppose B' is

$$B_{\sigma_1 \ldots \sigma_n \sigma_1^{-1} \ldots \sigma_n^{-1}}[\sigma_1', \ldots, \sigma_n', \sigma_1'^{-1}, \ldots, \sigma_n'^{-1}] \; .$$

Let B'' be $B'_{\sigma}[\sigma'']$, where $\sigma'' \notin A^{\pm} \cup (B')^{\pm}$. Then $\sigma''^{-1}B''$ is a copy of $\sigma^{-1}B$, and

$$B''_{\sigma''}[\sigma]^{\circ} = B' \; . \qquad (4)$$

Instead of (2) in the case (i), we have

$$\sigma(\sigma^{-1}B\tau) \cong (\sigma^{-1}B)\tau \qquad \text{(Theorem 13)}$$

$$\cong \sigma''^{-1}B''\tau \,. \qquad \text{(Theorem 15)} \qquad (5)$$

Therefore

$$A \cong A_{\sigma}[\sigma''^{-1}B''\tau] \,. \qquad \text{(Inf. 5b)} \qquad (6)$$

But every occurrence of σ'' in the right side of (5) can be replaced by σ, because of

$$\sigma''(\sigma''^{-1}B''\tau) \cong \sigma(\sigma^{-1}B\tau) \qquad \text{(Theorem 15)}$$

and Inf. 5a. Thus

$$(6) \cong (A_{\sigma}[\sigma''^{-1}B''\tau])_{\sigma''}[\sigma]^{\circ}$$

$$\cong (A_{\sigma}[B''\tau])_{\sigma''}[\sigma]^{\circ} \,. \qquad \text{(Theorem 11)} \qquad (7)$$

The right side of (7) is $A_{\sigma}[B'\tau]$ because of (4), namely $A \cong A_{\sigma}[B'\tau]$. Q.E.D.

<u>Theorem 22</u>. (Go-To leading to exits) If $\tau \notin A^{-}$ and $\sigma^{-1}\tau$ occurs in A, then

$$\vdash A \cong A_{\sigma}[\tau] \,. \qquad \text{(IVe†)}$$

<u>Proof</u>. $\sigma(\sigma^{-1}\tau) \cong \tau$ (Theorem 13, Theorem 11)

and Inf. 5b give the conclusion. Q.E.D.

<u>Theorem 23</u>. If $A^{+} \cap C^{-} = A^{-} \cap C^{+} = \phi$, $B^{+} \cap D^{-} = B^{-} \cap D^{+} = \phi$, $\vdash A \underset{R[C] \cup R[D] \cup X}{\cong} B$, and $\vdash C \underset{X}{\cong} D$, then

$$\vdash AC \underset{X}{\cong} BD \,. \qquad \text{(Ig†)}$$

<u>Proof</u>. (i) The case that

$$B^{+} \cap C^{-} = B^{-} \cap C^{+} = \phi \qquad (1)$$

is proved firstly. The first wff of the premise of the theorem implies

$$AC \underset{R[C] \cup R[D] \cup X}{\cong} BC \,. \qquad \text{(Inf. 6)} \qquad (2)$$

The second wff, $C \underset{X}{\cong} D$, implies

$$BC \underset{X}{\cong} BD \,. \qquad \text{(Inf. 7)} \qquad (3)$$

Thus

$$\vdash AC \underset{X}{\cong} BD \,. \qquad \text{(Theorem 2, (2), (3))}$$

(ii) If (1) does not hold, we consider the copies B' and C' such that $A^+ \cap C'^{-1}$, $A^- \cap C'^+$, $B'^+ \cap D^-$, $B'^- \cap D^+$, $B'^+ \cap C$, and $B'^- \cap C'^+$ are all ϕ. By theorem 15

$$B \cong B'$$

and

$$C \cong C' \,.$$

We carry out the following derivation.

$$\begin{aligned} AC &\cong AC' && \text{(Inf. 6)} \\ &\underset{X}{\cong} B'D && \text{(by (i) above)} \\ &\cong BD \,. && \text{(Inf. 7)} \end{aligned}$$

Namely

$$AC \underset{X}{\cong} BD \,. \qquad \text{Q.E.D.}$$

The above metatheorems show that wff $\mathfrak{J}$ is provable in the present formal system if it is provable in the previous system as noted at the beginning of this section. For the convenience of later use, theorems 11 and 12 will be modified as follows. (Proofs are essentially the same as before.)

Theorem 11. (Superfluous Labels) If $\sigma \notin A^-$, then $\vdash A \cong A_B[\sigma^{-1}B]$.

Theorem 12. (Disconnected Statements) If $A_B[\Lambda]^+ \cap B^- = \phi$ and B is preceded by a go-to in A, then

$$\vdash A \cong A_B[\Lambda] \,.$$

The first of the following theorems will be used in section 8, while the second is related to the notion of correctness. Theorem 18 says that two statements which are syntactically concatenations of a number of statements (loops may be contained semantically) are equivalent if the constituent statements are equivalent statement-wise, which fact is related to compilation. Moreover, this theorem gives an example of proving the equivalence of two statements which do not necessarily terminate.

Theorem 24. If

$$\vdash A_i \underset{X}{\cong} B_i ,$$

$$\vdash \sigma A_i \underset{X}{\cong} \sigma B_i$$

for each $\sigma \in A_i^-$, and

$$V[A_i] \subseteq X$$

for each $i \in [n]$, then

$$\vdash A_1 \ldots A_n \underset{X}{\cong} B_1 \ldots B_n ,$$

and

$$\vdash \sigma(A_1 \ldots A_n) \underset{X}{\cong} \sigma(B_1 \ldots B_n) \quad \text{for each } \sigma \in \bigcup_{i \in [n]} A_i^- .$$

Proof. Let C and D be

$$\tau_1 \tau_1^{-1} A_1 \tau_2 \tau_2^{-1} A_2 \ldots \tau_n \tau_n^{-1} A_n \tau_{n+1}$$

and

$$\tau_1 \tau_1^{-1} B_1 \tau_2 \tau_2^{-1} B_2 \ldots \tau_n \tau_n^{-1} B_n \tau_{n+1} ,$$

respectively, where $\tau_1, \ldots, \tau_{n+1}$ do not belong to $(A_1 \ldots A_n)^{\pm} \cup (B_1 \ldots B_n)^{\pm}$. Suppose $C^- = \{\sigma_1, \ldots, \sigma_n\}$, while $C^- = \bigcup_{i \in [n]} A_i^- + \{\tau_1, \ldots, \tau_n\}$ by definition. Then we notice the following.

$$\vdash \sigma \tau_i^{-1} A_i \underset{X}{\cong} \sigma \tau_i^{-1} B_i \tag{1}$$

for each $\sigma \in A_i^- + \{\tau_i\}$, because

$$\begin{aligned} \tau_i \tau_i^{-1} A_i &\cong A_i && \text{(Theorems 11 (extended), 13)} \\ &\underset{X}{\cong} B_i && \text{(premise of the theorem)} \\ &\cong \tau_i \tau_i^{-1} B_i , && \text{(Theorem 11, Theorem 13)} \end{aligned}$$

and, for each $\sigma \in A_i^-$,

$$\begin{aligned} \sigma \tau_i^{-1} A_i &\cong \sigma A_i && \text{(Theorem 11)} \\ &\underset{X}{\cong} \sigma B_i && \text{(premise of the theorem)} \\ &\cong \sigma \tau_i^{-1} B_i . && \text{(Theorem 11)} \end{aligned}$$

Therefore

$$\vdash \sigma\tau_i^{-1}A_i\tau_{i+1} \underset{X}{\tilde{\cong}} \sigma\tau_i^{-1}B_i\tau_{i+1} \qquad \text{(Inf. 6, (1))} \tag{2}$$

for each $\sigma \in A_i^- + \{\tau_i\}$. By theorem 16,

$$\vdash \sigma C \cong (\sigma\tau_i^{-1}A_i\tau_{i+1})'C \tag{3}$$

for each $\sigma \in A_i^- + \{\tau_i\}$, and

$$\vdash \sigma D \cong (\sigma\tau_i^{-1}B_i\tau_{i+1})'D \tag{4}$$

for each $\sigma \in B_i^- + \{\tau_i\}$. But

$$\begin{aligned}(\sigma\tau_i^{-1}A_i\tau_{i+1})' &\cong \sigma\tau_i^{-1}A_i\tau_{i+1} && \text{(Theorem 15)}\\ &\underset{X}{\tilde{\cong}} \sigma\tau_i^{-1}B_i\tau_{i+1} && \text{(by (2))}\\ &\cong (\sigma\tau_i^{-1}B_i\tau_{i+1})' , && \text{(Theorem 15)}\end{aligned}$$

so that

$$\vdash (\sigma\tau_i^{-1}A_i\tau_{i+1})' \underset{X}{\tilde{\cong}} (\sigma\tau_i^{-1}B_i\tau_{i+1})' . \tag{5}$$

We change the index i of Inf. 9 into j , define σ_j as $\tilde{\sigma}_j$, and substitute σ_j (we can simply use σ_j instead of $(1 \to \sigma_j, \delta^i)$ as shown in the proof of theorem 18), C , D , $(\sigma_j\tau_i^{-1}A_i\tau_{i+1})'$, $(\sigma_j\tau_i^{-1}B_i\tau_{i+1})'$, and C in place of D^i , A , B , A^i , B^i , and C , respectively. We note that

$$\vdash (\sigma\tau_i^{-1}A_i\tau_{i+1})' \underset{X}{\tilde{\cong}} (\sigma\tau_i^{-1}A_i\tau_{i+1})' . \qquad \text{(reflexivity)} \tag{6}$$

The wffs (3)-(6) constitute the premises of Inf. 9, and all the restrictions are apparently satisfied, so that

$$\vdash \sigma_k C \underset{X}{\tilde{\cong}} \sigma_k D \tag{7}$$

for each $\sigma_k \in C^-$. Therefore, for each $\sigma_k \in A^-$,

$$\begin{aligned}\sigma_k(A_1 \dots A_n) &\cong \sigma_k C && \text{(Theorems 11, 13)}\\ &\underset{X}{\tilde{\cong}} \sigma_k D && \text{(by (7))}\\ &\cong \sigma_k(B_1 \dots B_n) . && \text{(Theorems 11, 13)}\end{aligned}$$

Similarly,

$$
\begin{aligned}
A_1 \ldots A_n &\cong \tau_1 C \\
&\underset{X}{\cong} \tau_1 D \\
&\cong B_1 \ldots B_n \,.
\end{aligned}
$$
Q.E.D.

<u>Theorem 25</u>. (Verification Condition for Assignment Operator. Cf., Floyd (1967a)) Statement $x := f$ is partially correct w.r.t. p and q if and only if

$$\vdash_{\Gamma} p \supset q_x[f]^{\circ} \,.$$

<u>Proof</u>. We shall examine the conditions for p and q to satisfy

$$(p \to x := f, \Delta) \cong (p \to x := f; (q \to \Lambda, \Delta), \Delta) \,. \tag{1}$$

(See sections 4, 5; correctness, (1').)

$$
\begin{aligned}
(p \to x := f; (q \to \Lambda, \Delta), \Delta) &\cong (p \to (q_x[f]^{\circ} \to x := f, x := f; \Delta), \Delta) && \text{(Axiom 13)} \\
&\cong (p \wedge q_x[f]^{\circ} \to x := f, (p \wedge \neg\, q_x[f]^{\circ} \to \Delta, \Delta)) && \text{(Axioms 5, 11a)} \\
&\cong (p \wedge q_x[f]^{\circ} \to x := f, \Delta) \,. && \text{(Theorem 4)} \quad (2)
\end{aligned}
$$

Therefore (1) is equivalent to

$$(p \to x := f, \Delta) \cong (p \wedge q_x[f]^{\circ} \to x := f, \Delta) \,, \tag{3}$$

for which, obviously,

$$\vdash_{\Gamma} p \equiv p \wedge q_x[f]^{\circ} \,,$$

namely

$$\vdash_{\Gamma} p \supset q_x[f]^{\circ} \tag{4}$$

is necessary (see 3 below) and sufficient. Q.E.D.

<u>Remarks</u>

1. Formula of (4) is logically equivalent to Floyd's original formula (written in our notation):

$$\exists x_o (x = f_x[x_o]^{\circ} \wedge p_x[x_o]^{\circ}) \supset q \,,$$

provided that the equality axioms are satisfied. (Also, cf. Hoare (1969).)

2. We assumed the completeness of Γ (including the law of the excluded middle) in order to use Theorem 4.
3. The necessity is based on the meaning of formulas, which can be, however, improved as follows: We shall consider

$$\vdash (p \to \Lambda, \Delta) \cong \Lambda$$

as an assertion of the validity of formula p in the sense of predicate calculus, and denote it by

$$\vdash^* p \ .$$

Then we can prove $\vdash^* p \supset q_x[f]^\circ$ formally from wff (3) by the following derivation: Let r denote $p \wedge q_x[f]^\circ$, and A the statement $(r \to \Lambda, \Delta)$.

$$(p \to x := f, \Delta) \underset{\emptyset}{\cong} (p \to \Lambda, \Delta) \ . \qquad \text{(Axiom 8, Theorem 8)} \qquad (5)$$

$$(r \to x := f, \Delta) \underset{\emptyset}{\cong} (r \to \Lambda, \Delta) \qquad \text{(similarly)}$$

$$\underset{\emptyset}{\cong} (p \to \Lambda, \Delta) \ , \qquad \text{(by (5))}$$

so that

$$(p \to \Lambda, \Delta) \cong (r \to \Lambda, \Delta) \ . \qquad \text{(Inf. 3)} \qquad (6)$$

$$(p \to \Lambda, \Delta) \cong (p \wedge r \to \Lambda, (p \wedge \neg r \to \Delta, \Delta)) \qquad \text{(Theorem 4)}$$

$$\cong (p \to (r \to \Lambda, \Delta), \Delta) \qquad \text{(Axiom 11a)}$$

$$\cong (p \to (p \to \Lambda, \Delta), \Delta) \qquad \text{(by (6))}$$

$$\cong (p \wedge p \to \Lambda, (p \wedge \neg p \to \Delta, \Delta)) \qquad \text{(Axiom 11a)}$$

$$\cong (p \to \Lambda, \Delta) \ . \qquad \text{(Axiom 16b, Theorem 4)} \qquad (7)$$

Similarly

$$(\neg p \wedge \neg r \to \Lambda, \Delta) \cong (\neg p \to \Lambda, \Delta) \ . \qquad (8)$$

Therefore

$$(p \wedge r \vee \neg p \wedge \neg r \to \Lambda, \Delta) \cong (1 \to \Lambda, \Delta) \qquad \text{(Inf. 4, (7), (8))}$$

$$\cong \Lambda \ . \qquad (9)$$

But

$$(p \supset q_x[f]^\circ \to \Lambda, \Delta) \cong (p \equiv r \to \Lambda, \Delta) \qquad \text{(Axiom 16b)}$$

$$\cong (p \wedge r \vee \neg p \wedge \neg r \to \Lambda, \Delta) \ , \qquad \text{(similarly)} \qquad (10)$$

so that

$$\vdash (p \supset q_x[f]^\circ \rightarrow \Lambda, \Delta) \cong \Lambda . \qquad \text{(by (9), (10))}$$

(The sufficiency comes from Axiom 16b.)

4. Although the main reason that we introduced quantifiers into Algol-like statements (see section 2) is to include formulas of usual predicate calculus in conditional statements in connection with the notion of correctness, this syntactic generalization of Algol-like statements may not be essential. For, the study of Engeler (1967) seems to suggest that infinitary logic is frequently more appropriate than ordinary logic. It must be noted that the example given by Floyd (1967a) may be considered to be based upon infinitary logic. Also, the verification conditions for branch and join commands (the rest not being essential) can be stated and proved without using quantifiers, similarly to the above.

7. DECOMPOSITION OF STATEMENTS

Let V be a subset of $\mathscr{V}$ such that $\mathscr{V}-V$ contains infinite elements $w_0, w_1, \ldots,$ and L be a subset of $\mathscr{L}$ such that $\mathscr{L}-L$ contains infinite elements $\sigma_0, \sigma_1, \ldots$. By $\mathcal{A}_0$ is denoted the set of statements defined by induction as follows.

(d1) Λ belongs to $\mathcal{A}_0$.

(d2) For each $\sigma \in \mathscr{L}$, σ and σ^{-1} belong to $\mathcal{A}_0$.

(d3) For each $x \in \mathscr{V}$ and a fixed element w_0 of $\mathscr{V}-V$, $x := w_0$ and $w_0 := x$ belong to $\mathcal{A}_0$.

(d4) For each $\pi^{(n)} \in \mathscr{F}^{(n)}$ and $e_1, \ldots, e_{n-1}$ such that either $e_i \in \mathscr{F}^{(0)}$ or $e_i \in \mathscr{V}$ for each $i \in [n-1]$, $w_0 := \pi^{(n)} w_0 e_1 \ldots e_{n-1}$ belongs to $\mathcal{A}_0$.

(d5) For each $\rho^{(n)} \in \mathscr{P}^{(n)}$, $\sigma \in \mathscr{L}$, and $e_1, \ldots, e_{n-1}$ as above, $(\rho^{(n)} w_0 e_1 \ldots e_{n-1} \rightarrow \sigma, \Lambda)$ belongs to $\mathcal{A}_0$.

(e1) If A and B belong to $\mathcal{A}_0$, then AB belongs to $\mathcal{A}_0$. ($A^- \cap B^- = \phi$ should be satisfied. Otherwise, AB is not a statement.)

Let $\mathcal{A}_1$ be the set of statements consisting of all A such that $V[A] \subseteq V$, $A^\pm \subseteq L$, and that the logical symbols other than $\neg$ and $\vee$ do not occur in A .

We shall establish a function

$$\Phi : \mathcal{A}_1 \rightarrow \mathcal{A}_0 ,$$

which has the following characteristics.

1. Constructiveness: Φ is total and effectively defined.

2. Correctness: $\vdash A \underset{V}{\widetilde{=}} \Phi(A)$ for any $A \in \mathcal{A}_1$.

In other words, Φ is an algorithm that carries out a translation of $\mathcal{A}_1$ into $\mathcal{A}_0$, of which the latter consists of sequences of relatively simple statements. Moreover, we can formally prove that Φ always gives a statement equivalent to the original one insofar as the values of variables belonging to V and the destinations of exits are concerned. (Actually we prove the above also for each entry. Cf., proof of theorem 26.)

For the convenience of description, we introduce two sets of statements, as follows:

$$\mathcal{A}_2 = \{x := f \mid x \in \mathcal{V} \text{ and } V[f] \subseteq V\} .$$

$$\mathcal{A}_3 = \{(p \to \tau, \Lambda) \mid \tau \in \mathcal{L} \text{ and } V[p] \subseteq V\} .$$

Besides, $\mathcal{A}_1^*$, $\mathcal{A}_2^*$, and $\mathcal{A}_3^*$ will be used, whose elements differ from $\mathcal{A}_1$, $\mathcal{A}_2$, and $\mathcal{A}_3$, respectively, only in that some suffixes are added. (See definition of Θ below.)

<u>Definition of Φ</u>

Let Θ and Ψ be two functions as defined below. Then

$$\Phi(A) = \Psi(\Theta_0(A))$$

for each $A \in \mathcal{A}_1$.

1. <u>Definition of Θ</u>

We define the function

$$\Theta: \mathcal{A}_1 \times \mathfrak{n} \to \mathcal{A}_1^* ,$$

where the elements of $\mathcal{A}_1^*$ are statements whose symbols are possibly suffixed. For each A and each $\nu \in \mathfrak{n}$, $\Theta_\nu(A)$ denotes the image of (A, ν). Actually, however, Θ is extended so that, for each arithmetic expression f such that $V[f] \subseteq V$ and for each Boolean expression p such that $V[p] \subseteq V$, $\Theta_\nu(f)$ and $\Theta_\nu(p)$ are defined. Besides, two auxiliary functions

$$\lambda: \mathcal{Q}_1 \cup \{p \mid V[p] \subseteq V\} \to \mathfrak{n}$$

and

$$\mu: \{f \mid V[f] \subseteq V\} \to \mathfrak{n}$$

are defined.

Practical meaning of these functions are as follows.

$\mu(f)$: The number of required working storages to compute f .

$\Theta_\nu(f)$: The result of suffixing function symbols occurring in f so as to specify the allocation of working storages. (ν is irrelevant.)

$\lambda(p)$: The number of auxiliary labels to compute p , which is the number of occurrences of symbol $\neg$ in p .

$\mu(p)$: The number of required working storages to compute p .

$\Theta_\nu(p)$: The result of suffixing p to specify all the auxiliary labels using index greater than ν .

$\lambda(A)$ and $\Theta_\nu(A)$: Similar to $\lambda(p)$ and $\Theta_\nu(p)$.

Functions Θ , λ , and μ are defined simultaneously by induction on statements as follows.

<u>Atomic Statements</u>

(a1) $C = \Lambda$, σ , or σ^{-1}:
and
(a2)

$$\Theta_\nu(C) = C \quad \text{for each} \quad \nu .$$

$$\lambda(C) = 0 .$$

(a3) $C = x := f$, where $f = y$:

$$\mu(f) = 0 .$$

$$\Theta_\nu(f) = f \quad \text{for each} \quad \nu .$$

$$\Theta_\nu(C) = x := \Theta_\nu(f) . \tag{1}$$

$$\lambda(C) = 0 . \tag{2}$$

<u>Statements</u> (non-atomic)

(b1) $C = AB$:

$$\Theta_\nu(C) = \Theta_\nu(A)\Theta_{\nu+\lambda(A)}(B) .$$

$$\lambda(C) = \lambda(A) + \lambda(B) .$$

(b2) $C = x := f$, where $f = \pi^{(n)} f_0 \dots f_{n-1}$:

$$\mu(f) = M + m ,$$

where

$$M = \max_{0 \le i \le n-1} \mu(f_i) , \tag{3}$$

and m is the number of f_i such that $f_i \notin V$.

$$\Theta_\nu(f) = \pi^{(n)}_{M+1,\ldots,M+m}\, \Theta_\nu(f_0) \ldots \Theta_\nu(f_{n-1}) . \tag{4}$$

$\Theta_\nu(C)$ and $\lambda(C)$ are defined by (1) and (2) above.

(b3) $C = (p \to A,B)$, where $p = \rho^{(n)} f_0 \ldots f_{n-1}$:

$$\mu(p) = M + m ,$$

where M and m are defined by (3) and (4) above.

$$\Theta_\nu(p) = \rho^{(n)}_{M+1,\ldots,M+m}\, \Theta_\nu(f_0) \ldots \Theta_\nu(f_{n-1}) .$$

$$\lambda(p) = 0 .$$

(i) If A is τ and B is Λ , then

$$\Theta_\nu(C) = (\Theta_\nu(p) \to \tau, \Lambda) , \tag{5}$$

and

$$\lambda(C) = \lambda(p) . \tag{6}$$

(ii) If A is not of the form τ or B is not Λ , then

$$\Theta_\nu(C) = (\Theta_{\nu+\lambda(A)+\lambda(B)}(p) \to_{N+1,N+2} \Theta_\nu(A), \Theta_{\nu+\lambda(A)}(B)) , \tag{7}$$

where

$$N = \nu + \lambda(A) + \lambda(B) + \lambda(p) ,$$

and

$$\lambda(C) = N + 2 . \tag{8}$$

(c1) $C = (\neg\, p \to A,B)$:

$$\Theta_\nu(\neg\, p) = \neg_{\nu+1}\, \Theta_\nu(p) .$$

$$\lambda(\neg\, p) = \lambda(p) + 1 .$$

$\Theta_\nu(C)$ and $\lambda(C)$ are defined by (5)-(8) above. (Replace p by $\neg p$.)

(c2) $C = (p \vee q \to A, B)$:

$$\Theta_\nu(p \vee q) = \Theta_\nu(p) \vee \Theta_{\nu+\lambda(p)}(q) .$$

$$\lambda(p \vee q) = \lambda(p) + \lambda(q) .$$

$\Theta_\nu(C)$ and $\lambda(C)$ are defined by (5)-(8) above. (Replace p by $p \vee q$.)

2. Definition of Ψ

We define the function

$$\Psi: \mathcal{A}_1^* \cup \mathcal{A}_2^* \cup \mathcal{A}_3^* \to \mathcal{A}_0 .$$

By A^*, f^*, and p^* will be denoted $\Theta_\nu(A)$, $\Theta_\nu(f)$, and $\Theta_\nu(p)$, respectively, for certain values of ν. Thus, for instance, (b1) below, i.e.,

$$\Psi(A^*B^*) = \Psi(A^*)\Psi(B^*) ,$$

reads as follows:

Since $C = AB$, $\Theta_\nu(C)$ is of the form A^*B^*. Define $\Psi(A^*)\Psi(B^*)$ as $\Psi(\Theta_\nu(C))$.

Ψ is defined by induction as follows. (w_0 plays the role of an accumulator.)

Atomic Statements

(a1) $\Psi(\Lambda) = \Lambda$.

(a2) $\Psi(\sigma) = \sigma$.

$\Psi(\sigma^{-1}) = \sigma^{-1}$.

(a3) (i) $\Psi(w_0 := y) = w_0 := y$.

(ii) If $x \neq w_0$, then $\Psi(x := y)$ is defined by (1) below. (Substitute y in place of f .)

Statements (non-atomic)

(b1) $\Psi(A^*B^*) = \Psi(A^*)\Psi(B^*)$.

(b2) (i) $\Psi(w_0 := \pi^{(0)}) = w_0 := \pi^{(0)}$.

(ii) $\Psi(w_0 := \pi^{(n)}_{\alpha(1)\ldots\alpha(m)}\, f_0^* \ldots f_{n-1}^*) = C_{n-1} \ldots C_0 \,;\, w_0 := \pi^{(n)} w_0 u_1 \ldots u_{n-1}$, $(n \geq 1)$

where

$$u_i = \begin{cases} f_i & , \quad f_i \in V \\ w_{\alpha(\beta(i))} & , \quad f_i \notin V \end{cases} \qquad \text{for each } i \in [n-1] ,$$

C_0 is $w_0 := f_0$, and

$$C_i = \begin{cases} \Lambda & , \ f_i \in V \\ u_i := f_i & , \ f_i \notin V \end{cases} \qquad \text{for each } i \in [n-1] ,$$

$\beta(i)$ being defined by the following induction:

$$\beta(0) = 0 .$$

$$\beta(i+1) = \begin{cases} \beta(i) & , \ f_i \in V \\ \beta(i)+1 & , \ f_i \notin V . \end{cases}$$

(iii) If $x \neq w_0$, then

$$\Psi(x := f^*) = \Psi(w_0 := f^*)\, x := w_0 . \tag{1}$$

(b3) (i) $\Psi((\rho^{(0)} \rightarrow \tau, \Lambda)) = (\rho^{(0)} \rightarrow \tau, \Lambda)$.

(ii) $\Psi((\rho^{(n)}_{\alpha(1)\ldots\alpha(m)} f^*_0 \ldots f^*_{n-1} \rightarrow_{\gamma(1)\gamma(2)} \tau, \Lambda)$, $(n \geq 1)$,

where $C_0, \ldots, C_{n-1}, u_1, \ldots, u_{n-1}$ are the same as above (cf., (b2(ii))).

(iii) If A is not of the form τ or B is not Λ, then

$$\begin{aligned} &\Psi((p^* \rightarrow_{\gamma(1)\gamma(2)} A^*, B^*)) \\ &= \Psi((p^* \rightarrow_{\gamma(1)\gamma(2)} \sigma_{\gamma(1)}, \Lambda))\Psi(B^*)\sigma_{\gamma(2)}\sigma^{-1}_{\gamma(1)}\Psi(A^*)\sigma^{-1}_{\gamma(2)} . \end{aligned} \tag{2}$$

(c1) (i) $\Psi((\neg_\delta\ p^* \rightarrow \tau, \Lambda)) = \Psi((p^* \rightarrow \sigma_\delta, \Lambda))\tau\sigma^{-1}_\delta$.

(ii) If A is not of the form τ, or B is not Λ, then

$\Psi((\neg_\delta\ p^* \rightarrow_{\gamma(1)\gamma(2)} A, B))$ is defined by (2) above. (Substitute $\neg_\delta p^*$ in place of p^* .)

(c2) (i) $\Psi((p^* \vee q^* \rightarrow \tau, \Lambda)) = \Psi((p^* \rightarrow \tau, \Lambda))\Psi((q^* \rightarrow \tau, \Lambda))$.

(ii) If A is not of the form τ, or B is not Λ, then

$\Psi((p^* \vee q^* \rightarrow_{\gamma(1)\gamma(2)} A, B))$ is defined by (2) above. (Substitute $p^* \vee q^*$ in place of p^* .)

<u>Example</u>

We consider the statement

$$\underline{\text{if}}\ x < 0\ \underline{\text{then}}\ x := -x , \tag{1}$$

which was used as an example of compilation in Igarashi (1968).

Here, let us allow only binary $-$, and see how the statement

$$\underline{\text{if}}\ \ x<0\ \ \underline{\text{then}}\ \ x:=0-x\ , \tag{2}$$

namely

$$(x<0 \to x:=0-x,\Lambda) \tag{3}$$

in our notation, is treated.

Let A be $(\rho^{(1)}x \to x:=\pi^{(2)}\pi^{(0)}x,\Lambda)$. Then

$$\Theta_0(A) = A^* = (\rho^{(1)}x \to_{1,2} x:=\pi_1^{(2)}\pi^{(0)}x,\Lambda)$$

and

$$\Psi(A^*) = w_0 := x\,;\,(\rho^{(1)}w_0 \to \sigma_1,\Lambda)\sigma_2\sigma_1^{-1}\,;\,w_0 := \pi^{(0)}\,;\,w_0 := \pi^{(2)}w_0x\,;\,x := w_0\,;\,\sigma_2^{-1}\ .$$

Especially, we define $x<0$ as $\rho^{(1)}x$, 0 as $\pi^{(0)}$, and $x-y$ as $\pi^{(0)}xy$, so that A becomes (3).

For readability's sake, $\Phi(A)$, i.e., $\Psi(A^*)$ will be written in ALGOL 60 and listed with corresponding actions, symbols w_0, σ_1, and σ_2 being replaced by acc , L1 , and L2 , respectively.

	acc := x ;	load x	
	<u>if</u> acc < 0 <u>then</u> <u>go</u> <u>to</u> L1 ;	jump on minus L1	
	<u>go</u> <u>to</u> L2 ;	jump L2	
L1:		insert label L1	
	<u>acc</u> := 0 ;	load 0	(4)
	<u>acc</u> := <u>acc</u> − x ;	subtract x	
	x := <u>acc</u> ;	store x	
L2:		insert label L2	

Statement (4) is different only in trivial points from program $\mathfrak{B}$ (in the above paper) for which

$$\vdash(1)\ \underset{\{x\}}{\overset{=}{\ }}\ \mathfrak{B}$$

is proved as an example of derivation. That proof, for this <u>particular</u> pair of statements needed two pages of derivation (20 steps) preceded by one page (10 steps) for an auxiliary formula, being derived directly from the previous formal system. In the present paper, however, we shall prove, also formally, that

$$A \underset{V}{\cong} \Phi(A)$$

is valid for <u>every</u> $A \in \mathfrak{A}_1$, which implies that (2) $\underset{\mathcal{V}-\{acc\}}{\cong}$ (4).

8. FORMAL PROOF OF THE CORRECTNESS OF DECOMPOSITION

In this section we shall prove formally the following theorem which implies the validity of the transformation defined in the previous section.

<u>Theorem 26</u>. Let $(\mathcal{V},\varkappa,\mathfrak{R},\Gamma^o,J)$ be an interpretation such that $\pi_{\mathfrak{R}}$ is a total function for each $\pi \in \mathfrak{F}$ and that $\rho_{\mathfrak{R}}$ is a total predicate for each $\rho \in \mathfrak{P}$. Then

$$\vdash A \underset{V}{\cong} \Phi(A)$$

for any $A \in \mathfrak{A}_1$.

We shall prove the following lemmas first.

<u>Lemma 1</u>. If $x \notin V[f]$, then

$$\vdash x := f \,;\, y := g \underset{\mathcal{V}-\{x\}}{\cong} y := g_x[f]^o ,$$

and

$$\vdash x := f \,;\, (p \to \sigma, \Lambda) \underset{\mathcal{V}-\{x\}}{\cong} (p_x[f]^o \to \sigma, \Lambda) .$$

<u>Proof</u>. Choose z such that $z \neq x$.

$$\begin{aligned} x := f \,;\, y := g &\cong x := f \,;\, y := g \,;\, z := z && \text{(Axioms 2a, 6, etc.)} \\ &\cong x := f \,;\, y := g_x[f] \,;\, z := z && \text{(Axiom 7b)} \\ &\cong x := f \,;\, y := g_x[f] . && \text{(conversely)} \qquad (1) \end{aligned}$$

$$x := f \underset{\mathcal{V}-\{x\}}{\cong} \Lambda , \qquad \text{(Axiom 8)} \qquad (2)$$

so that, right multiplying both sides of (2) by $y := g_x[f]^o$, we obtain

$$x := f \,;\, y := g_x[f]^o \underset{\mathcal{V}-\{x\}}{\cong} y := g_x[f]^o . \qquad \text{(Inf. 6)} \qquad (3)$$

It must be noted that only $g_x[f]^\circ$, instead of an arbitrary $g_x[f]$, should be used because it must not contain x to use Inf. 6. By (1) and (3), the first wff is provable, while the latter can be proved in the same manner. Q.E.D.

Lemma 2. Let C and D denote $(p \to A, B)$ and $(p \to \tau_1, \Lambda)B\tau_2\tau_1^{-1}A\tau_2^{-1}$, respectively. Then

$$\vdash C \cong D ,$$

and

$$\vdash \sigma C \cong \sigma D$$

for each $\sigma \in A^- \cup B^-$.

Proof. Let $\hat{C}$ and $\hat{D}$ denote

$$\gamma^{-1}(p \to \alpha\alpha^{-1}A, \beta\beta^{-1}B)\delta$$

and

$$\gamma^{-1}(p \to \tau_1, \Lambda)\beta\beta^{-1}B\tau_2\tau_1^{-1}\alpha\alpha^{-1}A\tau_2^{-1}\delta ,$$

respectively, where α, β, γ, and δ do not belong to $C^\pm \cup D^\pm$. Then by theorems 11-13, $\vdash C \cong \hat{C}$, $\vdash \sigma C \cong \sigma\hat{C}$, $\vdash D \cong \hat{D}$, and $\vdash \sigma D \cong \sigma\hat{D}$, for any $\sigma \in A^- \cup B^-$. Let $\{\sigma_1, \ldots, \sigma_n\}$ be $A^- \cup B^- \cup \{\alpha, \beta, \gamma\}$.

$$\sigma_i\hat{D} \cong \begin{cases} (\sigma_i\alpha^{-1}A\tau_2^{-1}\delta)'\hat{D} & , \quad \sigma_i \in A^- \cup \{\alpha\} \\ (\sigma_i\beta^{-1}B\tau_2)'\hat{D} & , \quad \sigma_i \in B^- \cup \{\beta\} \\ (\gamma\gamma^{-1}(p \to \tau_1, \Lambda)\beta)'\hat{D} & , \quad \sigma_i = \gamma . \end{cases} \qquad \text{(Theorem 16, Theorem 17)}$$

But

$$\begin{aligned} (\sigma_i\beta^{-1}B\tau_2)'\hat{D} &\cong (\sigma_i\beta^{-1}B)'\tau_2\hat{D} \\ &\cong (\sigma_i\beta^{-1}B)'\delta\hat{D} , && \text{(Theorem 22)} \end{aligned}$$

and

$$\begin{aligned} (\gamma\gamma^{-1}(p \to \tau_1, \Lambda)\beta)'\hat{D} &\cong (p \to \tau_1, \Lambda)\beta\hat{D} \\ &\cong (p \to \alpha, \Lambda)\beta\hat{D} && \text{(Theorem 21)} \\ &\cong (p \to \alpha\beta, \beta)\hat{D} && \text{(Axiom 12a)} \\ &\cong (p \to \alpha, \beta)\hat{D} . && \text{(Theorem 12)} \end{aligned}$$

Therefore

$$\sigma_i\hat{D} \cong \begin{cases} (\sigma_i A)'\delta\hat{D} , & \sigma_i \in A^- \\ (\sigma_i B)'\delta\hat{D} , & \sigma_i \in B^- \\ A'\delta\hat{D} , & \sigma_i = \alpha \\ B'\delta\hat{D} , & \sigma_i = \beta \\ (p \to \alpha,\beta)\hat{D} , & \sigma_i = \gamma . \end{cases} \qquad (1)$$

Apparently (1) is provable if $\hat{C}$ is substituted in place of $\hat{D}$, so that

$$\sigma_i\hat{C} \cong \sigma_i\hat{D} \quad \text{for each} \quad \sigma_i \in A^- \cup B^- \cup \{\alpha,\beta,\gamma\} .$$

Therefore

$$\sigma C \cong \sigma\hat{C} \cong \sigma\hat{D} \cong \sigma D$$

for each $\sigma \in A^- \cup B^-$, and

$$C \cong \hat{C} \cong \gamma\hat{C} \cong \gamma\hat{D} \cong \hat{D} \cong D . \qquad \text{(Theorem 18)} \qquad \text{Q.E.D.}$$

<u>Lemma 3</u>. $\vdash \sigma(p \to A,B) \cong \sigma(q \to B,A)$ for each $\sigma \in A^- \cup B^-$.

<u>Proof</u>. Let τ be a label symbol such that $\tau \notin A^- \cup B^-$. Then

$$\sigma(p \to A,B)\tau \cong \sigma(p \to A\tau,B\tau) , \qquad \text{(Axiom 12b)} \qquad (1)$$

$$\sigma(q \to B,A)\tau \cong \sigma(q \to B\tau,A\tau) , \qquad \text{(Axiom 12b)} \qquad (2)$$

and by theorem 16

$$\sigma(p \to A\tau,B\tau) \cong \begin{cases} (\sigma A)'\tau(p \to A\tau,B\tau) , & \sigma \in A^- , \\ (\sigma B)'\tau(p \to A\tau,B\tau) , & \sigma \in B^- . \end{cases} \qquad (3)$$

Similarly

$$\sigma(q \to B\tau,A\tau) \cong \begin{cases} (\sigma A)'\tau(q \to B\tau,A\tau) , & \sigma \in A^- , \\ (\sigma B)'\tau(q \to B\tau,A\tau) , & \sigma \in B^- . \end{cases} \qquad (4)$$

Therefore, by theorem 18,

$$\sigma(p \to A\tau,B\tau) \cong \sigma(q \to B\tau,A\tau) , \qquad (5)$$

so that, using (1) and (2),

$$\sigma(p \to A,B)\tau \cong \sigma(q \to B,A)\tau .$$

Thus

$$\sigma(p \to A,B) \cong \sigma(q \to B,A) \ . \qquad \text{(Theorem 14)} \qquad \text{Q.E.D.}$$

Lemma 4. If the interpretation satisfies the premise of theorem 26, then

$$\vdash (p \to \tau,\Lambda)(q \to \tau,\Lambda) \cong (p \vee q \to \tau,\Lambda) \ .$$

Proof.

$$(p \to \tau,\Lambda)(q \to \tau,\Lambda) \cong (p \to \tau(q \to \tau,\Lambda),(q \to \tau,\Lambda)) \qquad \text{(Axiom 12a)}$$

$$\cong (p \to \tau,(q \to \tau,\Lambda)) \ . \qquad \text{(Theorem 12)} \quad (1)$$

$$(p \vee q \to \tau,\Lambda) \cong (\neg(\neg\, p \wedge \neg\, q) \to \tau,\Lambda) \qquad \text{(Axiom 16b)}$$

$$\cong (\neg\, p \wedge \neg\, q \to \Lambda,\tau) \qquad \text{(Axiom 10)}$$

$$\cong (\neg\, p \wedge \neg\, q \to \Lambda,(\neg\, p \wedge q \to \tau,\tau)) \qquad \text{(Theorem 4)}$$

$$\cong (\neg\, p \to (\neg\, q \to \Lambda,\tau),\tau) \qquad \text{(Axiom 11a)}$$

$$\cong (p \to \tau,(\neg\, q \to \Lambda,\tau)) \qquad \text{(Axiom 10)}$$

$$\cong (p \to \tau,(q \to \tau,\Lambda)) \ . \qquad \text{(Axiom 10)} \quad (2)$$

Statements (1) and (2) are identical. Q.E.D.

Proof of Theorem 26. We shall prove the following statements, which include the conclusion of the theorem, by induction.

1. For each $A \in \mathcal{O}_1$ such that A is neither of the form $x := f$ nor of the form $(p \to \tau,\Lambda)$,

$$\vdash A \underset{V}{\cong} \Psi(A^*)$$

and

$$\vdash \sigma A \underset{V}{\cong} \sigma\Psi(A^*)$$

for each $\sigma \in A^-$.

2. For each statement of the form $x := f$ belonging to $\mathcal{O}_1$ or $\mathcal{O}_2$,

$$\vdash x := f \underset{S[x:=f^*]}{\cong} \Psi(x := f^*) \ ,$$

where

$$S[x := f^*] = \{x\} \cup (\mathcal{V} - W[f^*] - \{w_0\}) \ ,$$

$W[f^*]$ being $\{w_i \mid i \text{ occurs in } f^* \text{ as suffix}\}$.

3. For each statement of the form $(p \to \tau, \Lambda)$ belonging to $\mathcal{A}_1$ or $\mathcal{A}_3$,

$$\vdash (p \to \tau, \Lambda) \underset{V}{\cong} \Psi((p^* \to \tau, \Lambda)) .$$

Since $V \subseteq S[x := f^*]$, these statements imply

$$\vdash A \underset{V}{\cong} \Psi(A^*)$$

for any $A \in \mathcal{A}_1$.

Atomic Statements

(a1), (a2(i)), (a2(ii)), and (a3(i)). $\Psi(A^*)$ is identical with A, so that the above statements are apparent.

(a3(ii)). $\Psi(w_0 := y) x := w_0$ is $w_0 := y ; x := w_0$, for which

$$w_0 := y ; x := w_0 \underset{\mathcal{V}-\{w_0\}}{\cong} x := y , \qquad \text{(Lemma 1)}$$

and

$$S[x := y] = \mathcal{V} - \{w_0\} .$$

Statements (non-atomic)

Hereafter the statements 1-3 will be used as the induction hypotheses.

(b1). By hypothesis 1,

$$A \underset{V}{\cong} \Psi(A^*) ,$$

$$\sigma A \underset{V}{\cong} \sigma\Psi(A^*) \quad \text{for each } \sigma \in A^- ,$$

$$B \underset{V}{\cong} \Psi(B^*) ,$$

and

$$\sigma B \underset{V}{\cong} \sigma\Psi(B^*) \quad \text{for each } \sigma \in B^- .$$

Therefore

$$AB \underset{V}{\cong} \Psi(A^*)\Psi(B^*)$$

and

$$\sigma AB \underset{V}{\cong} \sigma\Psi(A^*)\Psi(B^*) . \qquad \text{(Theorem 24)}$$

(b2(i)). Apparent because $\Psi(A^*)$ is identical with A.

(b2(ii)). Let $D^{(n)}$ be $\pi^{(n)} w_0 u_1 \ldots u_{n-1}$, $D_k^{(n)}$ be $\pi^{(n)} f_0 \ldots f_k u_{k+1} \ldots u_{n-1}$, and

$$T_k = \mathcal{V} - \bigcup_{i=0}^{k} W[f_i^*] - \{u_i \mid i \in [k] \text{ and } u_i \notin V\},$$

for $k = 0,\dots,n-1$. We prove first

$$C_k \dots C_0 D^{(n)} \underset{T_k}{\simeq} D_k^{(n)}, \tag{1}$$

for each k, by induction on k.

<u>Step</u> $k = 0$:

$$w_0 := f_0 \underset{S[w_0:=f_0^*]}{\simeq} C_0 . \qquad \text{(Hypothesis 2)} \tag{2}$$

We note that

$$R[D^{(n)}] = \{w_0, u_1, \dots, u_{n-1}\} \subseteq S[w_0 := f_0^*], \tag{3}$$

which is shown as follows.

$$S[w_0 := f_0^*] = \mathcal{V} - w[f_0^*] . \qquad \text{(by definition)}$$

But by definition of $W[f_0^*]$,

$$w_0 \notin W[f_0^*],$$

$$W[f_0^*] \cap V = \phi,$$

and, if $u_i \notin V$, then $u_i \notin W[f_0^*]$, $(i = 0,\dots,n-1)$. Thus $R[D^{(n)}] \cap W[f_0^*] = \phi$, so that (3) holds. By (2) and (3),

$$w_0 := f_0 ; D^{(n)} \underset{S[w_0:=f_0^*]\cup\{w_0\}}{\simeq} C_0 D^{(n)} \qquad \text{(Inf. 6)}$$

$$\simeq D_0^{(n)} . \qquad \text{(Axiom 7a)} \tag{4}$$

<u>Step</u> $k+1$: We use (1) as the supposition of induction. First, we prove the case that $u_{k+1} \notin V$.

$$C_{k+1}(C_k \dots C_0 D^{(n)}) \underset{T_k}{\simeq} C_{k+1} D_k^{(n)} . \qquad \text{(Inf. 7, (1))} \tag{5}$$

$$C_{k+1} \underset{S[u_{k+1}:=f_{k+1}^*]}{\simeq} u_{k+1} := f_{k+1} . \qquad \text{(Hypothesis 2)} \tag{6}$$

We note that

$$R[D_k^{(n)}] = (\bigcup_{i=0}^{k} V[f_i]) \cup \{u_{k+1},\ldots,u_{n-1}\} \subseteq S[u_{k+1} := f_{k+1}^*] \ , \tag{7}$$

which is shown similarly to the above, by

$$W[f_{k+1}^*] \cap V = \emptyset \ ,$$

$$V[f_i] \subseteq V \ ,$$

and, if $u_i \notin V$, then $u_i \notin W[f_{k+1}^*]$, $(i = 0,\ldots,n-1)$. Therefore

$$C_{k+1}D_k^{(n)} \underset{S[u_{k+1}:=f_{k+1}^*] \cup \{w_0\}}{\simeq} u_{k+1} := f_{k+1} \, ; \, D_k^{(n)} \qquad \text{(Inf. 6)} \tag{8}$$

$$\underset{\mathcal{V}-\{u_{k+1}\}}{\simeq} D_{k+1}^{(n)} \ . \qquad \text{(Lemma 1)} \tag{9}$$

But

$$T_k \cap (S[u_{k+1} := f_{k+1}^*] \cup \{w_0\}) \cap (\mathcal{V} - \{u_{k+1}\}) = T_k \cap (\{u_{k+1}\} \cup (\mathcal{V} - W[f_{k+1}])) \cap (\mathcal{V} - \{u_{k+1}\})$$
$$= T_k - W[f_{k+1}] - \{u_{k+1}\}$$
$$= T_{k+1} \ ,$$

so that

$$C_{k+1}\ldots C_0 D^{(n)} \underset{T_{k+1}}{\simeq} D_{k+1}^{(n)} \ . \qquad \text{((5), (8), (9), Theorem 2)} \tag{10}$$

Secondly, the case that $u_{k+1} \in V$ has to be proved. In this case, however, C_{k+1} is Λ , $D_{k+1}^{(n)}$ is identical with $D_k^{(n)}$, and $T_{k+1} = T_k$, so that (1) implies (10). Thus (1) has been proved.

Let $k = n - 1$.

$$C_{n-1}\ldots D_0 D^{(n)} \underset{T_{n-1}}{\simeq} D_{n-1}^{(n)} \ . \qquad \text{(by (1))} \tag{11}$$

Apparently

$$T_{n-1} = \mathcal{V} - W[f^*] = S[w_0 := f^*] \ ,$$

and $D_{n-1}^{(n)}$ is $w_0 := \pi^{(n)} f_0 \ldots f_{n-1}$, that is, $w_0 := f$, so that

$$C_{n-1}\ldots C_0 D^{(n)} \underset{S[w_0:=f^*]}{\simeq} w_0 := f \ . \qquad \text{(by (11))} \tag{12}$$

(b2(iii)). By (b2(ii)) above,

$$\Psi(w_0 := f^*) \underset{\mathcal{V}-W[f^*]}{\simeq} w_0 := f \ ,$$

so that

$$\Psi(w_0 := f^*)x := w_0 \underset{\{x\}\cup(\mathcal{V}-W[f^*])}{\simeq} w_0 := f \,;\, x := w_0 \qquad \text{(Inf. 6)}$$

$$\underset{\mathcal{V}-\{w_0\}}{\simeq} x := f \ . \qquad \text{(Lemma 1)}$$

Therefore

$$x := f \underset{\{x\}\cup(\mathcal{V}-W[f^*]-\{w_0\})}{\simeq} \Psi(x := f^*) \ . \qquad \text{(Theorem 2)}$$

(b3(i)). Apparent because $\Psi(A^*)$ is identical with A .

(b3(ii)). We have only to modify the proof of (b2(ii)) as follows. Let $D^{(n)}$ be $(\rho^{(n)}w_0u_1\ldots u_{n-1} \to \tau,\Lambda)$, $D_k^{(n)}$ be $(\rho^{(n)}f_0\ldots f_ku_{k+1}\ldots u_{n-1} \to \tau,\Lambda)$, and

$$T_k = \mathcal{V} - \bigcup_{i=0}^{k} W[f_i^*] - \{u_i \mid i \in [k] \text{ and } u_i \notin V\} - \{w_0\} \ ,$$

for $k = 0,\ldots,n-1$.

Then we can prove wff (1) above also for this case, using the axioms, theorems, etc. in the same manner. Letting $k = n-1$, we have

$$C_{n-1}\ldots D_0D^{(n)} \underset{\mathcal{V}-W[f^*]-\{w_0\}}{\simeq} (p \to \tau,\Lambda) \ . \qquad (13)$$

Therefore

$$\Psi((p^* \to \tau,\Lambda)) \underset{V}{\simeq} (p \to \tau,\Lambda) \ .$$

(b3(iii)). By (b3(ii)) above,

$$(p \to \sigma_{\gamma(1)},\Lambda) \underset{V}{\simeq} \Psi((p^* \to \sigma_{\gamma(1)},\Lambda)) \ . \qquad (14)$$

By hypothesis 1,

$$A \underset{V}{\simeq} \Psi(A^*) \ ,$$

$$\sigma A \underset{V}{\simeq} \sigma\Psi(A^*) \quad \text{for each } \sigma \in A^- \ .$$

$$B \underset{V}{\simeq} \Psi(B^*) \ ,$$

$$\sigma B \underset{V}{\simeq} \sigma\Psi(B^*) \quad \text{for each } \sigma \in B^- \ ,$$

and

$$\sigma_{\gamma(1)} \underset{V}{\cong} \sigma_{\gamma(1)} \quad \text{(reflexivity)}$$

Therefore

$$\Psi((p^* \to \sigma_{\gamma(1)}, \Lambda))\Psi(B^*)\sigma_{\gamma(2)}\sigma_{\gamma(1)}^{-1}\Psi(A^*)\sigma_{\gamma(2)}^{-1}$$

$$\underset{V}{\cong} (p \to \sigma_{\gamma(1)}, \Lambda)B\sigma_{\gamma(2)}\sigma_{\gamma(1)}^{-1}A\sigma_{\gamma(2)}^{-1} \quad \text{(Theorem 24)} \quad (15)$$

$$\cong (p \to A, B) . \quad \text{(Lemma 2)}$$

Thus

$$(p \to A, B) \underset{V}{\cong} \Psi((p^* \to_{\gamma(1)\gamma(2)} A^*, B^*)) . \quad (16)$$

By the same theorem and lemma,

$$\sigma(p \to A, B) \underset{V}{\cong} \sigma\Psi((p^* \to_{\gamma(1)\gamma(2)} A^*, B^*)) .$$

<u>(c1(i))</u>.

$$\Psi((p^* \to \sigma_\delta, \Lambda)) \underset{V}{\cong} (p \to \sigma_\delta, \Lambda) . \quad \text{(Hypothesis 3)} \quad (17)$$

Right multiplying both sides of (17) by $\tau\sigma_\delta^{-1}$, we have

$$\Psi((p^* \to \sigma_\delta, \Lambda))\tau\sigma_\delta^{-1} \underset{V}{\cong} (p \to \sigma_\delta, \Lambda)\tau\sigma_\delta^{-1} \quad \text{(Inf. 6)}$$

$$\cong (p \to \sigma_\delta\tau\sigma_\delta^{-1}, \tau\sigma_\delta'^{-1}) \quad \text{(Axiom 12a)}$$

$$\cong (p \to \Lambda, \tau) \quad \text{(Theorems 11-13)}$$

$$\cong (\neg\, p \to \tau, \Lambda) . \quad \text{(Axiom 10)}$$

Thus

$$\Psi((\neg_\delta\, p^* \to \tau, \Lambda) \underset{V}{\cong} (\neg\, p \to \tau, \Lambda) . \quad \text{(Theorem 2)}$$

<u>(c1(ii))</u>. By (c1(i)) above, (14) holds also in this case, so that the same proof as that of (b3(iii)) suffices. (Substitute $\neg p$ and $\neg_\delta\, p^*$ in place of p and p^* in (14), respectively.)

<u>(c2(i))</u>.

$$\Psi((p^* \to \tau, \Lambda)) \underset{V}{\cong} (p \to \tau, \Lambda) . \quad \text{(Hypothesis 3)} \quad (18)$$

$$\Psi((q^* \to \tau, \Lambda)) \underset{V}{\cong} (q \to \tau, \Lambda) . \quad \text{(similarly)} \quad (19)$$

Therefore, by theorem 24,

$$\Psi((p^* \to \tau,\Lambda))\Psi((q^* \to \tau,\Lambda)) \underset{V}{\cong} (p \to \tau,\Lambda)(q \to \tau,\Lambda)$$

$$\cong (p \vee q \to \tau,\Lambda) \ . \qquad \text{(Lemma 4)}$$

Thus

$$\Psi((p^* \vee q^* \to \tau,\Lambda)) \underset{V}{\cong} (p \vee q \to \tau,\Lambda) \ .$$

(c2(ii)). By (c2(i)) above, (14) holds also in this case, so that the same proof as that of (b3(iii)) suffices. (Substitute $p \vee q$ and $p^* \vee q^*$ in place of p and p^* in (14), respectively.) Q.E.D.

Acknowledgment

The writer acknowledges Professor J. McCarthy of Stanford University for his valuable suggestions regarding the reinforcement of the formalism. The writer also acknowledges Dr. J. W. de Bakker of Mathematisch Centrum for stimulating the refinement of the formalism and for giving interesting examples of derivation in his exposition (de Bakker (1969)). The writer thanks E. Ashcroft for his useful suggestions and critical reading of the manuscript.

PROVING CORRECTNESS OF IMPLEMENTATION TECHNIQUES

by

C. B. Jones and P. Lucas

1. INTRODUCTION

The existence of formal definitions of programming languages invites attempts to find a more systematic approach to the design of implementations. In particular, the work described below adopts the approach of isolating common parts of programming languages, modelling possible implementations on different levels of abstraction and exhibiting the correctness of such implementations.

The so-called Vienna Method is a collection of techniques based on abstract machines which interpret programs. These techniques are also used to define the implementation methods. The proof that an appropriately defined equivalence relation holds between the defining and implementation machines establishes the correctness of the latter.

From the subjects so far investigated, the block concept has been chosen to provide the example for this paper. The implementation method considered is the well-known display method.

All notation used is introduced in the next section, thus making the paper self-contained.

Section 3 is concerned with the example and includes a definition of the block concept and a correctness proof of the display method. Also included are some notes on the proof method used.

A review of other work done in Vienna on the description and justification of implementation methods is contained in section 4.

2. METHOD AND NOTATION

An attempt has been made to minimize the use of unconventional notation. However, some of the notation used in the Vienna definition work proved to be useful even for the simple example presented in this paper. To make this paper self-contained, this notation is introduced in section 2.1 below, as far as necessary; a more detailed exposition can be found in Lucas and Walk (1969).

2.1. Objects

Objects are either composite or elementary. A set of elementary objects is presupposed; natural numbers and certain sets of names are examples of elementary objects used in this paper (see section 3.1.2). Composite objects can be considered as finite tree structures with elementary objects at the terminal nodes. Branches are named with the restriction that two branches starting from the same node must not have the same name. Names of branches are called selectors. Subtrees of a given object are called components. Components directly attached to the root of a given tree structure are called immediate components.

The composite object with zero components is called the null object and is denoted by Ω .

Selectors are used to select components from given objects. Let s be a selector and x an object; then

s(x) denotes the immediate component of x whose name is s .

If there is no such component, including the case that x is an elementary object, then $s(x) = \Omega$.

The selection of components may be iterated, e.g., s1(s2(x)) denotes the s1 component of the s2 component of x .

Predicates over objects are defined in an obvious notation (see McCarthy (1962)). An object is said to be of a certain type if it satisfies a corresponding predicate.

In the sequel some objects will be used to represent mappings from certain finite sets into finite sets. By analogy with the terms used for mappings, one may talk about the domain $D(x)$ and the range $R(x)$ of a given object, where

$$D(x) \underset{Df}{=} \{s \mid s(x) \neq \Omega\}$$

$$R(x) \underset{Df}{=} \{s(x) \mid s(x) \neq \Omega\} .$$

For example, let

x =

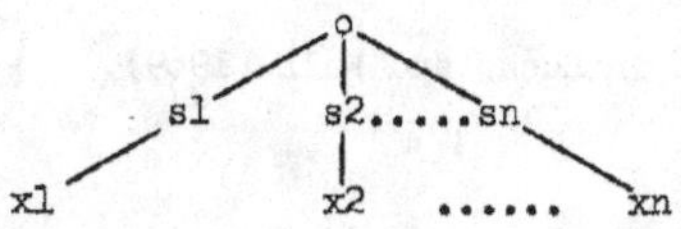

$$D(x) = \{s1, s2, \ldots, sn\} , \quad R(x) = \{x1, x2, \ldots, xn\} .$$

As a consequence $R(\Omega) = D(\Omega) = \{ \}$.

To remain in the present framework of objects, lists can be considered as objects whose immediate components are named by a subset of the natural numbers. However, in this paper elements of lists are referred to by the conventional subscript notation.

The following conventions are adopted for nonempty lists L :

$$\text{let} \quad L = \langle L_n, L_{n-1}, \ldots, L_1 \rangle$$

$$\ell(L) \underset{Df}{=} n \quad \text{length of the list ;}$$

$$top(L) \underset{Df}{=} L_n ;$$

$$rest(L) \underset{Df}{=} \langle L_{n-1}, \ldots, L_1 \rangle ;$$

$$rest(L,i) \underset{Df}{=} \langle L_i, L_{i-1}, \ldots, L_1 \rangle , \text{ for } 1 \leq i \leq n .$$

The following consequence is used:

(N1) for $L1 = rest(L2) \;\&\; 1 \leq i \leq \ell(L1)$: $rest(L1,i) = rest(L2,i)$.

2.2. Abstract Machines

For the present purpose, an abstract machine is defined by a set of initial states and a state transition function, $f(tx, \xi)$, which maps program texts, tx , and states into states.

A computation for a given initial state ξ^1 and a given program text tx is a sequence of states

$$\xi^1, \xi^2, \ldots, \xi^i, \xi^{i+1}, \ldots$$

such that

$$\xi^{i+1} = f(tx, \xi^i) .$$

The set of all possible states which the machine can assume is the set of all states occurring in computations of arbitrary initial states and program texts.

The state transitions of the underlying machine for section 3 will not be defined explicitly by a function but will only be constrained by certain postulates. This is because the paper is not about a specific language but a common part of many languages.

2.3. Twin Machines

The definitions of both the language part and its implementation are specified by abstract machines. For the purpose of proving certain equivalences between such machines, Lucas (1968) introduced the approach of combining the two machines into one, a so-called twin machine. Combining two machines into a twin machine means assembling the state components of the two machines into state components of the twin machine and defining the state transitions accordingly. The equivalence problem can then be formulated as a property of the states of the twin machine.

2.4. Notation Summary

logic:

$\neg$ not

& and

$\vee$ or

$\supset$ implication

$\forall$ universal quantifier

$\exists$ existential quantifier

sets:

$\in$ element of

$\notin$ not element of

$\cup$ union

$\cap$ intersection

$\subseteq$ subset or equal

$\{\ \}$ empty set

<u>conditional expressions</u>:

(prop $\rightarrow$ expr1,expr2) if prop then expr1 else expr2

<u>objects</u>:

$s(x)$ s component of x for: s is a selector, x is an object

I identity selector $(I(x) = x)$

Ω null object

$R(x)$ range of x (definition $\{s(x) \mid s(x) \neq \Omega\}$)

$D(x)$ domain of x (definition $\{s \mid s(x) \neq \Omega\}$)

$tp1 \Leftarrow tp_2$ tp1 immediately contained in tp2 (definition, see section 3.1.1)

<u>naming conventions</u>:

s- ... selector name

is- ... predicate name

<u>lists</u>: let $L = \langle L_n, L_{n-1}, \ldots, L_1 \rangle$

$\ell(L)$ length(=n)

top(L) L_n

rest(L) $\langle L_{n-1}, \ldots, L_1 \rangle$

rest(L,i) = $\langle L_i, L_{i-1}, \ldots, L_1 \rangle$

3. DEFINITION AND IMPLEMENTATION OF THE BLOCK CONCEPT

This section presents a definition of the block concept as first introduced into programming languages by ALGOL 60, Naur et al (1962), and a proof of correctness of a formulation of the stack mechanism used in the first complete implementations, Dijkstra (1960).[1]

The programming language under consideration is assumed to include the concept of blocks and procedures, and the possibility to pass procedures as arguments. Go to

---

[1] See also Van der Mey (1962).

statements and general call by name are not considered in order to avoid burdening the proofs.

The definition of the block concept is derived from the formal definitions of PL/I and ALGOL 60, Walk et al (1969), Lauer (1968). However, for this paper, the structure of programs being interpreted as well as the state and state-transitions of the abstract machine are specified only as far as is necessary to establish the subsequent proofs. This definition, therefore, illustrates the essentials of the block concept and the assumptions about the language necessary to guarantee the applicability of the stack mechanism.

The stack mechanism itself is also formulated as an abstract machine which, in the opinion of the authors, expresses the essential properties of this mechanism. No attempt is made to make the additional step from the general result to an actual machine implementation.

The proof of correctness is based on the twin machine, specified in section 3.1, which combines the defining machine and the stack mechanism.

The emphasis of the chapter is on the technique used to specify the block concept, its implementation, and on the proof of correctness. The end result has been known for many years but, apart from being an example of formally proving implementations correct, the proof indicates an argumentation which could be used informally in lecturing on the subject matter.

3.1. The Defining Model and its Implementation

3.1.1. The structure of programs

For the purpose of the present paper, it is sufficient to state that programs form a completely nested structure of blocks and procedure bodies. A program is itself a block which is referred to as the outermost block. More precisely, programs are objects of type is-block containing possibly nested components of type is-block or is-body.

Selectors[1] to components of programs are called text pointers (tp) . Consider a program tx and the set of all text pointers to blocks or procedure bodies of that

---

[1]Note that these are composite selectors, i.e., any selectors s1,s2,...,sn may be combined to form a composite selector $s1 \circ s2 \circ \ldots \circ sn$ where

$$s1 \circ s2 \circ \ldots \circ sn(x) \underset{Df}{=} s1(s2(\ldots(sn(x))\ldots)) \; .$$

program BP_{tx} . The text pointer to the entire program is I and $I \in BP_{tx}$, since programs are blocks.

The relation of tp1 being immediately contained in tp2 is written

$$tp1 \Leftarrow tp2 \quad \text{for} \quad tp1, \; tp2 \in BP_{tx} .$$

Since blocks and procedure bodies are completely nested, each tp1 , $tp1 \in BP_{tx}$ & $tp1 \neq I$ has a unique tp2 such that $tp2 \in BP_{tx}$ and $tp1 \Leftarrow tp2$. Thus for any given tp1, $tp1 \in BP_{tx}$, there is a unique chain of predecessors ending with I :

$$tp1 \Leftarrow tp2 \Leftarrow \ldots \Leftarrow I .$$

For each block or procedure body, there is a set of identifiers

$$ID(tx,tp) \quad \text{for} \quad \text{is-block}(tp(tx)) \vee \text{is-body}(tp(tx))$$

called the set of locally declared identifiers.

It is not necessary for the present purpose to state the attributes which can be declared with the identifiers nor how these attributes would be associated with the identifiers.

A pair (id,tp) is called an <u>occurrence</u> of the identifier id in tx if $tp(tx) = id$.

An occurrence is said to be immediately contained in a block or procedure body tp1 if its pointer part is contained in tp1 but is not contained in any block or procedure contained in tp1 .

3.1.2. <u>States of the twin machine and their significance</u>

The set of states of the twin machine (i.e., defining model) is the set of objects generated from the set of initial states by the state transitions (both initial states and state transitions are characterized in section 3.1.3). It would not be necessary to characterize the states in any other way. However, it seems advantageous to give a layout of the structure of the states and talk about the significance of the individual components of the states. Only certain components of the state are relevant for the present paper, therefore only the properties of these components need be fixed.

Like any other object, states are constructed from elementary objects and selectors, and these are now specified.

Elementary objects:

N natural numbers and zero

UN infinite set of names (referred to as unique names)[1]

TP set of text pointers

Selectors:

ID set of identifiers

UN set of unique names[1]

In addition, the following individual selectors are used: s-d, s-dn, s-U, s-tp, s-e, s-eo, s-epa . As mentioned in section 2, natural numbers can be considered as selectors for elements of lists.

The states of the twin machine satisfy the predicate is-state as defined below (in the style of McCarthy (1962)).

(S1) $\text{is-state}(\xi) \supset (\text{is-dump}(\text{s-d}(\xi)) \;\&\; \text{is-dendir}(\text{s-dn}(\xi)) \;\&\; \text{s-U}(\xi) \subseteq \text{UN})$;

(S2) $\text{is-dump}(d) \supset \text{is-de}(d_i)$ for $1 \leq i \leq \ell(d)$;

(S3) $\text{is-de}(de) \supset (\text{s-tp}(de) \in \text{TP} \;\&\; \text{is-env}(\text{s-e}(de)) \;\&\; \text{is-env}(\text{s-eo}(de)) \;\&\; \text{s-epa}(de) \in \text{N})$;

(S4) $\text{is-env}(e) \supset (D(e) \subseteq \text{ID}) \;\&\; (R(e) \subseteq \text{UN})$;

(S5) $\text{is-dendir}(dn) \supset (D(dn) \subseteq \text{UN}) \;\&\; (R(dn) \subseteq \text{DEN})$;

DEN is the set of all possible denotations; only procedure denotations are further specified;

(S6) $\text{is-proc-den}(den) \supset (den \in \text{DEN}) \;\&\; \text{s-tp}(den) \in \text{TP} \;\&\; \text{is-env}(\text{s-e}(den)) \;\&\; \text{s-epa}(den) \in \text{N})$.

Comments on the significance of the state components: Let ξ be a state (see S1):

$\text{s-d}(\xi)$ list called dump which corresponds to the dynamic sequence of block and procedure activations. The top element $\text{top}(\text{s-d}(\xi))$ corresponds to the most recent block or procedure activation;

$\text{s-dn}(\xi)$ denotation directory which associates unique names with their corresponding denotations, i.e., the denotation of a given unique name u is $u(\text{s-dn}(\xi))$;

$\text{s-U}(\xi)$ set of unique names used so far.

---

[1]Note the double role of unique names.

Let de be a dump element (see (S3)); such a dump element corresponds to a specific activation of a block or procedure.

s-tp(de) text pointer to the block or procedure;

s-e(de) environment which contains all referenceable identifiers of the block or procedure and their corresponding unique names for the specific activation. Thus if id is a referenceable identifier, then id(s-e(de)) yields its associated unique name;

s-eo(de) environment which contains all local identifiers of the block or procedure and their corresponding unique names for the activation;

s-epa(de) natural number which is the index of the dump element corresponding to the environmentally preceding block or procedure.[1]

Let den be a procedure denotation:

s-tp(den) pointer to the body of the procedure;

s-e(den) environment which resulted from the activation of the block which declared the procedure;

s-epa(den) the index of the dump element corresponding to the block activation in which the procedure was declared (environmentally preceding activation).

All s-e components are used exclusively by the defining model and all s-epa and s-eo components are used exclusively by the implementation model. The other components are common to both models.

The defining model is constructed with the copy rules of ALGOL 60 in mind (i.e., name conflicts of identifiers are resolved by suitable changes). The model deviates slightly from the copy rules in that new names are introduced for all identifiers and not only in cases where conflicts actually occur. Moreover the program text is not modified: instead the correspondence between identifiers and the new names introduced for them is recorded in the so-called environment. All information which may become associated with an identifier during computation (e.g., values of variables, procedures, etc.) is recorded in the denotation directory under the corresponding unique name. The set UN serves as the source for the unique names to be generated and the component U of the state contains all names already used. Procedure denotations

---

[1] Sometimes called statically or lexicographically preceding. The term environment chain is also used below in preference to the alternatives.

are composed of a text pointer to the body of the procedure and the environment which was current in the activation which introduced the procedure. Procedures are passed to parameters by passing their entire denotation.

The implementation model only keeps the local identifiers and their unique names for each activation. The so-called environment chain, established by the s-epa components of each dump element, permits reference to all referenceable identifiers. With this mechanism procedure denotations can be composed of the procedure body and index of the dump which corresponds to the activation which declared the procedure.

3.1.3. Initial states and state transitions

For convenience, some additional notational conventions are introduced.

Abbreviations for components of an arbitrary state:

$$d \underset{Df}{=} s\text{-}d(\xi) \quad \text{thus } d_i \text{ is the ith element of the dump of } \xi$$

$$dn \underset{Df}{=} s\text{-}dn(\xi)$$

$$U \underset{Df}{=} s\text{-}U(\xi) .$$

Components of the top element of the dump of ξ :

$$tp \underset{Df}{=} s\text{-}tp(top(d))$$

$$e \underset{Df}{=} s\text{-}e(top(d))$$

$$eo \underset{Df}{=} s\text{-}eo(top(d))$$

$$epa \underset{Df}{=} s\text{-}epa(top(d)) .$$

An arbitrary computation is a sequence of states:

$$\xi^1, \xi^2, \ldots, \xi^n, \xi^{n+1}, \ldots$$

where

ξ^1 ... arbitrary initial state

ξ^n ... state after n steps of the computation .

If possible, superscripts are avoided by using ξ , ξ' instead of ξ^n , ξ^{n+1} .

Finally, the above rules for abbreviating components of ξ may be analogously applied with superscripts, e.g.,

$$e' = s\text{-}e(top(d')) \quad \text{where} \quad d' = s\text{-}d(\xi') .$$

For the initial state of the twin machine, it is assumed that the outermost block has been activated and that this block has no declarations.[1] This means that the current text pointer tp points to the entire program, that there are no referenceable identifiers (empty environment and denotation directories) and that there are no used unique names.

An <u>initial state</u> ξ^1 has, therefore, the following properties:

(T1) $\ell(d^1)=1$

(T2) $tp^1=I$

(T3) $e^1=\Omega$

(T4) $eo^1=\Omega$

(T5) $epa^1=0$

(T6) $dn^1=\Omega$

(T7) $U^1=\{\ \}$.

The initial state may also depend upon the given input data for the program to be interpreted. However, this fact is irrelevant for the present discussion.

The following postulates on the state transitions are an attempt to characterize only block, procedure activation and the passing of procedures as parameters whilst avoiding restrictions of the considered class of programming languages with respect to other concepts and features they might contain. In particular, only a few assumptions are made about how the machine proceeds to interpret a given program text, e.g. no statement counter is mentioned in the postulates.

For a given program text tx , the machine starts a computation with a certain initial state. The possible transitions from one state to its successor state are constrained by the following postulates. For mathematical reasons, there is no condition under which the machine is said to stop. Instead, when, in a given state, the attempt is made to terminate the outermost block, this state is repeated infinitely many times, so that in the proof only infinite computations have to be considered.

Four cases of state transitions are distinguished:

1. block activation (including the treatment of locally declared identifiers);

[1] This is not a serious restriction because one can always embed programs in such a block.

2. procedure activation (including the passing of arguments, in particular, of procedures);
3. termination of blocks or procedures;
4. other state transitions.

Block and procedure activation have much in common so that is seems worthwhile to factor out these common properties.

The postulates refer to the transition from ξ to ξ' .

Common part of block and procedure activation

In either case, there is a text pointer tp-b to the block or procedure body to be activated. In the case of blocks, there is no specified way in which tp-b comes into existence, because the way in which interpretation proceeds is left, in general, unspecified. For procedure activations, tp-b is part of the denotation of the procedure identifier which caused the call. ID(tx,tp-b) yields the set of local identifiers declared in the block or the set of parameters of the procedure body. An auxiliary environment eo-b is constructed from the set of local identifiers by associating new unique names (which are at the same time recorded in the set of used names) with identifiers in the set. This auxiliary environment is then used as the new local environment and for the updating of the environment of the defining machine.

Given: tx , tp-b

Auxiliary environment eo-b :

(T8) $D(eo\text{-}b) = ID(tx, tp\text{-}b)$

(T9) $R(eo\text{-}b) \cap U = \{\ \}$.

State components:

(T10) $rest(d') = d$

(T11) $tp' = tp\text{-}b$

(T12) $eo' = eo\text{-}b$

(T13) for $u \in U$ & $is\text{-}proc\text{-}den(u(dn'))$: $u(dn') = u(dn)$

(T14) $U' = R(eo\text{-}b) \cup U$.

Comments:

ad (T9): This postulate only guarantees that the unique names used in the auxiliary environment are really new (not in the set of used names). This is sufficient for the proof, although to model ALGOL 60 or PL/I correctly, one must also guarantee that each identifier is associated with a different new name.

ad (T10): Notice that this also ensures that $\ell(d') = \ell(d) + 1$.

ad (T13): Guarantees that procedure denotations in the new state ξ' having old names have not been modified in any way. The postulate is sufficient in this form for the proof, although it permits deletion of existing procedure denotations upon activation. For a complete model of ALGOL 60 or PL/I, this would not be the case. No assumptions are made about denotations other than procedure denotations.

For block and procedure activations, it is necessary to create a new environment from a given one and the auxiliary environment containing the newly declared local identifiers of the block or procedure. The new environment is supposed to retain all the identifier-unique name pairs unless the identifiers are redeclared. This is achieved by the function "update" defined below.

Definition

(D1) update(e1,e2) = e3 such that

$$\mathrm{id}(e3) = \begin{cases} \mathrm{id} \in D(e2) \rightarrow \mathrm{id}(e2) \\ \mathrm{id} \notin D(e2) \rightarrow \mathrm{id}(e1) \end{cases}$$

for is-env(e1) and is-env(e2) .

The following immediate consequences will be useful:

(C1) $D(\text{update}(e1,e2)) = D(e1) \cup D(e2)$;

(C2) $R(\text{update}(e1,e2)) \subseteq R(e1) \cup R(e2)$.

The rest of the postulates can now be given.

Block activation

Given: tp-b , the text pointer to the block to be activated.

(T15) is-block(tp-b(tx)) ;

(T16) tp-b $\Leftarrow$ tp .

State components:

(T17) $e' = \text{update}(e, \text{eo-b})$

(T18) $\text{epa}' = \ell(d)$

(T19) for $u \in R(\text{eo-b})$ & is-proc-den(u(dn')) :

a) $\text{s-tp}(u(dn')) \Leftarrow \text{tp-b}$

is-body(s-tp(u(dn'))(tx))

b) $\text{s-e}(u(dn')) = e'$

c) $\text{s-epa}(u(dn')) = \ell(d')$.

Comments:

ad (T18): For blocks, the dynamically preceding activation is also the environmentally preceding one.

ad (T19): This postulate takes care of the procedure denotations that may have been introduced by local declarations within the block.

<u>Procedure activation</u>

Given: id-p , the identifier of the procedure to be activated.

(T20) $\text{id-p} \in D(e)$

(T21) is-proc-den(u-p(dn)) .

Abbreviations:

(T22) $\text{u-p} = \text{id-p}(e)$

$\text{tp-b} = \text{s-tp}(\text{u-p}(dn))$

$\text{e-p} = \text{s-e}(\text{u-p}(dn))$

$\text{epa-p} = \text{s-epa}(\text{u-p}(dn))$.

State components:

(T23) $e' = \text{update}(\text{e-p}, \text{eo-b})$

(T24) $\text{epa}' = \text{epa-p}$

(T25) for $u \in R(\text{eo-b})$ & is-proc-den(u(dn')) :

$\exists ul(ul \in R(e)$ & $u(dn') = ul(dn))$.

Comments:

ad (T24): The environmentally preceding activation is the one in which the procedure was declared and its denotation introduced.

ad (T25): Parameters are treated as the only local identifiers introduced by procedure activations. The postulate is concerned with passing procedures as arguments to parameters. No further assumptions are made on argument passing.

Termination of block and procedure activations

$\ell(d) > 1$:

(T26) $d' = rest(d)$

(T27) for $u \in U$ & is-proc-den(u(dn')) :

$u(dn') = u(dn)$

(T28) $U' = U$.

$\ell(d) = 1$:

(T29) $\xi' = \xi$.

Comments:

ad (T29): This case can be interpreted as an attempt to close the outermost block, i.e., the end of program execution. In this case, the state is repeated indefinitely in order to make all computations infinite.

Other state transitions

(T30) $d' = d$

(T31) for $u \in U$ & is-proc-den(u(dn')) :

$u(dn') = u(dn)$

(T32) $U' = U$.

3.2. Formulation of the Equivalence Problem

The equivalence problem in question is concerned with the reference to identifiers, more precisely, with the interpretation of the use of identifiers in the given program text.

It is assumed that reference to an identifier only ever involves reference to or change of the corresponding denotation in the denotation directory. Thus the environment in the defining model and the local environment and epa component of the implementation model are solely auxiliary devices to make the reference to the denotation directory possible. The unique names are irrelevant except for this purpose.

Therefore, the two reference mechanisms are considered to be equivalent if for any given state they yield the same unique name for any identifier, thus relating any identifier to the same entry in the denotation directory.

Reference mechanism of the defining model

The unique name which corresponds to a given referenceable identifier id for some state ξ is simply id(e) for $id \in D(e)$.

Reference mechanism 1 for the implementation model

The simplest way to get the unique name for a given referenceable identifier, using the implementation model, is to search along the environment chain to find the activation which introduced the identifier and its unique name. Two auxiliary functions[1] are introduced to accomplish this: the function s(id,d) which yields the number of steps along the environment chain to the activation which introduced the identifier; the function index(n,d) which yields the index of the dump element n steps back along the environment chain.

(D2) $s(id,d) = (id \in D(eo) \rightarrow 0 \text{ , } s(id,rest(d,epa)) + 1)$

(D3) $index(n,d) = (n = 0 \rightarrow \ell(d) \text{ , } index(n-1,rest(d,epa)))$

where $eo = s\text{-}eo(top(d))$ and $epa = s\text{-}epa(top(d))$.

The actual reference mechanism, for $id \in D(e)$, is

$$id(s\text{-}eo(d_{index(s(id,d),d)})) \text{ .}$$

Thus the first equivalence problem is, for $id \in D(e)$,

$$id(e) = id(s\text{-}eo(d_{index(s(id,d),d)})) \text{ .}$$

This result is proved as Theorem I in section 3.4.

Assumption

It is assumed that occurrences of identifiers are only interpreted if the block or body which immediately contains them is the current one (see section 3.1.1).

Under this assumption it turns out, as expected, that s(id,d) depends only on the given program text, i.e., the number of steps which will need to be made along the environment chain can be statically determined for each use of an identifier (see Theorem III).

Reference mechanism 2 of the implementation model

This mechanism differs from the previous one in the method by which the index of

---

[1] Recursive functions which rely on the environment chain are justified because Lemma 7 holds.

the required element is computed. A list called disp (for display) is introduced which contains indices of the dump elements of the environment chain. The display is used to determine the actual index for a given identifier.

Definitions

(D4) $\ell e(d) = (epa = 0 \rightarrow 0, \ell e(rest(d, epa)) + 1)$

where $epa = s\text{-}epa(top(d))$

(D5) $depth(id, d) = \ell e(rest(d, index(s(id, d), d)))$

(D6) $disp_i = index(\ell e(d) - i, d)$

for $0 \leq i \leq \ell e(d) - 1$.

The second reference mechanism, for $id \in D(e)$, is

$$id(s\text{-}eo(d_{disp_{depth(id,d)}})) .$$

The second equivalence problem is, for $id \in D(e)$,

$$id(e) = id(s\text{-}eo(d_{disp_{depth(id,d)}})) .$$

This result is proved as Theorem II in section 3.4.

The function depth is, under the above assumption, only a function of the program text (see Theorem III) as explained in Dijkstra (1960), who also introduced the display.

The display is usually kept as a state component and must be updated. It is sufficient, but not always necessary, to update the display every time the dump changes. The subject of optimization is discussed in Henhapl and Jones (1970).

The question may arise as to how relevant the presented implementation model is to actual implementations. Implementations as usually described (e.g., Randell and Russell (1964)) do not use unique names and denotations directories, but instead keep the information directly in the dump (stack). They use relative positions within each dump element instead of identifiers for access. Furthermore, procedure denotations, except for parameters, have not to be kept in the dump at all, since the text pointer can be computed statically and the index of the environmentally preceding activation can be computed in the same way as is already done for all other identifiers.

These deviations can be overcome by the following simple considerations. It is assumed that each identifier, when first introduced, is associated with a unique name different from the unique names chosen for all other identifiers. It follows that a unique name occurs at most once in the local environments of the dump. Therefore, the indirect step via the unique names is unnecessary. Since the only interesting entries of the denotation directory are those referred to by names occurring in some local environment of the dump, one can omit the denotation directory altogether and associate denotations directly with identifiers in the local environment. The introduction of relative positions instead of identifiers amounts to a trivial local change of names. The usual implementation where references are made via the integer pair (depth of nesting, relative position) is, therefore, derivable from the present model.

Apart from the final models being close to actual implementations, the lemmas and theorems give a clear insight into the structure of the state components and their relations.

3.3. Proof Principles

The basic proof principle in the subsequent proofs is induction on the number of steps of computations, i.e., a proof $P(\xi^n)$ for all n could, in general, take the form

basis: $P(\xi^1)$

induction step: a) $P(\xi^n) \supset P(\xi^{n+1})$, or

b) $\underset{m<n}{\forall} P(\xi^m) \supset P(\xi^n)$.

To prove the induction step, it is, in general, necessary to make four case distinctions according to the distinguished types of state transitions.

If the property to be proved is a property of the dump component $P(d^n)$ alone, it is valid to use induction on the length of the dump. To show this, the following two auxiliary lemmas are useful:

Lemma 1. (L1) For $1 \leq i < \ell(d^n)$: $\exists m(d^m = rest(d^n,i)\ \&\ 1 < m < n)$.

Lemma 2. (L2) $rest(d^n,1) = d^1$.

Both lemmas can be proved by induction on the steps of computations.[1]

---

[1] It is felt that inclusion of the detailed proofs at this point would detract from the development of the section. The interested reader may find them in Appendix I.

One can think of the set of all possible dumps as being generated from the dumps of the initial states and the state transitions corresponding to block and procedure activation only. This is valid because Lemma 1 shows that termination, for $\ell(d) > 1$, does not add new elements to the generated set, and from (T29) and (T30) it follows that the same is true for all other cases. From (T10) it follows that the rest of a dump generated by block and procedure activation is always identical to the dump of the preceding state. Therefore, the following induction principle is valid for proving say $P(d)$ for the specific machine and the dump component of its states.

basis: $P(d^1)$

induction step: a) $P(rest(d)) \supset P(d)$, or

b) $\forall_{i<\ell(d)} P(rest(d,i)) \supset P(d)$,

where, for the induction step, only the cases of block and procedure activations have to be considered.

Clearly the proof method also holds if $P(d)$ can be justified without appeal to the induction hypotheses. Furthermore, Lemma 1 states that for each instance of i, there exists a predecessor state which is identical to $rest(d,i)$. Therefore, from a given assertion $P(d)$, it is valid to conclude

$$\forall_{i<l(d)} P(rest(d,i)) .$$

3.4. Proofs[1]

Section 3.2 has introduced the two equivalence properties in which we are interested. These results, Theorems I and III, Th. III showing how certain indices can be statically computed and a number of lemmas used in the argument, are proved in this section.

Lemmas 3 and 4 show that the set of used unique names does not lose elements; procedure denotations whose names are in U cannot be changed to different procedure denotations. These results are required in Lemmas 5 and 6, where the inductive step based on procedure activation may rely on properties of any preceding element of the computation.

[1] Case distinctions are identified by underlining and indentation.

Lemma 3. (L3) For $1 \leq m < n$:

$$U^m \subseteq U^n .$$

Proof. The proof follows immediately from (T14), (T28), (T29), and (T32).

Lemma 4. (L4) For $1 \leq m < n$ & $u \in U^m$:

$$\text{is-proc-den}(u(dn^n)) \supset u(dn^n) = u(dn^m) .$$

Proof. The proof follows immediately from (L3), (T13), (T27), (T29), and (T31).

Lemma 5 shows that the range of the current environment is contained in U . The proof of the lemma is by induction on steps of the computation using Lemmas 1, 3, and 4 as well as the relevant state transition postulates.[1] Part (b) of the statement is included to make the induction for the case of procedure activation possible. The result is required in the proof of Lemma 6 to connect the fact that an identifier is in the domain of some environment component with the hypotheses of Lemma 4.

Lemma 5. (L5) a) $R(e) \subseteq U$;

b) for $u \in U$ & is-proc-den(u(dn)) : $R(\text{s-e}(u(dn))) \subseteq U$.

Lemma 6 states that any procedure denotation referenceable from some environment has an epa component not greater than the index of the dump containing that environment, and that the other components are those corresponding to the dump element which generated the denotation. The lemma is proved by induction on the steps of the computation using Lemmas 1, 4, and 5, as well as the relevant state transition postulates.[1] The result is required to establish the argument for the case of procedure activation in Lemmas 7, 8, and 10.

Lemma 6. (L6) For $u \in R(e)$ & is-proc-den(u(dn)) :

a) $1 \leq \text{s-epa}(u(dn)) \leq \ell(d)$;

b) $\text{s-e}(u(dn)) = \text{s-e}(d_{\text{s-epa}(u(dn))})$;

c) $\text{s-tp}(u(dn)) \Leftarrow \text{s-tp}(d_{\text{s-epa}(u(dn))})$.

Lemma 7 establishes that epa components point to an earlier element of the

[1]The proofs of Lemmas 5 and 6 have been put in Appendix I, because the reader who wishes to limit his reading of detailed proofs might find the subsequent ones more useful.

dump than that where they are contained. (The proof uses the result of section 3.3 that it is only necessary to consider block and procedure activations.) The lemma is required in order to show, in Theorem I and Lemma 9, that the preceding element of the environment chain is in the range of the induction hypotheses.

<u>Lemma 7</u>. (L7) For d such that $\ell(d) > 1$:

$$1 \leq epa < \ell(d) \ .$$

<u>Proof</u>. Proof for top(d') :

1	<u>top(d') generated by block activation</u>:	
2	$epa' = \ell(d)$	(T18)
3	$1 \leq epa' < \ell(d')$	2, (T10)
4	<u>top(d') generated by procedure activation</u>:	
5	$u\text{-}p \in R(e)$	(T20), (T22)
6	is-proc-den(u-p(dn))	(T21)
7	$1 \leq s\text{-}epa(u\text{-}p(dn)) \leq \ell(d)$	5, 6, (L6a)
8	$1 \leq epa' < \ell(d')$	(T24), (T22), 7, (T10)

Lemma 8 expresses the fundamental connection between the environment and search mechanisms; it shows that the environment component of an element of the dump differs from the environment component of the environmentally preceding dump element by exactly the local environment of the current element. (The proof uses the result of section 3.3 that it is only necessary to consider block and procedure activations.) The lemma makes it possible to prove Theorem I without the case distinction of the generating state transitions.

<u>Lemma 8</u>. (L8) For d such that $\ell(d) > 1$:

$$e = update(s\text{-}e(d_{epa}), eo) \ .$$

<u>Proof</u>. Proof for top(d') :

1	<u>top(d') generated by block activation</u>:	
2	$e' = update(e, eo\text{-}b)$	(T17)
3	$= update(e, eo')$	2, (T12)

4	$= \text{update}(s\text{-}e(d'_{\ell(d)}), eo')$	3, (T10), (N1)
5	$= \text{update}(s\text{-}e(d'_{epa'}), eo')$	4, (T18)
6	<u>top(d') generated by procedure activation</u>:	
7	$1 \leq epa\text{-}p \leq \ell(d)$	(T21), (T22), (T20), (L6a)
8	$e\text{-}p = s\text{-}e(d_{epa\text{-}p})$	(T21), (T22), (T20), (L6b)
9	$= s\text{-}e(d'_{epa\text{-}p})$	7, 8, (T10), (N1)
10	$e' = \text{update}(e\text{-}p, eo\text{-}b)$	(T23)
11	$= \text{update}(e\text{-}p, eo')$	10, (T12)
12	$= \text{update}(s\text{-}e(d'_{epa\text{-}p}), eo')$	9, 11
13	$= \text{update}(s\text{-}e(d'_{epa'}), eo')$	12, (T24)

Theorem I justifies the first result presented in section 3.2. (Its proof uses the induction on dumps discussed in section 3.3.) The result is further elaborated in Theorem III, where it is shown that an index can be computed from the text corresponding to the result of "s" , and is also used as a basis of Theorem II.

<u>Theorem I</u>. For $id \in D(e)$:

$$id(e) = id(s\text{-}eo(d_{\text{index}(s(id,d),d)})) .$$

<u>Proof</u>. Proof by induction on the dump:

<u>basis</u>:

1	$s\text{-}e(\text{rest}(d,1)) = e^1$	(L2)
2	vacuously true	1, (T3)

<u>induction step</u>:

3	$\ell(d) > 1$	
4	$id \in D(e)$	Hyp
5	$D(e) = D(s\text{-}e(d_{epa})) \cup D(eo)$	3, (L8), (C1)
6	<u>$id \in D(eo)$</u>:	
7	$id(e) = id(eo)$	3, (L8), 6, (D1)
8	$s(id,d) = 0$	6, (D2)
9	$\text{index}(s(id,d),d) = \ell(d)$	8, (D3)
10	$id(s\text{-}eo(d_{\text{index}(s(id,d),d)})) = id(eo)$	9
11	$id(e) = id(s\text{-}eo(d_{\text{index}(s(id,d),d)}))$	7, 10

12	$\underline{id \notin D(eo)}$:	
13	$id(e) = id(s\text{-}e(d_{epa}))$	3, (L8), 12, (D1)
14	$s(id,d) = s(id,rest(d,epa)) + 1$	12, (D2)
15	$index(s(id,d),d) = index(s(id,rest(d,epa)),rest(d,epa))$	14, (D3)
16	$1 \le epa < \ell(d)$	3, (L7)
17	$id \in D(s\text{-}e(d_{epa}))$	12, 5, 4
18	$id(s\text{-}e(d_{epa})) = id(s\text{-}eo(d_{index(s(id,rest(d,epa)),rest(d,epa))}))$	16,17,IH
19	$id(e) = id(s\text{-}eo(d_{index(s(id,d),d)}))$	4, 18, 13, 15

Lemma 9 establishes a relation which holds between the index and ℓe functions, since both follow exactly the same chain. The result is used in Theorem II to provide the link to the result of Theorem I.

Lemma 9. (L9) For $0 \le i \le \ell e(d)$:

$$\ell e(rest(d,index(i,d))) = \ell e(d) - i \; .$$

Proof. Proof by induction on $l(d)$:

basis:

1	$\ell(d) = 1$	
2	$epa = 0$	1, (L2), (T5)
3	$\ell e(d) = 0$	2, (D4)
4	$i = 0$	3
5	$rest(d,index(0,d)) = d$	(D3)
6	$\ell e(rest(d,index(i,d))) = \ell e(d) - i$	4, 5

induction step:

7	$\ell(d) > 1$	
8	$0 \le i \le \ell e(d)$	Hyp
9	$epa \ge 1$	7, (L7)
10	$\underline{i > 0}$:	
11	$\ell e(d) = \ell e(rest(d,epa)) + 1$	9, (D4)
12	$0 \le i - 1 \le \ell e(rest(d,epa))$	8, 10, 11
13	$\ell e(rest(rest(d,epa),index(i-1,rest(d,epa)))) = \ell e(rest(d,epa)) - (i-1)$	12, IH

14 $\ell e(rest(d,index(i,d))) = \ell e(d) - i$ 13, (D3), (D4)

15 $\underline{i=0}$:

16 $\ell e(rest(d,index(i,d))) = \ell e(d) - i$ 15, 4, 6

Theorem II justifies the second result presented in section 3.2. The result is further elaborated in Theorem III, where it is shown that depth can be computed from the text. Section 3.2 has pointed out that disp can be modelled by a state component.

<u>Theorem II</u>. For $id \in D(e)$:

$$id(e) = id(s\text{-}eo(d_{disp_{depth(id,d)}})) .$$

<u>Proof</u>.

1 $disp_{depth(id,d)} = index(\ell e(d) - \ell e(rest(d,index(s(id,d),d))),d)$ (D6), (D5)

2 $= index(s(id,d),d)$ 1, (L9)

3 $id \in D(e)$: Hyp

4 $id(e) = id(s\text{-}eo(d_{disp_{depth(id,d)}}))$ 3, Th. I, 2

Lemma 10 shows that the tp component of a dump element is the successor of the tp component of the environmentally preceding dump element. (The proof uses the result of section 3.3 that it is only necessary to consider block and procedure activations.) This lemma, extended over entire environment chains, is used in Theorem III.

<u>Lemma 10</u>. (L10) For d such that $\ell(d) > 1$:

$$tp \Leftarrow s\text{-}tp(d_{epa}) .$$

<u>Proof</u>. Proof for $top(d')$:

<u>$top(d')$ generated by block activation</u>:

1 $tp\text{-}b \Leftarrow tp$ (T16)

2 $\Leftarrow s\text{-}tp(d'_{\ell(d)})$ 1, (T10), (N1)

3 $\Leftarrow s\text{-}tp(d'_{epa'})$ 2, (T18)

4 $tp' \Leftarrow s\text{-}tp(d'_{epa'})$ 3, (T11)

<u>top(d') generated by procedure activation:</u>

5	$epa\text{-}p \leq \ell(d)$	(T20), (T22), (T21), (L6a)
6	$tp\text{-}b \Leftarrow s\text{-}tp(d_{epa\text{-}p})$	(T20), (T22), (T21), (L6c)
7	$\Leftarrow s\text{-}tp(d'_{epa\text{-}p})$	6, 5, (T10), (N1)
8	$\Leftarrow s\text{-}tp(d'_{epa'})$	7, (T24)
9	$tp' \Leftarrow s\text{-}tp(d'_{epa'})$	8, (T11)

Lemma 11 establishes that the relationship between the identifiers in ID(tx,tp) and in the component eo is preserved. (The proof uses the result of section 3.3 that it is only necessary to consider block and procedure activations.) The result is used in Theorem III.

<u>Lemma 11</u>. (L11) $D(eo) = ID(tx,tp)$.

<u>Proof</u>.

1	<u>basis</u>: $\ell(d) = 0$	
2	$D(e) = \{\ \}$	1, (L2), (T3)
3	$ID(tx,I) = \{\ \}$	Outermost block 3.1.1
4	$D(e) = ID(tx,tp)$	1, (L2), (T2), 2, 3
	<u>induction</u>: for top (d')	
5	$D(eo') = D(eo\text{-}b)$	(T12)
6	$= ID(tx,tp\text{-}b)$	5, (T8)
7	$= ID(tx,tp')$	6, (T11)

The functions "s" and "depth" are shown above as functions of the dump. However, this was done only to facilitate the above proofs, and Theorem III shows that the anticipated transition to functions of the text is possible. Values computed by a compiler could then be associated with the references. Since the result relates to the program texts, which have not been formally characterized, the argument is less formal than preceding proofs.

<u>Theorem III</u>. Equivalent functions to "s" and "depth" can be found which depend only upon the static text.

<u>Justification</u>.[1] Using the generalization principle given in section 3.3:

---

[1] The function index is used as a notational convenience in the following argument.

1 $\ell(d) > index(1,d) > index(2,d) > \ldots > index(n,d) = 1$ (L7), (D3)

2 $tp \Leftarrow s\text{-}tp(d_{index(1,d)}) \Leftarrow s\text{-}tp(d_{index(2,d)}) \Leftarrow \ldots \Leftarrow I$ 1, (L10), (T2)

Let $tp_1, tp_2, \ldots, I$ be the pointers to the blocks in which tp(tx) is nested; section 3.1.1 shows that

3 $tp \Leftarrow tp_1 \Leftarrow \ldots \Leftarrow I$

and also, since such chains are unique,

4 $tp_i = s\text{-}tp(d_{index(i,d)})$ 2, 3

5 $id \in D(s\text{-}eo(d_{index(i,d)})) \equiv id \in ID(tx, s\text{-}tp(d_{index(i,d)}))$ (L11)

6 $\equiv id \in ID(tx, tp_i)$ 5, 4

7 $s(id,d) = (id \in D(eo) \rightarrow 0, s(id, rest(d,epa)))$. (D2)

Given an occurrence (id,tp-id) , let tp-o be the pointer to the block or procedure body immediately containing it (see section 3.1.1). Section 3.2 (assumption) shows that occurrences are only referenced when the block or body which immediately contains them is the current one; then

8 $tp\text{-}o = tp$.

Let

9 $s'(id,tp) = (id \in ID(tx,tp) \rightarrow 0 , s'(id,tp'))$ where $tp \Leftarrow tp'$.

It can be shown by induction that

10 $s(id,d) = s'(id, tp\text{-}o)$ 7, 8, 6, 9

11 "s'" is the equivalent function to "s" which depends only upon the text.

A similar argument results in discovery of the required static function for depth.

3.5. Discussion

A number of aspects of the above example can now be discussed. The postulates on the state transitions, which form the basis of the above proof, are in some sense too restrictive. The condition that any "other statement" does not alter the stack or procedure denotations has been sufficient for constructing the proof but is clearly not necessary to achieve the same result. This is not a serious constraint, and where it need be violated (e.g., a sufficiently restricted type of procedure variable), the proof can be extended in a straightforward manner.

It is perhaps worth pointing out how the above example differs from the previous papers on blocks (see section 4). The definition of the state transitions by postulates avoids the argument from the control in Lucas (1968) and the necessity to give too much detail in the explicit state transition functions of Henhapl and Jones (1970). The result proved above goes further than that in Lucas (1968) and is made simpler than that in Henhapl and Jones (1970) by the idea of using the function "s" to compute the number of steps for "index" and showing, in a subsequent step, that it can be computed statically.

The proofs have always used a more or less explicit twin machine. However, the use of the derived proof principles (see section 3.3) appears to be a worthwhile extension and to be the kind of result which should be sought in future work.

The experience gained with this example suggests that, if the correctness proof is to be based on the equivalence of two abstract machines, there are certain essential steps in the argumentation. Although shorter proofs may be found, the authors foresee no conceptually simpler proof unless an appropriate, completely thought out, basis for the language definitions can be found.

4. OTHER RESULTS ON IMPLEMENTATIONS AND THEIR JUSTIFICATION

The interest in proofs of correctness of implementation methods was started at the Vienna Laboratory with Lucas (1968), and work has continued on the subject of blocks. However, implementation methods have been documented and justified for other aspects of languages, and this section reviews the main results.

It will set the context for the following discussion if it is mentioned that the emphasis, so far, has been to consider implementation problems as seen from their object time organization and to postpone consideration of the compile time processing.

4.1. Blocks

In the initial attempt to show the equivalence of two reference mechanisms (Lucas (1968)), the base reference method is the environment mechanism used in the current paper. The alternative method is a search for the first element of the static chain where the identifier occurs in the local environment component (i.e., a function similar to "s" above, but which returns the index of the dump). The problem considered is to prove equivalent the two mechanisms from definitions in the

"Vienna Method" notation. Thus the relations on which the proof is based are derived from the model. The proof itself, which introduced the twin machine concept used above, is divided in much the same way as Lemma 8 and Theorem 1 of the current paper.

Most of the remaining work on reference mechanisms is presented in Henhapl and Jones (1970). In that report, the possibility of block termination caused by go to statements is also considered. The base mechanism is the copy rule (see Naur et al (1962)), and display models, both of the type described above and of the type using unique static block names, are proved correct. A number of optimizations to the basic methods are also discussed.

Intermediate models are included so that, from a number of lemmas about one model, a theorem is proved which establishes the equivalence to the next model in a chain of equivalences. Some problems resulting from the choice of intermediate models are discussed in the summary of Henhapl and Jones (1970). A further comment on the steps made is that the direct proof of the equivalence of the display to the search models (using axioms of the text) appears less elegant than the approach adopted above. The definition of the state transitions by functions facilitates the derivation of relations required for the proofs, but still gives much unnecessary detail.

4.2. Loops

The problem of correctness proofs for optimized implementations of loop handling is discussed in Zimmermann (1970). This report begins by illustrating how most interpretations for loops can be represented by state transitions which

initialize (control variable, etc.);

execute body of the loop;

test (for repeat condition);

update (control variable, etc.);

exit.

Given two sets of such transition functions, a specific way of proving their equivalence is shown. This technique requires that relations are found which fulfil the role of Induction Hypotheses, but the proof itself consists only of proving certain lemmas which do not require induction. The technique has been used to show that assignment statements can be moved out of loops (providing certain constraints are satisfied) and that the evaluation of expressions, linear with respect to the control variable, can be handled in an efficient way.

The technique of factoring out the inductive reasoning for whole sets of proofs should be capable of generalization.

4.3. Storage

The properties of storage are defined in Walk et al (1969) by axioms. Thus the correctness of an implementation is established by showing that it satisfies the axioms; furthermore, finding any such model proves that the axioms are consistent. These questions are considered in Henhapl (1969), but more work is required, both on the axioms and relevant models. (The paper in the current volume, Bekić and Walk (1970), is a contribution to the former area.)

4.4. Expressions

Expression evaluation techniques are described extensively in the literature, a study of which has led to an extensive bibliography (Chroust (1970)). Work is now in progress on describing their object time operations in a uniform formal style. The value of progressing to correctness proofs is questionable, since the compile time aspects of many of the techniques would be a more likely source of error in actual implementations.

4.5. Bases for Proof Construction

A number of experiments have been made as to the possibility of basing the proofs on some other style of definition. However, apart from the minor changes made during the sequence of block papers, no completely thought out proposal has proved acceptable.

APPENDIX

This appendix contains the detailed proofs of Lemmas 1, 2, 5, and 6.

Lemma 1. (L1) For $1 \leq i < \ell(d^n)$,

$$\exists m(d^m = rest(d^n, i) \ \& \ 1 \leq m < n) .$$

Proof. Proof by induction on steps of the computation.

<u>basis</u>: ξ^1

1 lemma is vacuously true (T1)

<u>induction step</u>: $\xi^n \to \xi^{n+1}$

<u>block or procedure activation</u>:

2 $\underline{1 \leq i < \ell(d^{n+1})}$: Hyp

3 $\underline{1 \leq i < \ell(d^n)}$:

4 $\exists m(d^m = rest(d^n,i) \ \& \ 1 \leq m < n)$ 3, IH

5 $\exists m(d^m = rest(d^{n+1},i) \ \& \ 1 \leq m < n+1)$ 4, (T10), (N1)

6 $\underline{i = \ell(d^n)}$:

7 $d^n = rest(d^{n+1},i)$ 6, (T10)

8 $\exists m(d^m = rest(d^{n+1},i) \ \& \ i \leq m < n+1)$ 2, 5, 7

<u>termination</u>: case distinction according to state transition

9 $\underline{\ell(d^n) > 1}$:

10 $1 \leq i < \ell(d^{n+1})$ Hyp

11 $\exists m(d^m = rest(d^{n+1},i) \ \& \ 1 \leq m < n+1)$ 10, IH, (T26), (N1)

12 $\underline{\ell(d^n) = 1}$:

13 immediate from (T29) 12

<u>other state transitions</u>:

14 immediate from (T30)

<u>Lemma 2</u>. (L2) $rest(d^n,1) = d^1$.

<u>Proof</u>. Proof by induction on steps of the computation.

<u>basis</u>: ξ^1

1 lemma is identically true (T1)

<u>induction step</u>: $\xi^n \to \xi^{n+1}$

<u>termination</u>: case distinction according to state transition

2 $\underline{\ell(d^n) > 1}$:

3 $rest(d^{n+1},1) = d^1$ 2, IH, (T26), (N1)

4 $\underline{\ell(d^n) = 1}$:

5 immediate from (T29)

<u>all other state transitions (including activation)</u>:

6 $rest(d^{n+1},1) = rest(d^n,1)$ (T10), (T30)

7	$\text{rest}(d^{n+1},1) = d^1$	IH, 6

<u>Lemma 5</u>. (L5) a) $R(e) \subseteq U$;

b) for $u \in U$ & is-proc-den(u(dn)) : $R(\text{s-e}(u(dn))) \subseteq U$.

<u>Proof</u>. Proof by induction on the steps of the computation.

<u>basis</u>: ξ^1

1	a) $R(e^1) = \{\ \}$	(T3)
2	$R(e^1) \subseteq U$	1
3	b) vacuously true	(T7)

<u>induction step</u>: $\xi^1, \xi^2, \ldots, \xi \rightarrow \xi'$

<u>block activation</u>:

4	a) $R(e') \subseteq R(e) \cup R(\text{eo-b})$	(T17), (C2)
5	$\subseteq U \cup R(\text{eo-b})$	4, IHa
6	$\subseteq U'$	5, (T14)
7	b) $u \in U'$ & is-proc-den(u(dn'))	Hyp
	case distinction according to partitioning of U'	(T14)
8	<u>$u \in R(\text{eo-b})$</u>:	
9	$R(\text{s-e}(u(dn'))) = R(e')$	7, 8, (T19)
10	$R(\text{s-e}(u(dn'))) \subseteq U'$	9, 6
11	<u>$u \in U$</u>:	
12	$R(\text{s-e}(u(dn))) \subseteq U$	11, 7, (L4), IHb
13	$R(\text{s-e}(u(dn'))) \subseteq U$	12, 11, 7, (L4)
14	$R(\text{s-e}(u(dn'))) \subseteq U'$	13, (L3)

<u>procedure activation</u>:

15	a) $\text{u-p} \in U$	(T20), (T22), IHa
16	$R(\text{e-p}) \subseteq U$	15, (T21), IHb, (T22
17	$R(e') \subseteq R(\text{e-p}) \cup R(\text{eo-b})$	(T23), (C2)
18	$\subseteq U \cup R(\text{eo-b})$	17, 16
19	$\subseteq U'$	18, (T14)
20	b) $u \in U'$ & is-proc-den(u(dn'))	Hyp
	case distinction according to partitioning of U'	(T14)
21	<u>$u \in R(\text{eo-b})$</u>:	

22 $\exists ul(ul \in R(e)$ & $u(dn') = ul(dn))$ 20, 21, (T25)

let ul be such a name

23 $ul \in U$ 22, IHa

24 $R(s\text{-}e(ul(dn))) \subseteq U$ 23, 20, 22, IHb

25 $R(s\text{-}e(u(dn'))) \subseteq U$ 24, 22

26 $\subseteq U'$ 25, (L3)

27 $\underline{u \in U}$:

28 is-proc-den$(u(dn))$ 27, 20, (L4)

29 $R(s\text{-}e(u(dn))) \subseteq U$ 27, 28, IHb

30 $R(s\text{-}e(u(dn'))) \subseteq U'$ 29, 27, 20, (L4), (L3)

<u>termination</u>:

31 a) $\exists m(d^m = rest(d)$ & $1 \leq m < n)$ (L1)

let m be such an index

32 $d' = d^m$ 31, (T26)

33 $R(e') \subseteq U^m$ 32, IHa

34 $R(e') \subseteq U'$ 31, 33, (L3)

35 b) $u \in U'$ & is-proc-den$(u(dn'))$ Hyp

36 $u \in U$ 35, (T28)

37 $u(dn') = u(dn)$ 35, 36, (T27)

38 $R(s\text{-}e(u(dn))) \subseteq U$ 36, 35, 37, IHb

39 $R(s\text{-}e(u(dn'))) \subseteq U'$ 38, 37, (L3)

<u>other state transitions</u>:

a) and b) immediate from (T30), (T31), (T32)

<u>Lemma 6</u>. (L6) For $u \in R(e)$ & is-proc-den$(u(dn))$,

a) $1 \leq s\text{-}epa(u(dn)) \leq \ell(d)$;

b) $s\text{-}e(u(dn)) = s\text{-}e(d_{s\text{-}epa(u(dn))})$;

c) $s\text{-}tp(u(dn)) \Leftarrow s\text{-}tp(d_{s\text{-}epa(u(dn))})$.

<u>Proof</u>. Proof by induction on the steps of the computation.

<u>basis</u>: ξ^1

1 lemma is vacuously true (T3)

<u>induction step</u>: $\xi^1, \xi^2, \ldots, \xi \rightarrow \xi'$

2	$u \in R(e') \ \& \ \text{is-proc-den}(u(dn'))$	Hyp
	<u>block activation</u>:	
3	$R(e') \subseteq R(e) \cup R(eo\text{-}b)$	(T17), (C2)
	case distinction according to partitioning of $R(e')$	3
4	<u>$u \in R(eo\text{-}b)$</u>:	
5	$\text{s-epa}(u(dn')) = \ell(d')$	4, 2, (T19)
6	$1 \leq \text{s-epa}(u(dn')) \leq \ell(d')$	5
7	$\text{s-e}(u(dn')) = \text{s-e}(d'_{\text{s-epa}(u(dn'))})$	4, 2, (T19), 5
8	$\text{s-tp}(u(dn')) \Leftarrow \text{s-tp}(d'_{\text{s-epa}(u(dn'))})$	4, 2, (T19), (T11), 5
9	<u>$u \in R(e)$</u>:	
10	$u \in U$	9, (L5a)
11	$u(dn') = u(dn)$	10, 2, (L4)
12	$\text{s-epa}(u(dn)) \leq \ell(d)$	9, 2, 11, IHa
13	$1 \leq \text{s-epa}(u(dn')) \leq \ell(d')$	11, 12
14	$\text{s-e}(u(dn')) = \text{s-e}(d'_{\text{s-epa}(u(dn'))})$	IHb, 12, (N1)
15	$\text{s-tp}(u(dn')) \Leftarrow \text{s-tp}(d'_{\text{s-epa}(u(dn'))})$	IHc, 12, (N1)
	<u>procedure activation</u>:	
16	$R(e') \subseteq R(e\text{-}p) \cup R(eo\text{-}b)$	(T23), (C2)
	case distinction according to partitioning of $R(e')$	16
17	<u>$u \in R(eo\text{-}b)$</u>:	
18	$\exists ul(ul \in R(e) \ \& \ u(dn') = ul(dn))$	17, 2, (T25)
	let ul be such a name	
19	$\text{s-epa}(ul(dn)) \leq \ell(d)$	18, 2, IHa
20	$1 \leq \text{s-epa}(u(dn')) \leq \ell(d')$	18, 19
21	$\text{s-e}(u(dn')) = \text{s-e}(d'_{\text{s-epa}(u(dn'))})$	IHb, 19, (N1)
22	$\text{s-tp}(u(dn')) \Leftarrow \text{s-tp}(d'_{\text{s-epa}(u(dn'))})$	IHc, 19, (N1)
23	<u>$u \in R(e\text{-}p)$</u>:	
24	$u\text{-}p \in R(e) \ \& \ \text{is-proc-den}(u\text{-}p(dn))$	(T20), (T22), (T21)
25	$\text{epa-p} \leq \ell(d)$	24, IHa, (T22)
26	$\exists m(d^m = \text{rest}(d,\text{epa-p}) \ \& \ 1 \leq m < n)$	25, (L1)
	let m be such an index	

27	$e\text{-}p = e^m$	26, 24, IHb, (T22)
28	$u \in U^m$	27, (L5a)
29	$u(dn^m) = u(dn')$	28, 2, (L4)
30	$s\text{-}epa(u(dn^m)) \leq \ell(d^m)$	23, 27, 2, 29, IHa
31	$1 \leq s\text{-}epa(u(dn')) \leq \ell(d')$	29, 30, 26, 25
32	$s\text{-}e(u(dn')) = s\text{-}e(d'_{s\text{-}epa(u(dn'))})$	IHb, 30, (N1)
33	$s\text{-}tp(u(dn')) \Leftarrow s\text{-}tp(d'_{s\text{-}epa(u(dn'))})$	IHc, 30, (N1)
	<u>termination:</u> case distinction according to state transitions	
34	<u>$\ell(d) > 1$</u>:	
35	$\exists m(d^m = rest(d) \;\&\; 1 \leq m < n)$	(L1)
	let m be such an index	
36	$u \in R(e^m)$	2, 35
37	$u \in U^m$	36, (L5a)
38	$u(dn') = u(dn^m)$	37, 2, (L4)
39	$s\text{-}epa(u(dn^m)) \leq \ell(d^m)$	36, 2, 38, IHa
40	$1 \leq s\text{-}epa(u(dn')) \leq \ell(d')$	39, 38, 35
41	$s\text{-}e(u(dn')) = s\text{-}e(d'_{s\text{-}epa(u(dn'))})$	IHb, 39, (N1)
42	$s\text{-}tp(u(dn')) \Leftarrow s\text{-}tp(d'_{s\text{-}epa(u(dn'))})$	IHc, 39, (N1)
43	<u>$\ell(d) = 1$</u>:	
44	immediate from	(T29)
	<u>other state transitions:</u>	
45	immediate from	(T30), (T31), (T32)

ACKNOWLEDGEMENT

Much of the work done on the subject matter, but not represented in this paper, is due to Dr. W. Henhapl. The authors would like to take the opportunity of acknowledging the stimulus of their cooperation with him. The authors also gratefully acknowledge useful discussions with Dr. H. Bekić.

EXAMPLES OF FORMAL SEMANTICS*

by

Donald E. Knuth

Perhaps the most natural way to define the "meaning" of strings in a context-free language is to define attributes for each of the nonterminal symbols which arise when the strings are parsed according to the grammatical rules. The attributes of each nonterminal symbol correspond to the meaning of the phrase produced from that symbol. This point of view is expressed in some detail in Knuth (1968 a), where attributes are classified into two kinds, "inherited" and "synthesized". Inherited attributes are, roughly speaking, those aspects of meaning which come from the context of a phrase, while synthesized attributes are those aspects which are built up from within the phrase. There can be considerable interplay between inherited and synthesized attributes; the essential idea is that the meaning of an entire string is built up from local rules relating the attributes of each production appearing in the parse of that string. For each production in the context-free grammar, we specify "semantic rules" which define (i) all of the synthesized attributes of the nonterminal symbol on the left hand side of the production, and (ii) all of the inherited attributes of the nonterminal symbols on the righthand side of the production. The initial nonterminal symbol (at the root of the parse tree) has no inherited attributes. Potentially circular definitions can be detected using an algorithm formulated in Knuth (1968 a).

The purpose of this paper is to develop these ideas a little further and to present some additional examples of the "inherited attribute - synthesized attribute" approach to formal semantics. The first example defines the class of lambda expressions which have a reduced equivalent, in terms of a "canonical" reduced form. The

*The research reported here was supported in part by the Advanced Research Projects Agency of the Office of the Secretary of Defense (SD 183), and in part by IBM Corporation.

second example defines the simple programming language Turingol; this language was defined in Knuth (1968 a), in terms of conventional Turing machine quadruples, while the definition in this paper is intended to come closer to the fundamental issues of what computation really is, and to correspond more closely to problems which arise in the definition of large-scale contemporary programming languages.

The formal definitions in this paper are probably not in optimum form, but they seem to be a step in the right direction. It is hoped that the reader who has time to study these examples will be stimulated to develop the ideas further.

1. Lambda Expressions

Our first example of a formal definition concerns lambda expressions as discussed by Wegner (1968), restricting the set of variables to the forms x, x', x'', x''', etc. Informally, the lambda expressions we consider are either (i) variables standing alone; (ii) strings of the form λVE, where V is a variable (called a "bound variable") and E is a lambda expression; or (iii) strings of the form (E_1E_2), where E_1 and E_2 are lambda expressions. If E_1 has form (ii), (E_1E_2) denotes functional application, i.e., we may substitute E_2 for all "free" occurrences of V in E, making suitable changes to bound variables within E so that free variables of E_2 do not become bound. For example, $\lambda x(x'x)$ is a lambda expression in which x' is free but x is bound; it has the same meaning as $\lambda x''(x'x'')$ by renaming the bound variable, but $\lambda x'(x'x')$ has a different meaning. The lambda expression $(\lambda x'\lambda x(x'x)x)$ has the same meaning as $(\lambda x'\lambda x''(x'x'')x)$, by renaming a bound variable; and this has the same meaning as $\lambda x''(xx'')$, by substituting x for x'.

A lambda expression which contains no subexpressions of the form (λVEE_2) is called reduced. Some lambda expressions cannot be converted into an equivalent reduced form; the shortest example is $(\lambda x(xx)\lambda x(xx))$ which goes into itself under substitution. We say a lambda expression is reducible if it is equivalent to some reduced lambda expression. Our goal is to give a formal definition of the class of all reducible lambda expressions; this definition must make precise the notions of "free variables", "bound variables", "renaming", "substitution", etc. Fortunately, it is possible to create such a definition in a fairly natural way, using inherited and synthesized attributes.

Let E be a lambda expression. If E is reducible, our formal definition will define the meaning of E to be a string of characters which is a reduced lambda expression equivalent to E . The definition has the attractive property that two reducible lambda expressions are equivalent if and only if their meanings are exactly identical, character for character. (A proof of this assertion is beyond the scope of this paper, but can be based on the Church-Rosser theorem; cf. Wegner (1968).) The definition is iterative, in that the meaning of E might turn out to be the meaning of another lambda expression E_1 ; if E is irreducible, the process will never terminate, so we will obtain no meaning for E , but if E is reducible, the process will terminate in a finite number of steps. (Again the proof is beyond the scope of this paper, but uses well-known properties of lambda expressions.) It is recursively unsolvable to decide whether or not a given lambda expression is reducible, or if a given lambda expression is equivalent to the reduced form x , so an iterative procedure such as described below is probably the best we can do.

The formal definition involves some more or less standard notation. Let $\underline{N}$ be the set of nonnegative integers; $2^{\underline{N}}$ is the set of all subsets of $\underline{N}$. A <u>string</u> is a sequence of zero or more of the characters

$$x \quad \lambda \quad) \quad ' \quad ($$

and we let ϵ denote the empty string. The set of all strings is called $\underline{T}^*$. A function f is a set of ordered pairs $\{(x,f(x))\}$ whose first components are distinct; $\text{domain}(f) = \{x \mid (x,f(x)) \in f\}$. We write $\emptyset$ for the empty set or empty function;

$$f \uplus g = \{(x,g(x)) \mid x \in \text{domain}(g)\} \cup \{(x,f(x)) \mid x \in \text{domain}(f) \setminus \text{domain}(g)\}$$

denotes the function f "overridden" by the function g . If f is a function taking some subset of $\underline{N}$ into $2^{\underline{N}}$, and if $S \subseteq \underline{N}$, we write

$$\text{image of } S \text{ under } f = \cup\{f(x) \mid x \in S \cap \text{domain}(f)\} \cup \{x \mid x \in S \setminus \text{domain}(f)\} .$$

For example, if $S = \{2,3,4\}$ and $f = \{(2,\{1,4,5\}),(4,\{5,6\})\}$, then image of S under $f = \{1,3,4,5,6\}$.

If n is a nonnegative integer, "var(n)" denotes the string consisting of the letter x followed by n ' characters; thus, var(2) = x" . The number of characters is called the <u>index</u> of the variable.

Now we are ready for the formal definition itself; it is convenient to present the definition in a tabular format.

Terminal symbols: x λ) ' (

Nonterminal symbols: S E V

Start symbol: S

Inherited Attributes		
Name of Attribute	Type of Value	Significance
bound(E)	subset of <u>N</u>	indices of variables whose meaning is bound by the context of E
subst(E)	function from bound(E) into <u>T</u>*	specifies replacement text for substitutions
substf(E)	function from bound(E) into $2^{\underline{N}}$	specifies the indices of free variables in the corresponding replacement text
arg(E)	string	text (if any) used as argument in functional application
argf(E)	subset of <u>N</u>	indices of free variables in arg(E)

Synthesized Attributes		
Name of Attribute	Type of Value	Significance
meaning(S)	string	reduced text of lambda expressions (if it exists)
text(E)	string	string equivalent to E (includes substitutions and reductions)
free(E)	subset of <u>N</u>	indices of free variables occurring in E (before substitutions and reductions)
function(E)	<u>true</u> or <u>false</u>	is E explicitly a function?
reduced(E)	<u>true</u> or <u>false</u>	is E reduced?
index(V)	nonnegative integer	number of "primes" in the representation of this variable

Local Variables (used as abbreviations for brevity, in semantic rules 3.2)		
Name of Variable	Type of Value	Significance
mm	nonnegative integer	index chosen as new name of bound variable
rr	subset of <u>N</u>	indices of free variables in ss
ss	string	replacement text

Productions and Semantics				
Description	No.	Syntactic Rule	Example	Semantic Rules
Statement	1	$S \rightarrow E$	$(\lambda x(xx)\lambda x'(x'x))$	meaning(S): = <u>if</u> reduced(E) <u>then</u> text(E) <u>else</u> meaning(text(E)); bound(E): = subst(E): = substf(E): = argf(E): = ϕ ; arg(E): = ϵ .
Variable	2.1	$V \rightarrow x$	x	index(V): = 0 .
	2.2	$V_1 \rightarrow V_2'$	x'	index(V_1): = index(V_2)+1 .
Expression	3.1	$E \rightarrow V$	x'	function(E): = <u>false</u>; free(E): = {index(V)}; reduced(E): = <u>true</u>; text(E): = <u>if</u> index(V) $\in$ bound(E) <u>then</u> subst(E)(index(V)) <u>else</u> var(index(V)) .
	3.2	$E_1 \rightarrow \lambda V E_2$	$\lambda x'(x'x)$	function(E_1): = <u>true</u>; free(E_1): = free(E_2)$\setminus${index(V)}; reduced(E_1): = reduced(E_2); <u>if</u> arg(E_1) = ϵ , <u>then</u> (mm: = min{k $\in$ <u>N</u> | k $\notin$ image of free(E_1) under substf(E_1)} , ss: = var(mm) , rr: = {mm}) <u>else</u> ss: = arg(E_1) , rr: = argf(E_1) ; text(E_1): = <u>if</u> arg(E_1) = ϵ <u>then</u> "λ" ss text(E_2) <u>else</u> text(E_2); bound(E_2): = bound(E_1) $\cup$ {index(V)}; subst(E_2): = subst(E_1) $\uplus$ {(index(V),ss)}; substf(E_2): = subst(E_1) $\uplus$ {(index(V),tt)}; arg(E_2): = ϵ; argf(E_2): = ϕ .

Productions and Semantics (continued)				
Description	No.	Syntactic Rule	Example	Semantic Rules

	3.3	$E_1 \to (E_2E_3)$	(x'x)	function(E_1): = false;
				free(E_1): = free(E_2) ∪ free(E_3);
				reduced(E_1): = if function(E_2) then false else reduced(E_2) ∧ reduced(E_3);
				text(E_1): = if function(E_2) then text(E_2) else "(" text(E_2) text(E_3) ")" ;
				bound(E_2): = bound(E_3): = bound(E_1);
				subst(E_2): = subst(E_3): = subst(E_1);
				substf(E_2): = substf(E_3): = substf(E_1);
				arg(E_2): = text(E_3);
				argf(E_2): = image of free(E_3) under substf(E_3);
				arg(E_3): = ε;
				argf(E_3): = ∅ .

In rule 1, "meaning(text(E))" stands for meaning(S) in the derivation tree which arises when text(E) is parsed.

As an example of this formal definition, consider finding the "meaning" of (λx'λx(x'x)x) . We have the following parse tree, giving integer subscripts to the nonterminal symbols.

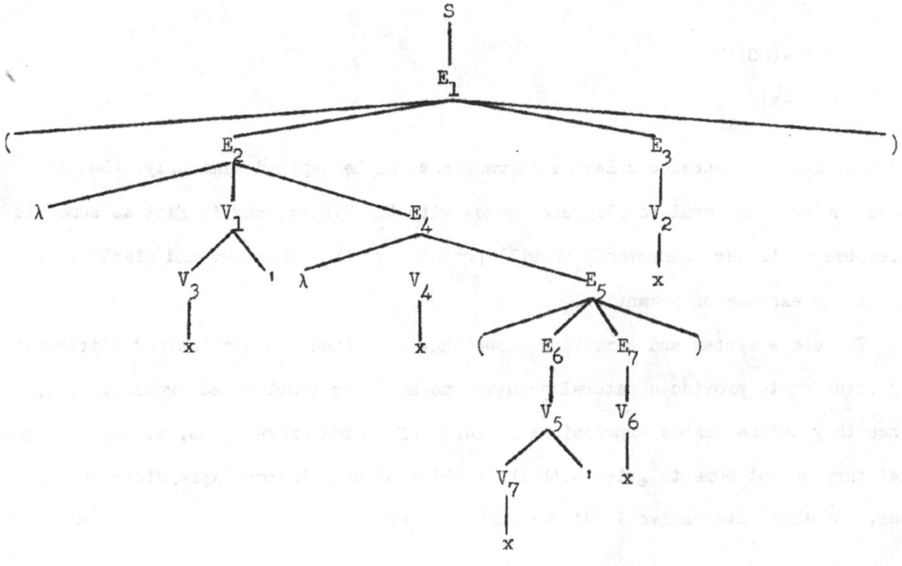

The semantic rules define the attributes as follows:

$$\text{index}(V_2) = \text{index}(V_3) = \text{index}(V_4) = \text{index}(V_6) = \text{index}(V_7) = 0 \ ;$$
$$\text{index}(V_1) = \text{index}(V_5) = 1 \ .$$

Node	bound	subst	substf	arg	argf	function	free	reduced	text
E_1	ϕ	ϕ	ϕ	ε	ϕ	<u>false</u>	{0}	<u>false</u>	λx'(xx')
E_2	ϕ	ϕ	ϕ	x	{0}	<u>true</u>	ϕ	<u>true</u>	λx'(xx')
E_3	ϕ	ϕ	ϕ	ε	ϕ	<u>false</u>	{0}	<u>true</u>	x
E_4	{1}	{(1,x)}	{(1,{0})}	ε	ϕ	<u>true</u>	{1}	<u>true</u>	λx'(xx')
E_5	{0,1}	F	G	ε	ϕ	<u>false</u>	{0,1}	<u>true</u>	(xx')
E_6	{0,1}	F	G	x'	{1}	<u>false</u>	{1}	<u>true</u>	x
E_7	{0,1}	F	G	ε	ϕ	<u>false</u>	{0}	<u>true</u>	x'

where $F = \{(0,x'),(1,x)\}$, $G = \{(0,\{1\}),(1,\{0\})\}$. Hence meaning(S)= meaning(λx'(xx')) , and we must parse λx'(xx') . A similar but much simpler derivation shows that meaning(λx'(xx')) = λx'(xx') .

Some of the semantic rules can be eliminated by making the syntax more complicated. For example, the class of reduced lambda expressions is defined by

$$S \to \lambda VS | N$$
$$N \to (NS) | V$$
$$V \to x | V' \ ,$$

and the class of nonreduced lambda expressions can be defined similarly. But it seems unwise in general to play such games with the syntax, and in fact as semantic rules become better understood, we will probably go the other way and simplify syntax at the expense of semantics.

The above syntax and semantics shows that inherited and synthesized attributes can interact to provide a natural solution to a rather complicated problem. But, since they define lambda expressions in terms of lambda expressions, it may be argued that they do not come to grips with the problem of what lambda expressions really mean. Another alternative is discussed in Section 4.

2. Turingol

A simple little language that describes Turing machine programs was introduced in Knuth (1968 a), where a semantic definition based on quadruples was given. The following example program gives the flavor of the "Turingol" language; it is a program designed to add unity to the binary integer which appears just left of the initially scanned square:

```
tape alphabet is blank, one, zero, point;
print "point";
go to carry;
test: if the tape symbol is "one" then
        {print "zero"; carry: move left one square; go to test};
print "one";
realign: move right one square;
        if the tape symbol is "zero" then go to realign.
```

It is worthwhile to search for a formal definition of Turingol which goes more deeply into the essential nature of computation itself, instead of assuming the knowledge of an artificial representation of Turing machines based on quadruples. The mapping from Turingol to quadruples is nontrivial and worthy of attention, but it is only part of the problem. Therefore, we shall now consider some approaches to the "total" problem of a Turingol definition.

One way to define Turingol, which we shall criticize later, is to introduce an intermediate language called TL/I ; we can define Turingol in terms of TL/I , and then we can define TL/I in terms of "conceptual computation". TL/I is a machine-like language, consisting essentially of sequential instructions whose operation codes are print, move, if, jump, and stop. The example TL/I program below is almost self-explanatory, so we shall turn immediately to the formal definition of Turingol.

It is convenient to let the symbol ν denote any positive integer, and to let the symbol σ stand for any string of alphabetic letters. These quantities could be syntactically defined, and we could make use of their attributes value(ν) , text(σ) , but, for simplicity, we may ignore such elementary operations, and we can identify numbers and letter strings with their representations. (In other words, we are assuming the existence of a "lexical scanning" mechanism, which must exist in some primitive form anyway to recognize the terminal symbols. We could in the same way have dispensed with index(V) in the lambda expression example above.)

The definition below involves "global" quantities, which may be regarded as attributes of the start symbol at the root of the parse tree, although their values are accessible at any node. All actions on global quantities can be reduced to a sequence of semantic rules relating appropriate local attributes, but it is simpler and more natural to abbreviate these rules using global quantities.

Global quantities may be global variables, global sets, or global counters. If ξ is a global variable and α is an expression, the notation

<u>define</u> $\xi \equiv \alpha$

stands for a <u>definition</u> of ξ. (A string of the language is "semantically erroneous" if its parse tree causes any global variable ξ to be defined more than once, or if any undefined global variable is used in an expression.)

If Σ is a global set, the notation

$\alpha \in \Sigma$

denotes <u>inclusion</u> of the value of expression α in the set Σ. If $\varkappa$ is a global counter, the notation

$\varkappa + \alpha$

denotes <u>increase</u> of the value of $\varkappa$ by the value of the integer expression α. Global sets start out empty, and global counters start out zero; when they appear in expressions, their value denotes the accumulated result of all inclusion or increasing operations specified in the entire parse tree. (Note that two inclusion or increasing operations can be done in any order.)

Here, finally, is a formal definition of Turingol in terms of TL/I :

Terminal symbols: σ . , : ;) (» « <u>tape</u> <u>alphabet</u> <u>is</u> <u>print</u> <u>move</u> <u>left</u> <u>right</u> <u>one</u> <u>square</u> <u>go</u> <u>to</u> <u>if</u> <u>the</u> <u>symbol</u> <u>then</u>

Nonterminal symbols: P S L D O

Start symbol: P

Inherited Attributes		
Name of Attribute	Type of Value	Significance
init(S), init(L)	positive integer	'address' of beginning of this statement or list

Synthesized Attributes		
Name of Attribute	Type of Value	Significance
fin(S), fin(L)	positive integer	'address' following this statement or list
index(D)	positive integer	number of symbols in declaration
d(O)	<u>left</u> or <u>right</u>	a direction

Global Variables		
Name of Attribute	Type of Value	Significance
label(σ), for all σ	positive integer	address associated with the identifier σ
symbol(σ), for all σ	positive integer	symbol number associated with the identifier σ

Global Counter		
Name of Attribute	Type of Value	Significance
nsymb	integer	number of symbols declared in this program

Global Set		
Name of Set	Type of Value	Significance
objprog	set of strings	TL/I program corresponding to the given Turingol program

Productions and Semantics

Description	No.	Syntactic Rule	Example	Semantic Rules
Declarations	1.1	D → tape alphabet is σ	tape alphabet is helen	index(D): = 1 ; nsymb + 1 ; define symbol(σ) ≡ 1 .
	1.2	$D_1 \rightarrow D_2, \sigma$	tape alphabet is helen, phyllis, pat	index(D_1): = index(D_2) + 1 ; nsymb + 1 ; define symbol(σ) ≡ index(D_1) .
Print statement	2.1	S → print "σ"	print "pat"	(init(S): print, symbol(σ)) ∈ objprog ; fin(S): = init(S) + 1 .
Move statement	2.2	S → move O one square	move left one square	(init(S): move, d(O)) ∈ objprog ; fin(S): = init(S) + 1 .
	2.2.1	O → left	left	d(O): = left .
	2.2.2	O → right	right	d(O): = right .
Go statement	2.3	S → go to σ	go to pieces	(init(S): jump, label(σ)) ∈ objprog ; fin(S): = init(S) + 1 .
Null statement	2.4	S →		fin(S): = init(S) .
Conditional statement	3.1	S_1 → if the tape symbol is "σ" then S_2	if the tape symbol is "phyllis" then print "pat"	(init(S): if, symbol(σ), fin(S_2)) ∈ objprog ; init(S_2): = init(S_1) + 1 ; fin(S_1): = fin(S_2) .
Labeled statement	3.2	$S_1 \rightarrow \sigma$: S_2	pieces: move left one square	define label(σ): = init(S_1) ; init(S_2): = init(S_1) ; fin(S_1): = fin(S_2) .
Compound statement	3.3	S → {L}	{print "pat"; go to pieces}	init(L): = init(S) ; fin(S): = fin(L) .
List of statements	4.1	L → S	print "pat"	init(S): = init(L) ; fin(L): = fin(S) .
	4.2	$L_1 \rightarrow L_2$;S	print "pat"; go to pieces	init(L_2): = init(L_1) ; init(S): = fin(L_2) ; fin(L_1): = fin(S) .
Program	5	P → D;L .	tape alphabet is helen, phyllis, pat; print "pat".	init(L): = 1 ; (fin(L): stop) ∈ objprog .

Example: The Turingol program for binary addition results in setting the global quantities

symbol(blank) = 1	nsymb = 4
symbol(one) = 2	label(carry) = 5
symbol(zero) = 3	label(test) = 3
symbol(point) = 4	label(realign) = 8 ,

and "objprog" is the (unordered) set of 11 strings

(1: print, 4)
(2: jump, 5)
(3: if, 2,7)
(4: print, 3)
(5: move, left)
(6: jump, 3)
(7: print, 2)
(8: move, right)
(9: if, 3, 11)
(10: jump, 8)
(11: stop) .

This set of strings is a TL/I program.

Now we can present a definition of TL/I . For this purpose, it is handy to extend context-free syntax slightly, allowing the production

$A \rightarrow$ set of B ,

where A and B are nonterminal symbols. This means that A can be, instead of a string (an ordered sequence), a set (unordered) of quantities having the form B .

A doubly-infinite tape, divided into squares and initially containing positive integers in each square, is manipulated by the actions of a TL/I program. There is a pointer which designates a square on the tape. One formal definition of TL/I is based on these concepts and an English language description of the operations to be done, as follows.

Nonterminal symbols: P S C

Terminal symbols: ν) (, : <u>if</u> <u>print</u> <u>jump</u> <u>move</u> <u>left</u> <u>right</u>

Start symbol: P

Inherited attribute: loc (a positive integer denoting the current position within the program)

Synthesized attributes: meaning (English language description of operations)

Global variables: action(ν) (English language description of operations starting at step ν)

Productions and Semantics				
Description	No.	Syntactic Rule	Example	Semantic Rules
Program	1	$P \rightarrow$ set of S	{(1: <u>print</u>,3), (2: <u>stop</u>)}	meaning(P): = "perform action(1)".
Statement	2	$S \rightarrow (\nu:C)$	(2: <u>stop</u>)	<u>define</u> action(ν) ≡ meaning(C) ; loc(C): = ν .
Command	3.1	$C \rightarrow$ <u>if</u>, ν_1, ν_2	<u>if</u>, 2, 7	meaning(C): = "if the tape square pointed to contains ν_1 , then perform action(loc(C) + 1) ; otherwise perform action(ν_2)" .
	3.2.1	$C \rightarrow$ <u>move</u>, <u>left</u>	<u>move</u>, <u>left</u>	meaning(C): = "move the pointer one square left, then perform action(loc(C) + 1)" .
	3.2.2	$C \rightarrow$ <u>move</u>, <u>right</u>	<u>move</u>, <u>right</u>	meaning(C): = "move the pointer one square right, then perform action(loc(C) + 1)" .
	3.3	$C \rightarrow$ <u>print</u>, ν	<u>print</u>, 3	meaning(C): = "erase the number on the square pointed to, replace it by ν, and then perform action(loc(C) + 1)".
	3.4	$C \rightarrow$ <u>jump</u>, ν	<u>jump</u>, 5	meaning(C): = "perform action(ν)" .
	3.5	$C \rightarrow$ <u>stop</u>	<u>stop</u>	meaning(C): = "stop" .

This example has some more or less undesirable properties, if not outright errors, although we can rectify the situtation in several interesting ways. If we take the definition literally as it stands, most TL/I programs will have "infinite" meanings; i.e., meaning(P) will never be defined in a finite number of steps, and we need to consider a limiting process. Thus, the meaning of our binary addition example comes out to be

> "perform "erase the number of the square pointed to, replace it by 4, and then perform "perform "move the pointer one square left, then perform "perform "if the tape square pointed to contains 2, then perform "re- place ..."; otherwise perform "erase the number on the square pointed to, replace it by 2, and then perform "move ...""""""""".

Being infinite, this doesn't really constitute an English sentence, nor does it read too well! It is essentially an infinite branching structure:

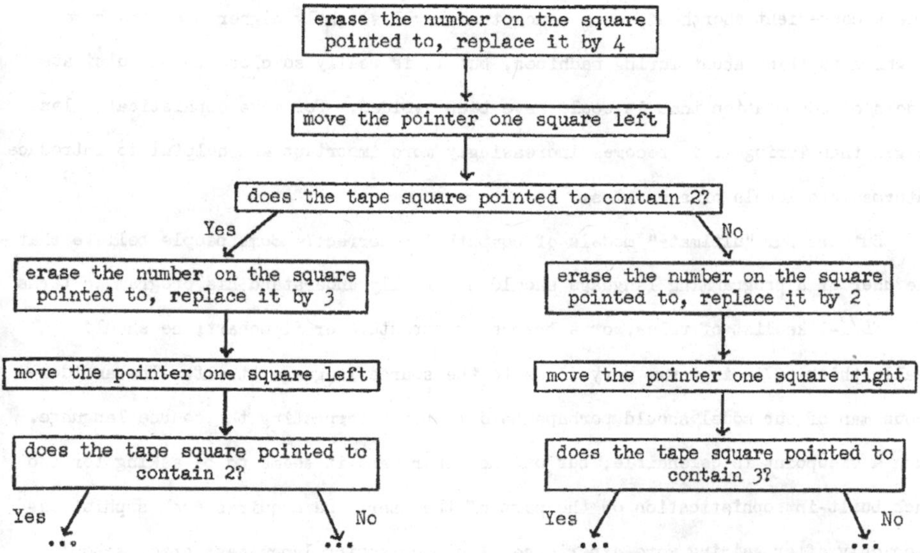

Instead of this infinite branching structure, we can take another point of view; rather than expanding the "action" parts of the meanings, we can consider the set of defined actions as a table which constitutes the meaning. Our example program then means "perform action(1)", where action(1) = "erase the number on the square pointed to, replace it by 4, and then perform action(2)";

action(2) = "perform action(5)";

⋮

action(11) = "stop".

We can now imagine a man who performs the process specified by a TL/I program, given a doubly-infinite tape and a set of defined actions as above; with one hand, he points to the action he is currently doing, while his other hand points to a place on the tape (and holds a pencil and an eraser). This is the "ambidextrous man" model of computation.

At this point, we can make some observations about the two-level definition of Turingol that appears above. Is it really necessary to introduce something like TL/I, or should we have gone directly to, say, the infinite branching structure or the ambidextrous-man model? A glance at the definitions shows that, indeed, we could have done things in one step. The introduction of TL/I serves only to provide a convenient shorthand, or a conceptual level slightly higher than the base, in which to think about Turing machines, but it is really so close to our ultimate models of computation that it could have been avoided. For more sophisticated languages than Turingol, it becomes increasingly more important and helpful to introduce intermediate levels of semantics.

But are our "ultimate" models of computation correct? Some people believe that the user of a programming language should not really understand his program in terms of a TL/I-like list of rules, or a branching structure or flowchart; he should really think of it in terms very close to the source language itself. The ambidextrous man of our model should perhaps be directly interpreting the source language. Such a viewpoint is defensible, but on the other hand it seems to be asking for too much built-in sophistication on the part of the user. He acquires such sophistication only after gaining more experience with programming languages; grade school children can understand simple machine languages, but they are not ready for Algol. Perhaps that is the reason many computer science educators are reporting that introductory courses in programming are usually more successful if the students are first taught a simple machine-like language before they learn algebraic languages. They need to understand the underlying principles of computation (what computers do) before seeing "problem-oriented languages". Therefore, it is likely that the models discussed above aren't too primitive. Furthermore, as a practical reality, a person programming well in some current language (FORTRAN, COBOL, ALGOL, PL/I, SNOBOL, etc.) should perhaps think of his program in some terms related to its actual machine representation, so that he knows what different constructions really "cost" him.

If the models aren't too primitive, are they too sophisticated? For example, positive integers should perhaps be defined in terms of Peano's postulates, etc.; maybe all the concepts should be further formalized in terms of set theory or category theory. This takes things from a domain children can understand into a more formal area which is able to support mechanical proof procedures. In this paper, our concern is with finding a natural conceptual basis for definitions; the basis must correspond to the way we actually think about computation, otherwise the related formalisms are not likely to be fruitful. Suitable formalisms will correspond to the natural conceptual basis rather closely, so we need not choose a more primitive formalism.

3. A Digression

Definition of programming language semantics by means of synthesized and inherited attributes is intended to correspond closely to the way people understand that language; the problem of producing compilers for that language is not a main goal, for it is possible to understand the meaning of a language without having to understand how to write a compiler for it. The success of context-free grammars as a model for syntax is based on its natural intuitive appeal (since the syntactic tree structures form a first approximation to the semantic structures), not on the fact that parsing algorithms can be devised for such grammars. A grammar is "declarative" rather than "imperative"; it expresses the essential relationships between things without implying that these relationships have been deduced using any particular algorithm. In general, we want to avoid any preoccupation with bits, advancing pointers, building and unbuilding lists, when such things have little or nothing to do with the intrinsic meaning of the language we are defining. On the other hand, once a "natural" mode of definition has been found, the next step should be to make practical use of it in the automation of software production; it is a happy circumstance when an intuitive description of a system can be almost automatically transformed into a practical working model based on that system. Much work remains to be done on the question of whether formal definitions such as those of this paper can be converted automatically to decent software programs; the following example may be useful as a test case for such techniques.

Consider the problem of writing an assembler for TL/I , converting a TL/I program into a sequence of bits suitable for interpretation by instructions on a microprogrammed computer. To make the problem interesting, we shall assume that we want to compress the length of the code, letting the number of bits that represent addresses and symbols be a parameter. The following "formal semantics" specifies this transformation precisely, in a problem-oriented fashion.

Let Memory(n,k) stand for the sequence of k bit positions

Memory(n) Memory(n+1) ... Memory(n+k-1) ,

and let Binary(n,k) stand for the sequence of k bits representing $(n \bmod 2^k)$ in binary notation. Let length(α) denote the length of string α . TL/I can now be defined as follows:

Nonterminal symbols, terminal symbols, start symbol: As before.

Synthesized attributes: code(C), a string of bits representing a coded instruction.

length(C), the number of bits in code(C).

Global variables: Memory(ν), a bit representing part of the encoded program.

loc(ν), a positive integer representing the first bit location of an instruction.

a , a positive integer representing the size of address specifications.

b , a nonnegative integer representing the size of symbol specifications.

nsymb, number of symbols (computed by the Turingol definition).

Global counters: addrs, the number of address fields in the program.

bits, the number of bits in non-address fields of the program.

Productions and Semantics	
Syntactic Rule	Semantic Rules
$P \to$ set of S	<u>define</u> $loc(1) \equiv 1$; $a := \min\{k \in \underline{N} \mid bits + k \cdot addrs \le 2^k\}$; $b := \min\{k \in \underline{N} \mid nsymb \le 2^k\}$.
$S \to (\nu : C)$	$Memory(loc(\nu), length(C)) := code(C)$; <u>define</u> $loc(\nu+1) \equiv loc(\nu) + length(C)$.
$C \to$ <u>if</u> , ν_1 , ν_2	$code(C) := 00\ Binary(\nu_1 - 1, b)\ Binary(loc(\nu_2), a)$; $bits + (b+2)$; $addrs + 1$; $length(C) := 2 + b + a$.
$C \to$ <u>move</u>, <u>left</u>	$code(C) := 010$; $bits + 3$; $length(C) := 3$.
$C \to$ <u>move</u>, <u>right</u>	$code(C) := 011$; $bits + 3$; $length(C) := 3$.
$C \to$ <u>print</u> , ν	$code(C) := 10\ Binary(\nu - 1, b)$; $bits + (2+b)$; $length(C) := 2 + b$.
$C \to$ <u>jump</u> , ν	$code(C) := 11\ Binary(loc(\nu), a)$; $bits + 2$; $addrs + 1$; $length(C) := 2 + a$.
$C \to$ <u>stop</u>	$code(C) := 11\ Binary(0, a)$; $length(C) := 2 + a$; $bits + 2$; $addrs + 1$.

Note that these rules specify a three pass process (first we count the addrs , then we can compute a and the loc's , then we can fill in the addresses) in a compact "declarative" manner.

4. Information Structures

The above definitions have adhered to old fashioned ways to represent information (integers, functions, sets, etc.), while Computer Science suggests that we ought to use some slightly different models and develop their formalisms further. A wide variety of applications (see, for example, Knuth (1968 b), Chapter 2) suggests that it is useful to represent the information in the real world, and its structural interrelationships, by means of things called "nodes". Each node consists of several "fields", which contain values; the values may be integers, strings, sets, etc., but (more importantly) the values may be references (i.e., pointers or links) to other nodes. The idea of references can be and has been formalized in various ways in terms of classical concepts, but recent experience suggests the usefulness of regarding references themselves as primitives. This often frees us from making arbitrary but conceptually irrelevant choices when we are representing information; for example, index sets are often used in mathematics when they really don't belong, and integers were used in our definition of lambda expressions and Turingol above, although we really wanted only unique labels and a notion of order.

Let us, therefore, consider making semantic definitions in terms of the proper data structures. A study of the Turingol-to-TL/I example shows that we should replace the set of location-instruction pairs in TL/I by a string of nodes (i.e., an ordered sequence of nodes). Each node corresponds to an instruction; jump and if nodes contain references to other nodes. We can concatenate strings of nodes just like strings of letters; so, for example, we can do away with the inherited attribute "init(S)" . The semantics for rule 4.2, $L_1 \to L_2;S$, becomes simply "meaning(L_1):= meaning(L_2) meaning(S)" . In this way, we obtain a more appealing (and more simple) formal definition, because all attributes are synthesized except for those which are implicitly present in global quantities. This idea of node strings containing pointers between the nodes, instead of absolute addresses which have to be determined by strict sequence rules, has been very successful in some studies recently conducted by the author on an experimental compiler-generating language.

Instead of making a complete listing of Turingol's semantics from the string-of-nodes point of view, it is perhaps even more interesting to consider the slightly more complicated problem of translating Turingol into a "self-explanatory flowchart". We may regard the meaning of a Turingol program as a set of nodes whose structure is that of a flowchart, easily readable by any ambidextrous man who wants to perform the algorithm. We use the notation

$$\underline{\text{new}}(\xi_1 := \eta_1 ; \xi_2 := \eta_2 ; \ldots ; \xi_m := \eta_m)$$

to denote the creation of a new node with m fields; the field named ξ_j contains the value η_j , for $1 \leq j \leq m$. The value of new(...) is a reference to this node. It is interesting to compare the Turingol definition below with the definition above, since the inherited-vs.-synthesized roles of "init" and "fin" are reversed!

Nonterminal symbols, terminal symbols, start symbol: As before.

Inherited attributes: fin(S), fin(L) , reference to node which follows statement or list.

Synthesized attributes:

init(S), init(L) , reference to node which begins statement or list;

index(D) , positive integer, the number of symbols in declaration;

d(O) , "left" or "right", a direction.

Global variables:

label(σ), for all σ , reference to the node corresponding to the label identifier σ .

symbol(σ), for all σ , positive integer, the symbol number associated with the identifier σ .

Fields of nodes: The COMMAND field contains strings of words and numbers explaining what to do when reaching this node; the YES, NO, and NEXT fields contain references to other nodes.

All nodes generated by the new operation constitute the "meaning" of a Turingol program.

Productions and Semantics		
No.	Syntactic Rule	Semantic Rules
1.1	D → tape alphabet is σ	index(D): = 1; define symbol(σ) ≡ 1 .
1.2	$D_1 \rightarrow D_2$,σ	index(D_1): = index(D_2) + 1 ; define symbol(σ) ≡ index(D_1) .
2.1	S → print "σ"	init(S): = new(COMMAND: = "Erase the number on the square pointed to, and replace it by symbol(σ); then go on to the NEXT node.";NEXT: = fin(S)) .
2.2	S → move O one square	init(S): = new(COMMAND: = "Move the pointer one square to the d(O); then go on to the NEXT node." ; NEXT: = fin(S)).
2.2.1	O → left	d(O): = left.
2.2.2	O → right	d(O): = right.

Productions and Semantics (continued)		
No.	Syntactic Rule	Semantic Rules
2.3	$S \to$ <u>go</u> <u>to</u> σ	init(S): = <u>new</u>(COMMAND: = "Go on to the NEXT node." ; NEXT := label(σ)) .
2.4	$S \to$	init(S): = <u>new</u>(COMMAND: = "Go on to the NEXT node." ; NEXT := fin(S)) .
3.1	$S_1 \to$ <u>if</u> <u>the</u> <u>tape</u> <u>symbol</u> <u>is</u> "σ" <u>then</u> S_2	init(S_1): = <u>new</u>(COMMAND: = "If the tape square pointed to contains symbol(ν), then go on to the YES node, otherwise go on to the NO node." ; YES: = init(S_2) ; NO := fin(S_1)) ; fin(S_2) := fin(S_1) .
3.2	$S_1 \to \sigma : S_2$	init(S_1): = init(S_2) ; fin(S_2): = fin(S_1) ; <u>define</u> label(σ) := init(S_1) .
3.3	$S \to \{L\}$	init(S): = init(L) ; fin(L): = fin(S) .
4.1	$L \to S$	init(L): = init(S) ; fin(L): = fin(S) .
4.2	$L_1 \to L_2 ; S$	init(L_1): = init(L_2) ; fin(L_2): = init(S) ; fin(S): = fin(L_1) .
5	$P \to D ; L$	<u>new</u>(COMMAND: = "Start at the NEXT node." ; NEXT: = init(L)) . fin(L): = <u>new</u>(COMMAND: = "Stop.").

This definition will produce the following flowchart from the binary addition example:

(The flowchart contains three "go on" nodes which seem redundant, although there are cases such as "loop: <u>go</u> <u>to</u> loop" which show that they cannot be eliminated entirely.)

symbol(blank) ≡ 1 , symbol(one) ≡ 2 , symbol(zero) ≡ 3 , symbol(point) ≡ 4 .

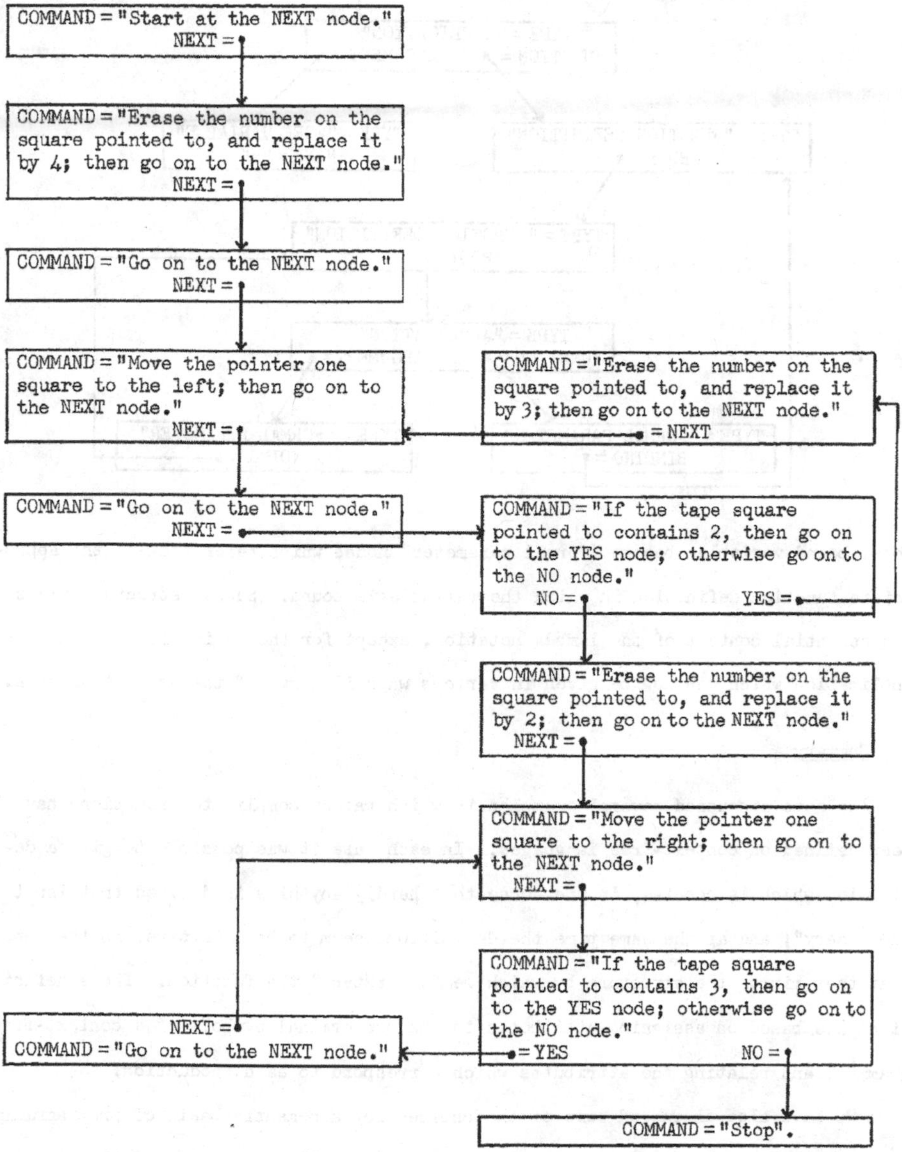

We could also consider lambda expressions again from the standpoint of appropriate information structure. It is an easy exercise to define the semantics so that, for example, the lambda expression $(\lambda x'\lambda x(x'x)x)$ becomes the structure:

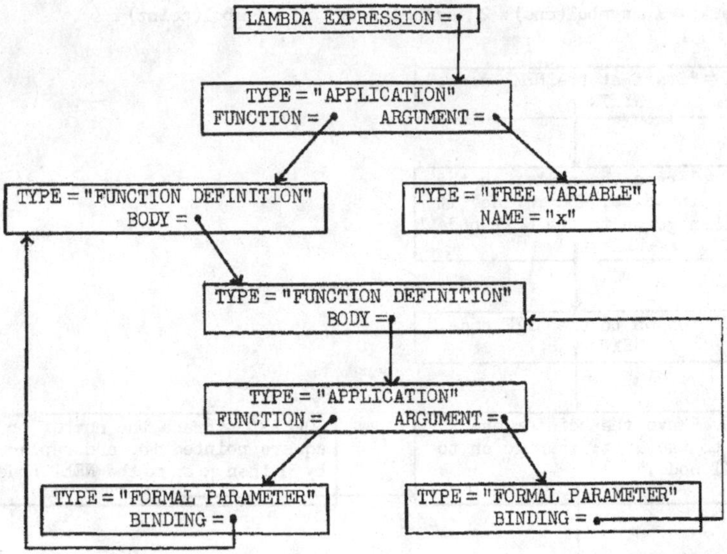

Thus, bound variables become "formal parameter" nodes which refer back to the appropriate function definition in which the variable is bound. Such a structure gives the essential content of the lambda notation, except for the definition of functional application which can now be given in various ways in terms of the node structures.

5. Summary

We have discussed several examples in which rather complicated functions have been defined on context-free languages. In each case it was possible to give a definition which is concise, in the sense that hardly anything is defined that isn't "necessary"; and at the same time the definitions seem to be intuitive, in the sense that they mirror the structure by which we "understand" the function. These definitions are based on assigning attributes to the nonterminal symbols of a context-free grammar, and relating the attributes which correspond to each production.

We have also discussed some of the choices for a semantic basis of programming languages. If we are interested in information-theoretic properties of algorithms, we may prefer an infinite branching structure as a computational model; if we are interested in representations of algorithms which are analogous to real live computer programs, we may prefer an "ambidextrous man" (essentially an automaton) model; if we are interested in the underlying structure, we may prefer a flowchart model. Other models are also possible. Whatever the model, formal definition via attributes seems to be helpful.

At the present time these ideas are being used and extended by Wayne Wilner, to construct a formal definition of a major programming language, SIMULA 67. The complexity of this language (over 300 productions in the syntax) makes the semantics slightly less transparent than the examples in this paper, but, in fact, the definition turns out to be simpler than anticipated.

Perhaps the main direction for future work which is suggested by the examples of this paper is to devise a suitable "context-free grammar" for arbitrary node structures instead of just strings. Attribute-definition on such grammars may lead to a very natural declarative language for problem solving in terms of relevant structures.

Abstract. A technique of formal definition, based on relations between "attributes" associated with nonterminal symbols in a context-free grammar, is illustrated by several applications to simple, yet typical, problems. First we define the basic properties of lambda expressions, involving substitution and renaming of bound variables. Then a simple programming language is defined using several different points of view. The emphasis is on "declarative" rather than "imperative" or "algorithmic" forms of definition.

EXPERIENCE WITH INDUCTIVE ASSERTIONS

FOR PROVING PROGRAMS CORRECT*

by

Ralph L. London

Introduction

A most effective method for proving that many computer programs are correct is the inductive assertion method. It is described with examples in Floyd (1967a), King (1969), Knuth (1968b, pp. 14-20), and Naur (1966); further examples of its use may be found in Good and London (1970), London (1970 c,d), and London and Halton (1969). The basic idea of the method is that one makes assertions concerning the progress of the computation at certain points in the code. The proof consists of verifying that each assertion is true each time control reaches that assertion, under the assumption that the previously encountered assertions are true.

London (1970 b,d) discusses the advantages and feasibility of proving programs. The present paper puts forth a notation or framework for stating assertions and gives some suggestions for its effective use, including in verifying the assertions. Two running computer systems for producing correctness proofs are summarized. Previously proved programs are analyzed to gain further insight into proofs using inductive assertions. The emphasis is on obtaining practical results.

A Notation for Assertions

Two related abilities are needed in stating assertions. First, one needs simple formulas relating variables. For example, the crucial assertions needed to prove the Extended Euclidean algorithm in Knuth (1968b, pp. 14-16) are

*This work is supported by the National Science Foundation, Grant GJ-583, and the Mathematics Research Center, United States Army, under Contract Number DA-31-124-ARO-D-462.

$$am + bn = d \tag{1}$$

and

$$a'm + b'n = d \tag{2}$$

while the key to proving a subtle exponential routine in King (1969, pp. 183-189) is the assertion

$$ZX^Y = A^B \ . \tag{3}$$

Second, one needs the ability to state the partial computation that has been accomplished or what is true over a range of values. The key assertions in proving a program in Floyd (1967a, p. 20) which sums the first N elements of the array A are, using the summation operator Σ ,

$$S = \sum_{j=1}^{I} A[j] \tag{4}$$

and the closely related

$$S = \sum_{j=1}^{I-1} A[j] \ . \tag{5}$$

For a program to sort an array M of N elements into ascending order, the proof in London (1970 c) uses the important assertion

$$M[p] \leq M[p+1] \quad \text{for} \quad I+1 \leq p \leq N-1 \tag{6}$$

or, using the and operator $\wedge$,

$$\bigwedge_{p=I+1}^{N-1} M[p] \leq M[p+1] \ . \tag{7}$$

The second type of assertions may also be expressed using three dots $(p = I+1, \ldots, N-1)$, using quantifiers $(\forall p(I+1 \leq p \leq N-1))$, or using other convenient symbolism.

All the above assertions, except (6), are members of the class of assertions defined by the pattern

expression	relation	operator {bounds or other modifications}	expression	(8)

where expression is arithmetic or Boolean; relation is $=, >, <, \geq, \leq, \neq, \in, \notin$; and operator is $\Sigma, \Pi, \wedge, \vee, \max, \min$. (Certain of the elements of (8) may be missing.

For example, "operator" is missing in (1) and both the first "expression" and "relation" are missing in (7).)

The class of assertions may be increased by using the same pattern but allowing additional kinds of expressions, relations, and operators that are appropriate to the particular program being proved. Furthermore, a single assertion may be composed of several assertions, each in the form of (8), joined by the four logical connectives: and, or, not, imply.

Using the Notation to State Assertions

The pattern (8) is intended as an informal statement of the type of assertions that are needed and are useful. Exactly what is needed in a particular example will, of course, depend on that example. However, the idea is to assert, in sufficiently general terms using the program variables (especially the loop control variables), what holds invariantly. The idea may also be expressed as the need to write an induction hypothesis as if a proof were to be done by ordinary mathematical induction. More concretely, what is required is an assertion describing the incremental processing accomplished by the ith execution of a loop, or more generally, the total processing accomplished by the first through the ith execution, as in (4) and (7), or even (3).

Certain necessary assertions may have to be more complex than the ones giving the final result. For example, compare (3) with the result $Z = A^B$. Assertions at an innermost loop, say, are usually more involved than the final ones.

It is difficult to be more explicit. King (1969, p. 113) states, "... the creation of predicates [assertions] remains an art learned by experience." Experience, of course, includes the study of existing proofs as well as giving proofs oneself. Naur (1966, p. 313), calling assertions by the descriptive term General Snapshots, advises, "... the values of variables given in a General Snapshot normally at best can be expressed as general, mathematical expressions or by equivalent formulations. I have to say 'at best' because in many cases we can only give certain limits on the value, and I have to admit 'equivalent formulations' because we do not always have suitable mathematical notation available."

That the notation (8) for assertions is useful in practice is seen by considering the assertions in successful proofs. The last section of this paper profiles ten inductive assertion proofs. All of the assertions in these proofs can be stated in the form of (8). While some of the assertions as originally written do not fit (8), each can be rewritten in a natural way to conform to (8).

In addition, all assertions in the twelve examples in King (1969, Appendices II and III) fit into (8) provided the quantifier notation is replaced as in (7). King's assertions are written as what he calls super Boolean expressions, a notion he precisely defines in BNF notation. They are essentially ordinary Algol Boolean expressions augmented by $\forall$ and $\exists$ to bind simple integer variables. Super Boolean expressions cover much but not all of (8).

In writing assertions, a convenient device is to give a name to a complicated assertion and to include provision for a variable "parameter". One then uses this name instead of the actual assertion. For example, in London (1970 c) the assertion

$$\bigwedge_{k=2s}^{n} M[k+2] \geq M[k] \qquad (9)$$

was needed. This set of inequalities was named $A(s)$, where s is effectively a formal parameter. The actual assertions made were then $A(i_o)$, $A(i)$, $A(i_o+1)$, $A(i+1)$, $A(n+2+1)$, and $A(2)$. This device is also used extensively in London and Halton (1969) for simplified reference to rather complicated assertions.

Another device, that of introducing new names, arises because the program variables (or names) appearing in the assertions always refer to the current values of the variables at the point in the program where the assertion is made. The original starting values may be lost, so sometimes there is need for additional variables to complete the proof. Thus, in the exponential routine that computes Z as A^B, the given inputs A and B are renamed X and Y; it is X and Y (but not A and B) that are altered by the program in the course of the computation. (Cf. an Algol procedure with parameters called by value.) The point is that both the original values A and B and the current values X and Y are needed in the proof as in assertion (3).

It may be necessary to create new names if the program does not do so explicitly. One convenient way is to use a subscript zero on the variable to denote the original value. In London (1970 c) this was done for a simple variable and also for the array name M . The latter allowed the assertion that the current array M is a permutation of the original array M_o (thus showing that no array elements were lost in the sorting).

The names may be created by a suitable assertion, for example,

$$X = A \text{ and } Y = B \text{ .} \tag{10}$$

In London and Halton (1969), the proof of Algorithm (91) includes an assertion in a loop which creates the needed names by using the current value of the loop control variable as a subscript.

Unfortunately, one needs assertions beyond the key or crucial assertions (1)-(7) to complete the respective proofs. These additional assertions cover certain necessary details which are nevertheless important and all too easy to overlook. Examples include ensuring that array references are within bounds, that certain variables are always integers and that bounds on variables hold. Each such assertion of detail may well be made at several points in the program and perhaps as a separate assertion from the crucial assertions. (More than one assertion is often profitably made at a point.)

The number of assertions that appear in a proof of a given program will vary with different tastes, styles, and techniques of proof. Generally one would expect an assertion at the exit(s) to state the results and an assertion at the entrance(s) giving initial assumptions, if any. Furthermore, each closed path (loop) in the program must contain at least one assertion. The same assertion may serve for more than one loop. Otherwise one is, of course, free to place additional assertions at selected points to aid in verifying conveniently all the assertions. While certain programs can be verified easily with the minimum number of assertions necessary, this is not recommended general strategy, especially for large programs. The other extreme--with assertions at every possible point--is likewise to be avoided in general.

The usually challenging and creative job of stating assertions does not complete the proof. The significant task of verifying the assertions remains just as an ordinary induction hypothesis must be proved and not merely stated. What must be done is the subject of the next section.

Verifying the Assertions

Assume at least one assertion at each entrance and at each exit--if necessary the trivial assertion TRUE. Several assertions at one point are here considered as one conjoined assertion. Consider all paths of control between assertions that start and end with an assertion, that contain no other assertions and that contain no program statement more than once. There are a finite number of such paths, since every loop contains an assertion (and there are only a finite number of assertions and statements). Each path leads to a "verification condition", a conjecture which must be proved to be a theorem. A proven verification condition verifies the assertion at the end of the path, but only for that particular path. Note that an assertion is often at the end of more than one path and, accordingly, requires more than one verification. If all verification conditions are proved, then the program is correct. Failure to prove a verification condition does not necessarily mean the program is incorrect, since improper assertions may have been chosen.

There are several ways to construct verification conditions. Two of the ways, forward accumulation and backward substitution, are illustrated below by an example from King (1969, pp. 20ff.). Full details of these and other algorithms are in King (1969) and Good (1970), including the proof that each leads to valid proofs, the equivalences between them and the relative advantages and disadvantages of their use.

Suppose assertion B is to be verified along a path that begins with assertion A . Forward accumulation starts with A , accumulates terms for intermediate statements, changes notation as needed, and ends at B . Backward substitution starts with B , makes substitutions directly in B , and ends at A .

The construction of verification conditions is shown for the example which computes z as x*y . The assertions are enclosed in braces. Paths are identified by pairs of assertion numbers with the second number denoting the assertion to be verified.

{assertion 1: true}

1. w←y ;

2. $z \leftarrow 0$;

L: {assertion 2: $z+w*x=x*y$}

3. if $w=0$, then {assertion 3: $z=x*y$} return z ;

4. $z \leftarrow x+z$;

5. $w \leftarrow w-1$;

go to L ;

Verification conditions (vc) by forward accumulation:

1-2: vc: $\text{true} \wedge w=y \wedge z=0 \supset z+w*x=x*y$

2-3: vc: $z+w*x=x*y \wedge w=0 \supset z=x*y$

2-2: derived in stages (sample notation: "z" is current value, "z1" is old value)

after 3: $z+w*x=x*y \wedge w \neq 0$

after 4: $z1+w*x=x*y \wedge w \neq 0 \wedge z=x+z1$

after 5: $z1+w1*x=x*y \wedge w1 \neq 0 \wedge z=x+z1 \wedge w=w1-1$

vc: $z1+w1*x=x*y \wedge w1 \neq 0 \wedge z=x+z1 \wedge w=w1-1 \supset z+w*x=x*y$

Verification conditions by backward substitution:

1-2: start with assertion 2: $z+w*x=x*y$

after 2: $0+w*x=x*y$

vc: $\text{true} \supset y*x=x*y$

2-3: start with assertion 3: $z=x*y$

after 3: $w=0 \supset z=x*y$

vc: $z+w*y=x*y \supset (w=0 \supset z=x*y)$

2-2: start with assertion 2: $z+w*x=x*y$

after 5: $z+(w-1)*x=x*y$

after 4: $x+z+(w-1)*x=x*y$

after 3: $w \neq 0 \supset x+z+(w-1)*x=x*y$

vc: $z+w*x=x*y \supset (w \neq 0 \supset x+z+(w-1)*x=x*y)$

(All the verification conditions can be shown to be theorems. Having done so, it has been proved that _if_ assertion 3 is reached, then $z=x*y$ as required. To show that assertion 3 _is_ reached, i.e., the program terminates, requires the additional assumption $y \geq 0$. Termination is usually shown separately, although inductive assertions can be stated and verified for termination as well.)

From this example it can be seen that for a given program, the number and complexity of the verification conditions (and the ease of their proofs) are determined by the assertions and the interconnection of statements between assertions. More points with assertions will usually give simpler verification conditions but more of them. For example, the number of verification conditions might be increased, but each one made simpler, by stating assertions at a point between two other assertions. Indeed, there may be no loss. King (1969, p. 24) gives an extreme example of a loop-free program where adding assertions decreases the number of verification conditions.

To construct and to prove the verification conditions involves several types of information: the semantic properties of the programming language in which the program is written; the properties of the expressions, relations, and operators of the assertions; and the semantics of the problem domain. A human prover can usually, without undue difficulty, select from this information and organize it into verifications of the assertions. Although he often does not explicitly construct verification conditions, this concept, plus an accurate statement of all the axioms and properties used, would serve to formalize his verifications.

In his verifications he does not state the use of every axiom and elementary theorem of arithmetic, or those of symbol manipulation more generally; neither does he state properties of the assertions and the problem domain. To do so would render already tedious proofs essentially impossible. The result is a rigorous but not formal proof and, therefore, a proof not subject to machine proof-checking.

An alternative to human-constructed proofs is, of course, to build an automatic program verification system. One such effort is described in the next section and is compared with a partially automatic system.

Automatic Proof Construction

The separate approaches of King (1969) and Good (1970) demonstrate that the computer itself can be of significant help in proving programs correct. King's prototype Program Verifier, running on an IBM 360 Model 65, uses the inductive assertion method on programs written in a simple Algol-like programming language, restricted to integers, but including one-dimensional arrays. The assertions are all written by a human as super Boolean expressions, noted earlier. Powerful formula simplification and manipulation routines are used to construct the (backward) verification conditions and, using an automatic theorem prover, to prove them. Specialized techniques for integers are also used.

Good's system, running on a Burroughs B5500, requires man-machine cooperation to complete a proof. It uses the inductive assertion method on a programming language similar to King's but without arrays or procedures and with declarations. The human uses a teletype to make all the assertions which can be any text string. Verification conditions are produced by the system; it looks for instances of program variable names in an assertion and substitutes appropriately. The human supplies the proof of the verification conditions, again as a text string, which the system accepts unquestioningly. The system provides significant additional help to the human by keeping track of which assertions remain to be verified along which paths, by allowing proofs to be changed, by retrieving assertions or proofs, by giving the complete or partial proof at the end, etc.

King's verifier has proved some interesting but small programs. The class of programs to which the verifier is applicable is also restricted. Some needed improvements he himself cites are increasing the expressive power of the assertions, keeping the expressions generated of reasonable size, increasing the power of the simplification process, adding more sophisticated features to the programming language and providing a genuine open-ended theorem prover. King defends his claim (p. 115): "we see no major hurdle which prevents the system from being continually upgraded by removing the limitations." Doing so will be far from trivial, but a good foundation does exist.

Good's system was tried on only a few small problems. The assertions given as text strings worked surprisingly well. They generated useable, intelligible, and easily provable verification conditions. This seemingly absurd way to state assertions shows real promise.

In writing a proof, however, there are problems for the human in identifying and referencing assertions, parts of the program, parts of the already completed proof, and other objects, all of which he needs. These are caused partly by the fact that while the human would like to view his program as a series of linear statements, the system forces him to view it as a graph which it created from the original program. The similar graph in King's verifier causes no problem because no interaction is involved.

With some effort by the human, then, proofs can be completed. But, as expected, human-factors problems were uncovered; some suggested improvements remain to be implemented and tested. There are also problems remaining in allowing the human to modify his assertions or, even worse, to modify his program. The difficulty is to allow this without completely invalidating the partially completed proof. In short, Good has made a significant start in the realization of a man-machine program proving system as well as identifying some of the remaining problems.

Both King (1969) and Good (1970) contain numerous additional insights and suggestions for understanding and using the inductive assertion method. In particular, Good has provided a formal basis for further consideration of the correctness problem that unifies some of the results obtained by others.

It is important to note that King and Good agree on the value of both approaches. King, in referring to the manual proof of Good and London (1968), says (p. 166), "Computer assistance to this tedious work appears to be as necessary for them [the authors] as human assistance is for our theorem prover. Some workable compromise between the efforts of man and machine seems to offer the most hope." Good (p. 164) suggests "combining the algebraic simplification and automatic theorem prover approach of King with the man-machine interactive approach described here."

The proposed assertion notation (8), when sufficiently formalized, may be suitable for use with mechanical theorem provers. Properties of the assertions, stated as axioms, would be used along with, say, the axioms of arithmetic and the semantics of the programming language in generating and proving the verification conditions. Both Rutledge (1968, pp. 13-14) and King (1969) suggest the need and desirability of so doing.

To illustrate, consider the simple example of finding the largest element L in the array A[1: N] where $1 \leq N$, an integer. This is a slight modification of an example given by Naur (1966).

$\{$assertion 0: true$\}$

1. $L \leftarrow A[1]$;

2. $I \leftarrow 2$;

   B: $\{$assertion 1: $L = \max_{j=1}^{I-1} A[j]\}$

3. if $I > N$, then $\{$assertion 3: $L = \max_{j=1}^{N} A[j]\}$ return L ;

4. if $A[I] > L$, then $L \leftarrow A[I]$;

   $\{$assertion 2: $L = \max_{j=1}^{I} A[j]\}$

5. $I \leftarrow I + 1$;

6. go to B ;

There are two properties or axioms of the max operator which are needed:

$$\max_{j=1}^{1} A[j] = A[1] \tag{11}$$

and

$$L = \max_{j=1}^{I-1} A[j] \wedge A[I] > L \supset A[I] = \max_{j=1}^{I} A[j]$$
$$\vee\ L = \max_{j=1}^{I-1} \wedge\ A[I] \leq L \supset L = \max_{j=1}^{I} A[j] \ . \tag{12}$$

(12) gives the two cases of the ith element being the new maximum or not. (12) would justify the induction step of an ordinary induction and (11) the basis step. Either formally or informally, assertion 1 can be verified on path 0-1 by (11) and on path 2-1 using statement 5. Assertion 2 can be verified on both paths 1-2 by (12), using one case on each path. Assertion 3 can be verified on path 1-3 by assertions not shown involving bounds on the integers I and N .

Naur's proof, not surprisingly, involves essentially the same assertions and verifications as given above. However, in place of the max operator, he uses a verbal definition. In addition, his verifications use the facts expressed by (11) and (12), but in words--perfectly appropriate for his informal, but nevertheless, rigorous presentation.

The ability to introduce needed additional notation into assertions and to give axioms describing properties of the notation seems necessary and appropriate in proving by automatic means a wider class of programs. Many useful operators have properties analogous to (11) and (12) as basis and induction steps, although not all properties are conveniently expressed in this format. Yet inductive definitions should work well with inductive assertions. Indeed, the real issue is not desirability, but feasibility. This remains an open question.

Profiles of Specific Inductive Assertion Proofs

Another way to understand how proofs are constructed by the inductive assertion method is to study proofs given by the method. Some reasonably representative inductive assertion proofs (constructed by hand) are listed in Table 1 together with some descriptors of the proofs. The result is a profile of the proofs. The entries are only approximate, since in compiling this table some rather arbitrary decisions had to be made. Nevertheless, the data are of interest.

All of the proofs in Table 1 work forward to verify the assertions. Verification conditions are not formally generated although, of course, the verifications could be recast into this form, often straightforwardly and directly.

Definitions of terms in Table 1:

1. Key assertions: Those assertions which express the induction hypothesis or the critical relationship needed. Assertions of detail are excluded.
2. Assumptions: A verification by initial assumption.
3. Recopy unchanged: A verification consisting of quoting a previous assertion without alteration.
4. Non-trivial calculation: A verification requiring some manipulation of assertions and statements.
5. Program logic: A verification which uses only the logic or semantics of the programming language on previous assertions.
6. Problem features: A verification which uses the features, semantics, or interpretation of the problem domain.

Terms 2, 3, and 4 provide one classification of the total verifications; terms 5 and 6 are another classification.

The proofs, identified in Table 1 by letters, are as follows:

A, B, and C: Variations of the exponential example of King (1969, pp. 183-189). A is his code and his assertions, B has an additional assertion, and C contains one unnecessary code statement. The proofs were written by London.

Descriptor \ Proof	Exponential A	Exponential B	Exponential C	TREESORT 3 D(siftup)	TREESORT 3 E(body)
Problem statement	1 line	1 line	1 line	$\frac{1}{4}$ page	2 lines
Lines of code	9	9	10	12	10
Flowchart boxes	6	6	7	10	5
Total assertions	4	5	14	34	21
Key assertions	2-50	3-60	6-43	13-38	15-71
Size of proof (pages)	$\frac{1}{2}$	$\frac{1}{2}$	1	$2\frac{2}{3}$	$1\frac{1}{2}$
Total verifications	7	9	21	43	28
Assumptions	1-14	1-11	1-5	2-5	0-0
Recopy unchanged	1-14	2-22	9-43	22-51	0-0
Non-trivial calculation	5-71	6-67	11-52	19-44	28-100
Program logic	5-71	6-67	14-67	36-84	17-61
Problem features	2-29	3-33	7-33	7-16	11-39

Descriptor \ Proof	Asymptotic Series F(97)	Asymptotic Series G(91)	Asymptotic Series H(100)	Interval Arithmetic I	OUTSIDE-ACE J
Problem statement	1 line	$\frac{1}{3}$ page	$\frac{2}{3}$ page	$2\frac{1}{4}$ pages	$\frac{1}{2}$ page
Lines of code	6	11	37	19	9
Flowchart boxes	5	10	29	17	5
Total assertions	5	10	24	10	6
Key assertions	3-60	6-60	15-63	9-90	5-83
Size of proof (pages)	$\frac{1}{4}$	$\frac{1}{2}$	3	$6\frac{1}{2}$	1
Total verifications	6	13	34	10	8
Assumptions	1-17	1-8	1-3	0-0	0-0
Recopy unchanged	3-50	3-23	12-35	0-0	1-12
Non-trivial calculation	2-33	9-69	21-62	10-100	7-88
Program logic	4-67	9-69	25-74	3-30	5-63
Problem features	2-33	4-31	9-26	7-70	3-37

TABLE 1. Proof Profiles

Note: M-N means M instances and N% of the total. For statements count as one flowchart box.

D and E: TREESORT 3 is the algorithm certified correct in London (1970 c). D proves the procedure siftup. E proves the body of the algorithm in which siftup is called.

F, G, and H: Three of the six algorithms proved in London and Halton (1969). F is algorithm (97), G is (91), and H is (100).

I: The interval arithmetic code in Good and London (1970), an excerpt typical of Good and London (1968).

J: OUTSIDEACE, Example 3 of London (1970 d).

The A and B proofs are quite similar. The extra verifications caused by the added assertion do not alter the profiles significantly and, indeed, the two proofs are essentially the same. It is strictly an individual preference whether a case analysis is expressed as an assertion, as in B , or whether it is handled in the verifications, as in A . Proof C has more assertions than either A or B , the difference being that the same assertions are repeated. This leads to shorter verifications with twice as much straight recopying but with the same breakdown between program logic and problem features.

The TREESORT proofs D and E each have a large number of assertions, since that seemed the most convenient way to ensure an accurate proof. In retrospect some assertions could probably be eliminated although perhaps at some cost in the clarity of the proof. Note in D that half of the verifications are done merely by recopying an assertion. Moreover, a large fraction of the verifications follow from the program logic alone. In E , there are calls on the procedure siftup. This explains the absence of recopying. It also explains the apparent discrepancy between the number of non-trivial calculations on the one hand and the large fraction in program logic, since the effect of procedure calls was considered to be program logic.

Proofs F and J are for simple programs each of which could be proved with fewer assertions than shown. Proof J is somewhat more complex than F because J's verifications involve translating the encoding of the problem domain back to the problem domain. This could be changed by using assertions directly involving the encoding. The encoding could then be translated separately once and for all at the end.

Program I contains no loop and is essentially a decision tree. Thus the assertions on the branches are all necessary and are all verified with non-trivial calculations that depend mainly on the problem features. The size of the problem statement and the size of the proof include reformulating the original problem to a form ready for implementation.

Proofs G and H are involved with loops. Program G is a single loop while program H is three simple loops, two of which are within a fourth outer loop. When the loops are separated and assertions made and verified noting this structure, the resulting profiles of G and H are similar to each other and to A and B .

Examples 1 and 2 in London (1970 d) can also be done by inductive assertions. Both examples involve a linear series of N tests for bidding bridge hands. The goal is to show that each hand satisfies precisely one test from the series. Let H_i be the set of hands passing the ith test. The subgoal of at most one test can be asserted by

$$\bigwedge_{i \neq j} (H_i \cap H_j = \text{null}) \ , \tag{13}$$

and the subgoal of at least one test by

$$\bigvee_{i=1}^{N} H_i = \text{all hands} \ . \tag{14}$$

The verification of assertions (13) and (14) is precisely the same case analysis proofs as in the examples. Each proof profile consists of one assertion proved by a non-trivial calculation depending heavily on the problem features--an extreme profile indeed.

The profiles may well vary with different proof styles and techniques. The proofs also probably contain additional information, and there is need for other useful descriptors. These profiles should be of value in assessing the ability to automate program proving.

Conclusion

This paper has been an informal presentation of observations and experiences using inductive assertions to prove programs correct. Practical rather than theoretical results have been stressed. If computer scientists find the suggestions useful and applicable and if provably correct programs result, then the overall aim of the paper will have been attained.

Acknowledgement

Allen Newell suggested to me the idea of a proof profile and some of the descriptors.

MATHEMATICAL THEORY OF PARTIAL CORRECTNESS*

by

Zohar Manna

Introduction

We normally distinguish between two classes of algorithms: deterministic algorithms and non-deterministic algorithms. A deterministic algorithm defines a single-valued (partial) function, while a non-deterministic algorithm defines a many-valued function. Therefore, while there are only a few properties of interest (mainly termination, correctness, and equivalence) for deterministic algorithms, there are many more (determinacy, for example) for non-deterministic algorithms.

In this work, we show that it is possible to express all properties regularly observed in such algorithms in terms of the 'partial correctness' property (i.e., the property that the final results of the algorithm, if any, satisfy some given input-output relation).

This result is of special interest since 'partial correctness' has already been formalized in predicate calculus for many classes of deterministic algorithms, such as flowchart programs (Floyd (1967 a) and Manna (1969 a)), functional programs (Manna and Pnueli (1970)), and Algol-like programs (Ashcroft (1970)); and also for certain classes of non-deterministic algorithms, such as choice flowchart programs (Manna (1970)) and parallel flowchart programs (Ashcroft and Manna (1970)). Thus the formalization in predicate calculus of all other properties of such algorithms is straightforward.

---

*The research reported here was supported in part by the Advanced Research Projects Agency of the Office of the Secretary of Defense (SD-183).

A preliminary version of this work was presented under the title "Second-Order Mathematical Theory of Computation" at the ACM Symposium on Theory of Computing (May 1970). All propositions stated in this work without proof are proved formally in that paper.

Similarly, Manna and McCarthy (1970) have formalized 'partial correctness' of functional programs in partial function logic; therefore, again, the formalization of all other properties of functional programs in partial function logic is straightforward.

Other papers related to this work are those of Burstall (1970), Cooper (1969 a, 1969 b) and Manna (1969 b).

1. Deterministic Algorithms

An algorithm P (with input variable x and output variable z) is said to be **deterministic** if it defines a single-valued (partial) function $z=P(x)$ mapping D_x (the input domain) into D_z (the output domain). That is, for every $\xi \in D_x$, $P(\xi)$ is either undefined or defined with $P(\xi) \in D_z$.

Examples: In the sequel we shall discuss the following four deterministic algorithms for computing $z=x!$ where $D_x = D_z = \{\text{the non-negative integers}\}$.

(a) Flowchart programs: Consider the flowchart programs P_1 (Figure 1) and P_2 (Figure 2). Here $(y_1,y_2) \leftarrow (y_1-1, y_1 \cdot y_2)$, for example, means that y_1 is replaced by y_1-1 and y_2 is replaced by $y_1 \cdot y_2$, simultaneously. The computation of P_1 for input $x=3$ for example, is

$$(y_1,y_2): (3,1) \to (2,3) \to (1,3\cdot 2) \to (0,3\cdot 2\cdot 1),$$

while the computation of P_2 for input $x=3$ is

$$(y_1',y_2'): (0,1) \to (1,1) \to (2,1\cdot 2) \to (3,1\cdot 2\cdot 3).$$

(b) Functional programs: Consider the functional programs

$$P_3:\ z=F(x) \ \underline{\text{where}}\ F(y) \Leftarrow \underline{\text{if}}\ y=0\ \underline{\text{then}}\ 1\ \underline{\text{else}}\ y\cdot F(y-1);$$

and

$$P_4:\ z=F(x,0)\ \underline{\text{where}}\ F(x,y') \Leftarrow \underline{\text{if}}\ y'=x\ \underline{\text{then}}\ 1\ \underline{\text{else}}\ (y'+1)\cdot F(x,y'+1).$$

Here '$\Leftarrow$' stands for 'is defined recursively by' (see McCarthy (1963a)). The computation of P_3 for $x=3$ is

$$z:\ F(3) \rightarrow 3\cdot F(2) \rightarrow 3\cdot 2\cdot F(1) \rightarrow 3\cdot 2\cdot 1\cdot F(0) \rightarrow 3\cdot 2\cdot 1\cdot 1\ ,$$

while the computation of P_4 for $x=3$ is

$$z:\ F(3,0) \rightarrow 1\cdot F(3,1) \rightarrow 1\cdot 2\cdot F(3,2) \rightarrow 1\cdot 2\cdot 3\cdot F(3,3) \rightarrow 1\cdot 2\cdot 3\cdot 1\ .$$

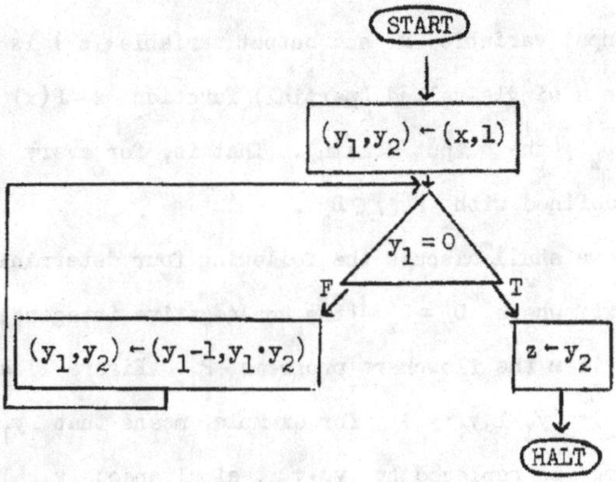

FIGURE 1: The flowchart program P_1 for computing $z=x!$

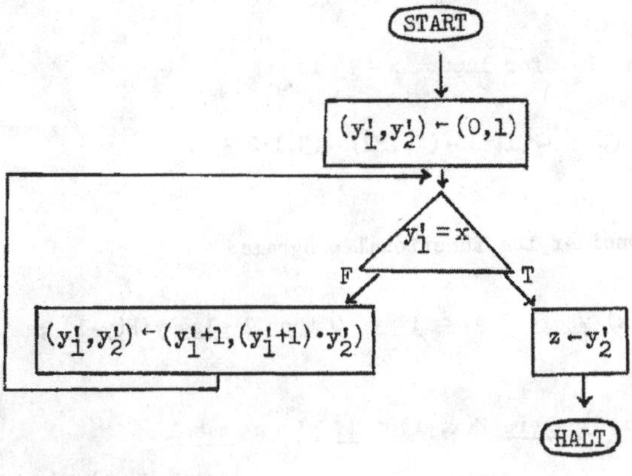

FIGURE 2: The flowchart program P_2 for computing $z=x!$

Let $\psi(x,z)$ be a total predicate over $D_x \times D_z$ (called the <u>output precidate</u>), and let $\xi \in D_x$. We say that

1. (i) $P(\xi)$ is <u>partially correct</u> with respect to ψ if either $P(\xi)$ is undefined, or $P(\xi)$ is defined and $\psi(\xi,P(\xi)) = T$;

   (ii) $P(\xi)$ is <u>totally correct</u> with respect to ψ if $P(\xi)$ is defined and $\psi(\xi,P(\xi)) = T$.

Let P_1 and P_2 be any two comparable deterministic algorithms, i.e., algorithms with the same input domain D_x and the same output domain D_z. We say that

2. (i) $P_1(\xi)$ and $P_2(\xi)$ are <u>partially equivalent</u> if either $P_1(\xi)$ or $P_2(\xi)$ is undefined, or both $P_1(\xi)$ and $P_2(\xi)$ are defined and $P_1(\xi) = P_2(\xi)$;

   (ii) $P_1(\xi)$ and $P_2(\xi)$ are <u>totally equivalent</u> if both $P_1(\xi)$ and $P_2(\xi)$ are defined and $P_1(\xi) = P_2(\xi)$.

3. (i) $P_1(\xi)$ is an <u>extension</u> of $P_2(\xi)$ if whenever $P_2(\xi)$ is defined, then so is $P_1(\xi)$ and $P_1(\xi) = P_2(\xi)$;

   (ii) $P_1(\xi)$ and $P_2(\xi)$ are <u>equivalent</u> if either both $P_1(\xi)$ and $P_2(\xi)$ are undefined, or both $P_1(\xi)$ and $P_2(\xi)$ are defined and $P_1(\xi) = P_2(\xi)$.

Our main purpose in this section is to show that all these properties can be expressed in terms of partial correctness as described by the following propositions:

(a) $P(\xi)$ is defined if and only if $P(\xi)$ is not partially correct w.r.t. F (false);

(b) $P(\xi)$ is totally correct w.r.t. ψ if and only if $P(\xi)$ is not partially correct w.r.t. $\sim\psi$;

(c) $P_1(\xi)$ is partially equivalent to $P_2(\xi)$ if and only if $\forall\psi\, [P_1(\xi)$ is partially correct w.r.t. ψ OR $P_2(\xi)$ is partially correct w.r.t. $\sim\psi]$;

(d) $P_1(\xi)$ is totally equivalent to $P_2(\xi)$ if and only if $\forall\psi\, [P_1(\xi)$ is not partially correct w.r.t. ψ OR $P_2(\xi)$ is not partially correct w.r.t. $\sim\psi]$;

(e) $P_1(\xi)$ is an extension of $P_2(\xi)$ if and only if $\forall\psi\, [P_1(\xi)$ is partially correct w.r.t. ψ IMPLIES $P_2(\xi)$ is partially correct w.r.t. $\psi]$; and finally

(f) $P_1(\xi)$ is equivalent to $P_2(\xi)$ if and only if $\forall\psi\, [P_1(\xi)$ is partially correct w.r.t. ψ IF AND ONLY IF $P_2(\xi)$ is partially correct w.r.t. $\psi]$.

The proof of propositions (a) and (b) is straightforward. Propositions (c)-(e) are best proved by considering the contra-positive relations:

(c') $P_1(\xi)$ is not partially equivalent to $P_2(\xi)$ (i.e., both $P_1(\xi)$ and $P_2(\xi)$ are defined and $P_1(\xi) \neq P_2(\xi)$) if and only if $\exists \psi [P_1(\xi)$ is not partially correct w.r.t. ψ AND $P_2(\xi)$ is not partially correct w.r.t. $\sim\psi]$;

(d') $P_1(\xi)$ is not totally equivalent to $P_2(\xi)$ (i.e., either $P_1(\xi)$ or $P_2(\xi)$ is undefined, or both $P_1(\xi)$ and $P_2(\xi)$ are defined and $P_1(\xi) \neq P_2(\xi)$) if and only if $\exists \psi [P_1(\xi)$ is partially correct w.r.t. ψ AND $P_2(\xi)$ is partially correct w.r.t. $\sim\psi]$; and

(e') $P_1(\xi)$ is not an extension of $P_2(\xi)$ (i.e., either $P_2(\xi)$ is defined and $P_1(\xi)$ is undefined, or both $P_1(\xi)$ and $P_2(\xi)$ are defined and $P_1(\xi) \neq P_2(\xi)$) if and only if $\exists \psi [P_1(\xi)$ is partially correct w.r.t. ψ AND $P_2(\xi)$ is not partially correct w.r.t. $\psi]$.

Proposition (f) follows directly from (e) since

(f') $P_1(\xi)$ is equivalent to $P_2(\xi)$ if and only if $P_1(\xi)$ is an extension of $P_2(\xi)$ and $P_2(\xi)$ is an extension of $P_1(\xi)$.

Suppose for a given deterministic algorithm P (mapping integers into integers) we wish to formalize properties such as being total and monotonically increasing (i.e., $x > x' \Rightarrow P(x) > P(x')$). Unfortunately, our definitions of partial and total correctness are not general enough to include such simple properties. We therefore introduce more general notions of partial and total correctness.

Let P_i $(1 \leq i \leq n)$ be n deterministic algorithms with input variables x_i , output variables z_i , input domains D_{x_i} , and output domains D_{z_i} , respectively. Let $\Psi(x_1,z_1,\ldots,x_n,z_n)$ be any total predicate over $D_{x_1} \times D_{z_1} \times \ldots \times D_{x_n} \times D_{z_n}$. We say that

4. (i) $P_1(\xi_1),\ldots,P_n(\xi_n)$ are partially correct w.r.t. Ψ if either at least one of the $P_i(\xi_i)$ is undefined, or each $P_i(\xi_i)$ is defined and $\Psi(\xi_1,P_1(\xi_1),\ldots,\xi_n,P_n(\xi_n)) = T$.

   (ii) $P_1(\xi_1),\ldots,P_n(\xi_n)$ are totally correct w.r.t. Ψ if each $P_i(\xi_i)$ is defined and $\Psi(\xi_1,P_1(\xi_1),\ldots,\psi_n,P_n(\xi_n)) = T$.

Note that for $n = 1$ we obtain properties 1(i) and 1(ii) as special cases of properties 4(i) and 4(ii), respectively. For $n = 2$ and $\Psi(x_1,z_1,x_2,z_2)$: $x_1 = x_2 \supset z_1 = z_2$ we obtain properties 2(i) and 2(ii) as special cases of 4(i) and 4(ii), respectively. For $n = 2$ and $\Psi(x_1,z_1,x_2,z_2) : x_1 > x_2 \supset z_1 > z_2$ where P_1 and P_2 are identical to P , we obtain the above monotonicity property.

It is interesting that these general notions of correctness can still be expressed just by means of the usual partial correctness, as described below.

(g) $P_1(\xi_1),\ldots,P_n(\xi_n)$ are partially correct w.r.t. Ψ if and only if

$\exists\psi_1\ldots\exists\psi_n\{$ $P_1(\xi_1)$ is partially correct w.r.t. ψ_1

AND $P_2(\xi_2)$ is partially correct w.r.t. ψ_2

⋮

AND $P_n(\xi_n)$ is partially correct w.r.t. ψ_n

AND $\forall y_1\ldots\forall y_n[\psi_1(\xi_1,y_1)$ AND...AND $\psi_n(\xi_n,y_n)$ IMPLIES $\Psi(\xi_1,y_1,\ldots,\xi_n,y_n)]\}$;

(h) $P_1(\xi_1),\ldots,P_n(\xi_n)$ are totally correct w.r.t. Ψ if and only if

$\forall\psi_1\ldots\forall\psi_n\{$ $P_1(\xi_1)$ is partially correct w.r.t. ψ_1

AND $P_2(\xi_2)$ is partially correct w.r.t. ψ_2

⋮

AND $P_n(\xi_n)$ is partially correct w.r.t. ψ_n

IMPLIES $\exists y_1\ldots\exists y_n[\psi_1(\xi_1,y_1)$ AND...AND $\psi_n(\xi_n,y_n)$ AND $\Psi(\xi_1,y_1,\ldots,\xi_n,y_n)]\}$.

Proposition (g) is straightforward. (Hint: to prove that the left-hand side implies the right-hand side, choose ψ_i in such a way that $\psi_i(\xi_i,\eta_i)=T \Longleftrightarrow P_i(\xi_i)$ is defined and $\eta_i=P_i(\xi_i)$.) Proposition (h) follows from (g) since:

(h') $P_1(\xi_1),\ldots,P_n(\xi_n)$ are totally correct w.r.t. Ψ if and only if

$P_1(\xi_1),\ldots,P_n(\xi_n)$ are not partially correct w.r.t. $\sim\Psi$.

2. Formalization of Partial Correctness of Deterministic Algorithms

The above propositions imply that if one knows, for example, how to formalize partial correctness of a given deterministic algorithm in predicate calculus, the formalization of all other properties of the algorithm in predicate calculus is straightforward. As a matter of fact, partial correctness has already been formalized in predicate calculus for many classes of deterministic algorithms.

In this section we illustrate the flavor of such formalizations.

(A) Flowchart Programs and Predicate Calculus

Let us consider, for example, a flowchart program P of the form described in Figure 3, with a given output predicate $\psi(x,z)$ over $D_x \times D_z$. Here, input(x) maps D_x into D_y , test(x,y) is a predicate over $D_x \times D_y$, operator(x,y) maps $D_x \times D_y$ into D_y , and output(x,y) maps $D_x \times D_y$ into D_z .

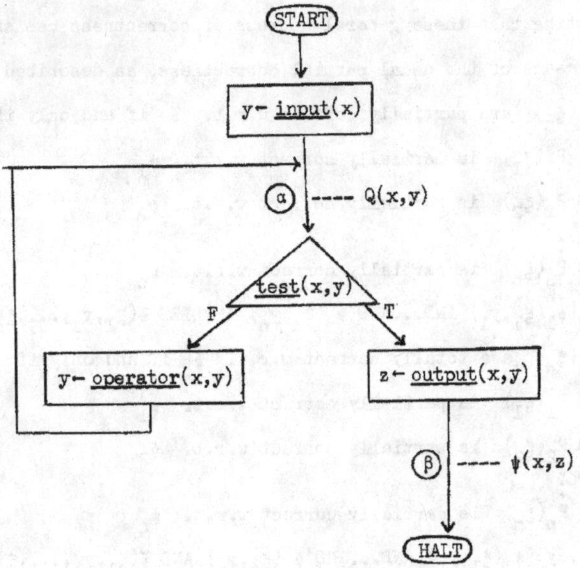

FIGURE 3: The flowchart program P

We associate the (unspecified) predicate variable $Q(x,y)$ with arc α and the given output predicate $\psi(x,z)$ with arc β, and construct the following formula $W_P(x,\psi)$:

$$\exists Q\{Q(x,\underline{input}(x)) \quad \text{---------- initialization}$$
$$\wedge\ \forall y[Q(x,y) \wedge \sim\underline{test}(x,y) \supset Q(x,\underline{operator}(x,y))] \quad \text{---------- induction}$$
$$\wedge\ \forall y[Q(x,y) \wedge \underline{test}(x,y) \supset \psi(x,\underline{output}(x,y))]\} \quad \text{---------- conclusion}$$

or, equivalently,

$$\exists Q\{Q(x,\underline{input}(x))$$
$$\wedge\ \forall y[Q(x,y) \supset \text{IF } \underline{test}(x,y) \text{ THEN } \psi(x,\underline{output}(x,y))$$
$$\text{ELSE } Q(x,\underline{operator}(x,y))]\}\ .$$

Here, IF A THEN B ELSE C stands for $(A\supset B) \wedge (\sim A\supset C)$. Note that $D\supset$ IF A THEN B ELSE C is logically equivalent to $(D\wedge A\supset B) \wedge (D\wedge\sim A\supset C)$.

The key result is that for any given input $\xi \in D_x$, $P(\xi)$ is partially correct w.r.t. ψ <u>if and only if</u> $W_P(\xi,\psi)$ is true (Manna (1969 a)).

Example 1: For the flowchart program P_1 (Figure 1), it follows that:[1] $P_1(\xi)$ is partially correct w.r.t. $z=x!$ if and only if $W_{P_1}(\xi, z=x!)$ is true, where $W_{P_1}(\xi, z=x!)$ is

$$\exists Q\{Q(\xi,\xi,1) \wedge \forall y_1 \forall y_2[Q(\xi,y_1,y_2) \supset \text{IF } y_1=0 \text{ THEN } y_2=\xi! \text{ ELSE } Q(\xi,y_1-1,y_1\cdot y_2)]\} .$$

Note that for $Q(\xi,y_1,y_2)$ being the predicate $y_2\cdot y_1!=\xi!$, the formula is true.

Example 2: For the flowchart program P_2 (Figure 2), it follows similarly that: $P_2(\xi)$ is partially correct w.r.t. $z=x!$ if and only if $W_{P_2}(\xi, z=x!)$ is true, where $W_{P_2}(\xi, z=x!)$ is

$$\exists Q\{Q(\xi,0,1) \wedge \forall y_1' \forall y_2'[Q(\xi,y_1',y_2') \supset \text{IF } y_1'=\xi \text{ THEN } y_2'=\xi! \text{ ELSE } Q(\xi,y_1'+1,(y_1'+1)\cdot y_2')]\} .$$

Note that for $Q(\xi,y_1',y_2')$ being the predicate $y_2'=y_1'!$, the formula is true.

(B) Functional Programs and Predicate Calculus

Consider, for example, a functional program P of the form

$z=F(x,\underline{\text{input}}(x))$ <u>where</u>

$F(x,y) \Leftarrow$ <u>if</u> <u>test</u>(x,y) <u>then</u> <u>output</u>(x,y)

<u>else</u> <u>operator1</u>$(x,y,F(x,$<u>operator2</u>$(x,y)))$,

with a given output predicate $\psi(x,z)$ over $D_x \times D_z$. Here, <u>input</u>(x) maps D_x into D_y, <u>test</u>(x,y) is a predicate over $D_x \times D_y$, <u>output</u>(x,y) maps $D_x \times D_y$ into D_z, <u>operator1</u> maps $D_x \times D_y \times D_z$ into D_z, and <u>operator2</u> maps $D_x \times D_y$ into D_y.

We associate the predicate variable $Q(x,y,z)$ with $F(x,y)$, and construct the following formula $W_P(x,\psi)$:

$$\exists Q\{\forall z[Q(x,\underline{\text{input}}(x),z) \supset \psi(x,z)] \wedge \forall y[\text{IF } \underline{\text{test}}(x,y) \text{ THEN } Q(x,y,\underline{\text{output}}(x,y)) \text{ ELSE } \forall t[Q(x,\underline{\text{operator2}}(x,y),t) \supset Q(x,y,\underline{\text{operator1}}(x,y,t))]]\}.$$

[1] Here, $D_x = D_z = \{\text{the non-negative integers}\}$, $y=(y_1,y_2)$, and $D_y = \{\text{all pairs of non-negative integers}\}$.

Here, Q represents F in such a way that for every triple $(\xi,\eta,\zeta)\in D_x\times D_y\times D_z$, if $F(\xi,\eta)=\zeta$, then $Q(\xi,\eta,\zeta)=T$, but not necessarily conversely.

The key result is that for any given input $\xi\in D_x$, $P(\xi)$ is partially correct w.r.t. ψ <u>if and only if</u> $W_P(\xi,\psi)$ is true (Manna and Pnueli (1970), see also Park (1970)).

<u>Example 3</u>: For the functional program P_3 :

$$z=F(x) \ \underline{\text{where}}$$
$$F(y) \Leftarrow \underline{\text{if}} \ \ y=0 \ \ \underline{\text{then}} \ \ 1 \ \ \underline{\text{else}} \ \ y\cdot F(y-1) \ ,$$

it follows that: $P_3(\xi)$ is partially correct w.r.t. $z=x!$ if and only if $W_{P_3}(\xi\,,z=x!)$ is true, where $W_{P_3}(\xi\,,z=x!)$ is

$$\exists Q\{\forall z[Q(\xi,z)\supset z=\xi!]$$
$$\wedge\ \forall y[\text{IF } \ y=0 \ \text{ THEN } \ Q(y,1) \ \text{ ELSE } \ \forall t[Q(y-1,t)\supset Q(y,y\cdot t)]]\}\ .$$

Note that for $Q(y,z)$ being the predicate $z=y!$ the formula is true.

<u>Example 4</u>: For the functional program P_4 :

$$z=F(x,0) \ \underline{\text{where}}$$
$$F(x,y')\Leftarrow \underline{\text{if}} \ \ y'=x \ \ \underline{\text{then}} \ \ 1 \ \ \underline{\text{else}} \ \ (y'+1)\cdot F(x,y'+1)\ ,$$

it follows that: $P_4(\xi)$ is partially correct w.r.t. $z=x!$ if and only if $W_{P_4}(\xi\,,z=x!)$ is true, where $W_{P_4}(\xi\,,z=x!)$ is

$$\exists Q\{\forall z[Q(\xi,0,z)\supset z=\xi!]$$
$$\wedge\ \forall y'[\text{IF } y'=\xi \text{ THEN } Q(\xi,y',1) \text{ ELSE } \forall t[Q(\xi,y'+1,t)\supset Q(\xi,y',(y'+1)\cdot t)]]\}\ .$$

Note that for $Q(\xi,y',z)$ being the predicate $z\cdot y'!=\xi!$, the formula is true.

The formulas constructed here are independent of the syntax of the language in which the algorithms are expressed, and, therefore, we can use our results to formalize in predicate calculus the equivalence of algorithms defined by different languages. From proposition (f) it follows, for example, that for every input ξ, $P_1(\xi)$ and $P_3(\xi)$ are equivalent if and only if $\forall\psi[W_{P_1}(\xi,\psi)\equiv W_{P_3}(\xi,\psi)]$ is true.

The reader should realize that the flowchart program P (Figure 3) can be represented equivalently (see McCarthy (1962)) by the functional program P' :

$z = F(x, \underline{input}(x))$ <u>where</u>

$F(x,y) \Leftarrow \underline{if}\ \underline{test}(x,y)\ \underline{then}\ \underline{output}(x,y)\ \underline{else}\ F(x, \underline{operator}(x,y))$.

However, $W_{P'}(x,\psi)$ is

$$\exists Q\{\forall z[Q(x,\underline{input}(x),z) \supset \psi(x,z)] \wedge \forall y[\text{IF } \underline{test}(x,y) \text{ THEN } Q(x,y,\underline{output}(x,y)) \text{ ELSE } \forall t[Q(x,\underline{operator}(x,y),t) \supset Q(x,y,t)]]\} ;$$

while $W_P(x,\psi)$ was

$$\exists Q\{Q(x,\underline{input}(x)) \wedge \forall y[Q(x,y) \supset \text{IF } \underline{test}(x,y) \text{ THEN } \psi(x,\underline{output}(x,y)) \text{ ELSE } Q(x,\underline{operator}(x,y))]\}.$$

Although both $W_P(x,\psi)$ and $W_{P'}(x,\psi)$ essentially formalize partial correctness of $P(x)$ w.r.t. ψ , they seem to be quite different. Intuitively, the difference between the two formalizations is that $Q(x,y)$ in $W_P(x,\psi)$ represents all current values of (x,y) at arc α during the computation of P, while $Q(x,y,z)$ in $W_{P'}(x,\psi)$ represents the final value of z when computation of P starts at arc α with initial values (x,y) .

(C) Functional Programs and Partial Function Logic

Consider again a functional program P of the form

$z = F(x, \underline{input}(x))$ <u>where</u>

$F(x,y) \Leftarrow \underline{if}\ \underline{test}(x,y)\ \underline{then}\ \underline{output}(x,y)$
$\qquad \underline{else}\ \underline{operator1}(x,y,F(x,\underline{operator2}(x,y)))$,

with a given output predicate $\psi(x,z)$.

We construct the following formula $\tilde{W}_P(x,\psi)$:

$$\exists F\{[*F(x,\underline{input}(x)) \supset \psi(x,F(x,\underline{input}(x)))] \wedge \forall y[F(x,y) \overset{*}{=} \underline{if}\ \underline{test}(x,y)\ \underline{then}\ \underline{output}(x,y)\ \underline{else}\ \underline{operator1}(x,y,F(x,\underline{operator2}(x,y)))]\} .$$

Here, "$\exists F$" stands for "there exists a partial function F mapping $D_x \times D_y$ into D_z such that..."; "$*F(x,\underline{input}(x))$" stands for the total predicate (mapping D_x

into $\{T,F\}$) "$F(x,\underline{input}(x))$ is defined"; and $\stackrel{*}{=}$ is just the natural extension of the usual equality relation, defined as follows: $A \stackrel{*}{=} B$ if and only if either both expressions A and B are defined and represent the same element (of D_z , in this case) or both expressions are undefined.

The key result is that for every given $\xi \in D_x$, $P(\xi)$ is partially correct w.r.t. ψ __if and only if__ $\widetilde{W}_P(\xi,\psi)$ is true (Manna and McCarthy (1970)).

__Example 5__: For the functional program P_4 :

$$z = F(x,0) \ \underline{\text{where}}$$

$$F(x,y') \Leftarrow \underline{\text{if}} \ y' = x \ \underline{\text{then}} \ 1 \ \underline{\text{else}} \ (y'+1)\cdot F(x,y'+1) \ ,$$

it follows that: $P_4(\xi)$ is partially correct w.r.t. $z = x!$ if and only if $\widetilde{W}_{P_4}(\xi , z = x!)$ is true, where $\widetilde{W}_{P_4}(\xi , z = x!)$ is

$$\exists F\{[*F(\xi,0) \supset F(\xi,0) = \xi!]$$
$$\wedge \ \forall y'[F(\xi,y') \stackrel{*}{=} \underline{\text{if}} \ y' = \xi \ \underline{\text{then}} \ 1 \ \underline{\text{else}} \ (y'+1)\cdot F(\xi,y'+1)]\} \ .$$

Note that for $F(\xi,y')$ being the partial function

$$F(\xi,y') = \begin{cases} \xi!/y'! & \text{if } \ y' \le \xi \\ \text{undefined} & \text{if } \ y' > \xi \end{cases}$$

the formula is true.

3. Non-Deterministic Algorithms

One natural extension of our study is obtained by considering non-deterministic algorithms rather than deterministic algorithms.

An algorithm P (with input variable x and output variable z) is said to be __non-deterministic__ if it defines a many-valued function $P(x)$, mapping elements of D_x (the input domain) into __subsets__ of D_z (the output domain); that is, for every $\xi \in D_x$, $P(\xi)$ is a (possibly empty) subset Z of D_z , where each $\zeta \in Z$ is the final result of some computation of P with input ξ .

__Examples__: We first describe three non-deterministic programs for computing $z = x!$, making use of the deterministic programs P_1-P_4 introduced in Section 1.

(a) Parallel flowchart program: In Figure 4 we have described a simple parallel flowchart program P_5 for computing $z = x!$. The program includes a 'BEGIN-END' block which consists of two branches, the left branch being the body of program P_1 and the right branch being the body of program P_2 , after changing the test statements to $y_1 = y_1'$ in both.

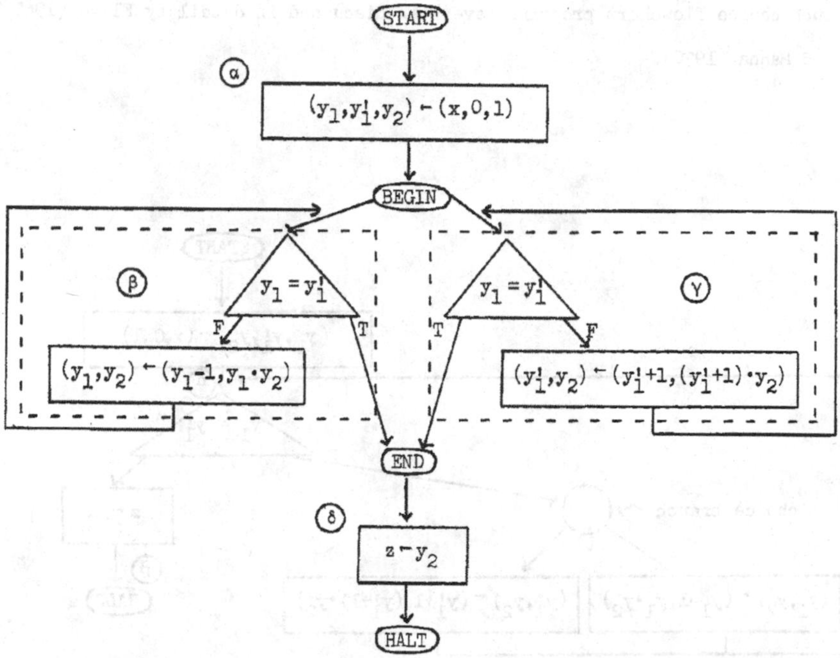

FIGURE 4: The parallel flowchart program P_5 for computing $z = x!$

The program is executed as follows. First statement α is executed. Entering the block either the statements in β or the statements in γ are executed, chosen arbitrarily. The execution proceeds asynchronously, i.e., between the execution of two consecutive β's , we may execute an arbitrary number of γ's ; and conversely, between the execution of two consecutive γ's we may execute an arbitrary number of β's . β and γ cannot be executed at the same time. Therefore, one can consider execution to be performed with a single processor switching between the two branches. We exit from the block and execute statement δ when either of the two branches reaches the END node. Such parallel programs are discussed in detail in Ashcroft and Manna (1970).

(b) Choice flowchart program: In Figure 5 we have described a choice flowchart program for computing $z = x!$. A branch of the form  is called a <u>choice branch</u>. It means that upon reaching the choice branch during execution of the program, we are allowed to proceed with either branch, chosen arbitrarily. Such choice flowchart programs have been discussed in detail by Floyd (1967 b) and Manna (1970).

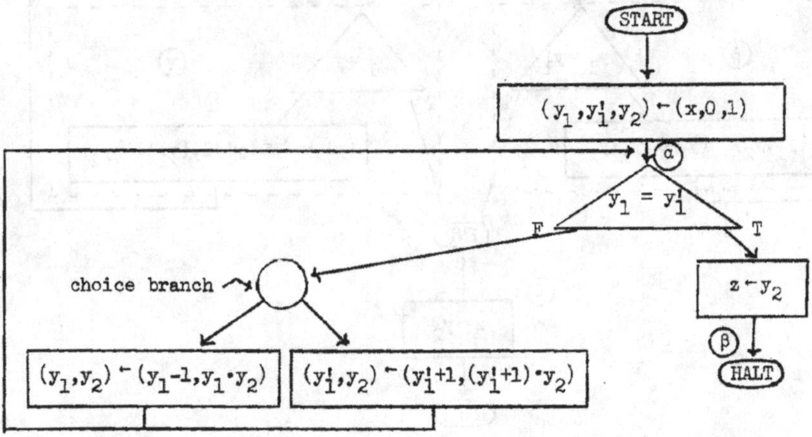

FIGURE 5: The choice flowchart program P_6 for computing $z = x!$

Note that for any given input x both P_5 and P_6 yield the same set of computations. For $x = 3$, for example, there are exactly 8 different possible executions of each program, described by the following tree (each triple represents the current value of (y_1, y_1', y_2)) :

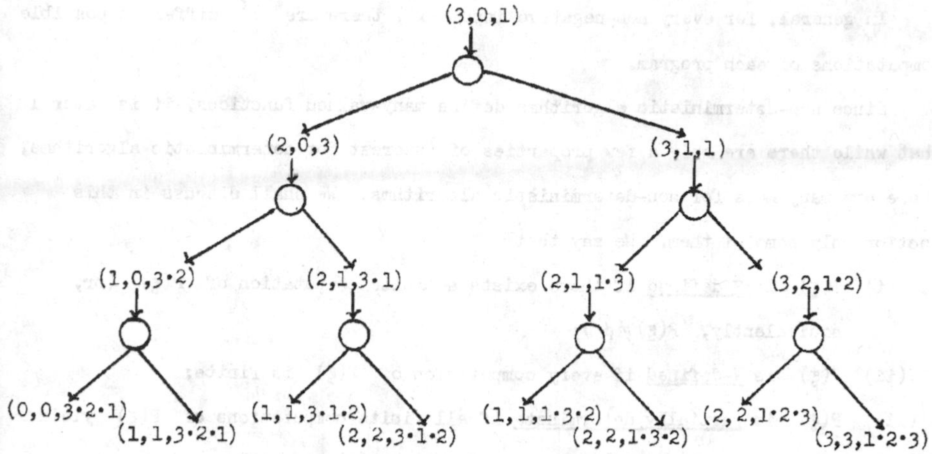

(c) Choice functional program: Consider the following choice functional program P_7:

$z = F(x,0)$ <u>where</u>

$F(y,y') \Leftarrow$ <u>if</u> $y=y'$ <u>then</u> 1 <u>else</u> <u>choice</u>$(y \cdot F(y-1,y'),(y'+1) \cdot F(y,y'+1))$.

The <u>choice</u> function here has the same meaning as the choice branch in P_6 ; it corresponds to McCarthy's (1963) <u>amb</u> (ambiguous) function. For $x=3$, P_7 also yields 8 different possible computations as described by the following tree:

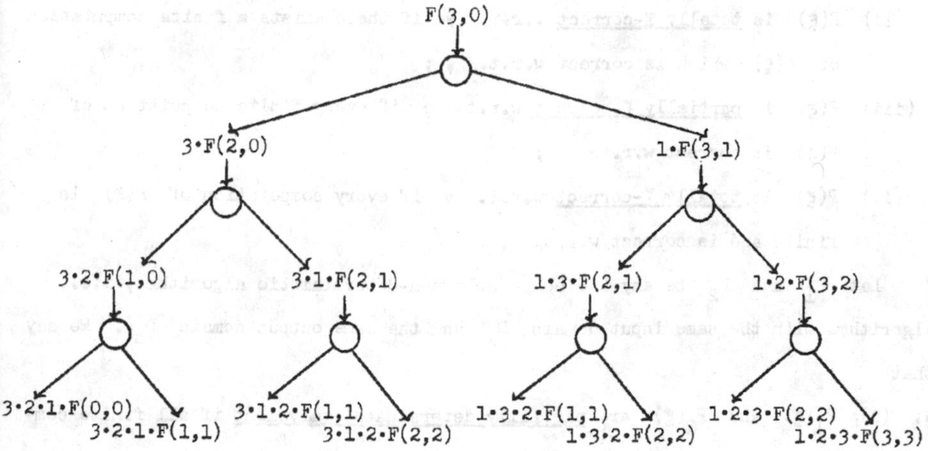

In general, for every non-negative input x, there are 2^x different possible computations of each program.

Since non-deterministic algorithms define many-valued functions, it is natural that while there are only a few properties of interest for deterministic algorithms, there are many more for non-deterministic algorithms. We shall discuss in this section only some of them. We say that

1. (i) $P(\xi)$ is <u>$\exists$-defined</u> if there exists a finite computation of $P(\xi)$ (or, equivalently, $P(\xi) \neq \phi$);

   (ii) $P(\xi)$ is <u>$\forall$-defined</u> if every computation of $P(\xi)$ is finite;

   (iii) $P(\xi)$ is <u>partially determinate</u> if all finite computations of $P(\xi)$ yield the same final result (or, equivalently, $P(\xi)$ is either empty or a singleton);

   (iv) $P(\xi)$ is <u>totally determinate</u> if all computations of $P(\xi)$ are finite and yield the same final result.

Let $\psi(x,z)$ be a total predicate over $D_x \times D_z$, and let $\xi \in D_x$. A finite computation of $P(\xi)$ is said to be <u>correct w.r.t. ψ</u> if for its final value ζ, $\psi(\xi,\zeta) = T$. We say that

2. (i) $P(\xi)$ is <u>partially $\exists$-correct</u> w.r.t. ψ if either there exists an infinite computation of $P(\xi)$, or there exists a finite computation of $P(\xi)$ which is correct w.r.t. ψ;

   (ii) $P(\xi)$ is <u>totally $\exists$-correct</u> w.r.t. ψ if there exists a finite computation of $P(\xi)$ which is correct w.r.t. ψ;

   (iii) $P(\xi)$ is <u>partially $\forall$-correct</u> w.r.t. ψ if every finite computation of $P(\xi)$ is correct w.r.t. ψ;

   (iv) $P(\xi)$ is <u>totally $\forall$-correct</u> w.r.t. ψ if every computation of $P(\xi)$ is finite and is correct w.r.t. ψ.

Let P_1 and P_2 be any two comparable non-deterministic algorithms, i.e., algorithms with the same input domain D_x and the same output domain D_z. We say that

3. (i) $P_1(\xi)$ and $P_2(\xi)$ are <u>partially determinate-equivalent</u> if all finite computations of $P_1(\xi)$ and $P_2(\xi)$ yield the same final result (or,

equivalently, $P_1(\xi) \cup P_2(\xi)$ is either empty or a singleton);

(ii) $P_1(\xi)$ and $P_2(\xi)$ are <u>totally determinate-equivalent</u> if all computations of $P_1(\xi)$ and $P_2(\xi)$ are finite and yield the same final result.

4. (i) $P_1(\xi)$ <u>partially extends</u> $P_2(\xi)$ if, for every finite computation of $P_2(\xi)$, there exists a finite computation of $P_1(\xi)$ that yields the same final value (or, equivalently, $P_1(\xi) \supseteq P_2(\xi)$);

(ii) $P_1(\xi)$ <u>totally extends</u> $P_2(\xi)$ if $P_1(\xi)$ partially extends $P_2(\xi)$, and if there exists an infinite computation of $P_2(\xi)$, then there is also an infinite computation of $P_1(\xi)$.

5. (i) $P_1(\xi)$ and $P_2(\xi)$ are <u>partially equivalent</u> if $P_1(\xi)$ partially extends $P_2(\xi)$ and conversely (or, equivalently, $P_1(\xi) = P_2(\xi)$);

(ii) $P_1(\xi)$ and $P_2(\xi)$ are <u>totally equivalent</u> if $P_1(\xi)$ totally extends $P_2(\xi)$ and conversely.

Our main purpose in this section is to show that all these properties can be expressed in terms of the two notions of partial correctness, namely partial $\exists$-correctness and partial $\forall$-correctness.

(a) $P(\xi)$ is $\exists$-defined if and only if $P(\xi)$ is not partially $\forall$-correct w.r.t. F (false);

(b) $P(\xi)$ is $\forall$-defined if and only if $P(\xi)$ is not partially $\exists$-correct w.r.t. F (false);

(c) $P(\xi)$ is partially determinate if and only if $\forall\psi[P(\xi)$ is partially $\forall$-correct w.r.t. ψ OR $P(\xi)$ is partially $\forall$-correct w.r.t. $\sim\psi]$;

(d) $P(\xi)$ is totally determinate if and only if $\forall\psi[P(\xi)$ is not partially $\exists$-correct w.r.t. ψ OR $P(\xi)$ is not partially $\exists$-correct w.r.t. $\sim\psi]$;

(e) $P(\xi)$ is totally $\exists$-correct w.r.t. ψ if and only if $P(\xi)$ is not partially $\forall$-correct w.r.t. $\sim\psi$;

(f) $P(\xi)$ is totally $\forall$-correct w.r.t. ψ if and only if $P(\xi)$ is not partially $\exists$-correct w.r.t. $\sim\psi$;

(g) $P_1(\xi)$ and $P_2(\xi)$ are partially determinate-equivalent if and only if $\forall\psi[P_1(\xi)$ is partially $\forall$-correct w.r.t. ψ OR $P_2(\xi)$ is partially $\forall$-correct w.r.t. $\sim\psi]$;

(h) $P_1(\xi)$ and $P_2(\xi)$ are totally determinate-equivalent if and only if $\forall\psi\,[P_1(\xi)$ is not partially $\exists$-correct w.r.t. ψ OR $P_2(\xi)$ is not partially $\exists$-correct w.r.t. $\sim\psi]$;

(i) $P_1(\xi)$ partially extends $P_2(\xi)$ if and only if $\forall\psi\,[P_1(\xi)$ is partially $\forall$-correct w.r.t. ψ IMPLIES $P_2(\xi)$ is partially $\forall$-correct w.r.t. $\psi]$;

(j) $P_1(\xi)$ totally extends $P_2(\xi)$ if and only if $\forall\psi\,[P_2(\xi)$ is partially $\exists$-correct w.r.t. ψ IMPLIES $P_1(\xi)$ is partially $\exists$-correct w.r.t. $\psi]$;

(k) $P_1(\xi)$ and $P_2(\xi)$ are partially equivalent if and only if $\forall\psi\,[P_1(\xi)$ is partially $\forall$-correct w.r.t. ψ IF AND ONLY IF $P_2(\xi)$ is partially $\forall$-correct w.r.t. $\psi]$;

(ℓ) $P_1(\xi)$ and $P_2(\xi)$ are totally equivalent if and only if $\forall\psi\,[P_1(\xi)$ is partially $\exists$-correct w.r.t. ψ IF AND ONLY IF $P_2(\xi)$ is partially $\exists$-correct w.r.t. $\psi]$.

Propositions (a), (b), (e), and (f) are straightforward by definition. Propositions (c), (d), (g), (h), (i), and (j) are best proved by considering the corresponding contra-positive relations. Propositions (k) and (ℓ) follow from propositions (i) and (j), respectively.

4. Formalization of Partial Correctness of Non-Deterministic Algorithms

For a given non-deterministic program P and an output predicate $\psi(x,z)$, we would like to construct two formulas $W^{\exists}(x,\psi)$ and $W^{\forall}(x,\psi)$ in predicate calculus, such that for every given input value $\xi \in D_x$:

(i) $P(\xi)$ is partially $\exists$-correct w.r.t. ψ if and only if $W^{\exists}(\xi,\psi)$ is true, and

(ii) $P(\xi)$ is partially $\forall$-correct w.r.t. ψ if and only if $W^{\forall}(\xi,\psi)$ is true.

Then, using the formulas $W^{\exists}(x,\psi)$ and $W^{\forall}(x,\psi)$, the formalization of all other properties of P in predicate calculus is straightforward.

We shall illustrate the construction of $W^{\exists}(x,\psi)$ and $W^{\forall}(x,\psi)$ for the choice flowchart program P_6 (Figure 5) and the choice functional program P_7 . The main idea behind this formalization is that the effect of the choice branch is represented by an '$\vee$' connective in $W^{\exists}(x,\psi)$, while it is represented by an '$\wedge$' connective in $W^{\forall}(x,\psi)$.

To construct $W_{P_6}^{\exists}(x, z=x!)$ and $W_{P_6}^{\forall}(x, z=x!)$, associate the predicate variable $Q(x,y_1,y_1',y_2)$ with arc α in Figure 5 and the predicate variable $z=x!$ with arc β. Then $W_{P_6}^{\exists}(x, z=x!)$ is

$$\exists Q\{Q(x,x,0,1) \wedge \forall y_1 \forall y_1' \forall y_2 [Q(x,y_1,y_1',y_2) \supset \text{IF } y_1 = y_1' \text{ THEN } y_2 = x! \\ \text{ELSE } [Q(x,y_1-1,y_1',y_1 \cdot y_2) \vee Q(x,y_1,y_1'+1,(y_1'+1)\cdot y_2)]]\} .$$

$W_{P_6}^{\forall}(x, z=x!)$ is similar with the '$\vee$' connective replaced by '$\wedge$'.

The reader can verify easily that for every non-negative integer ξ, both formulas $W_{P_6}^{\exists}(\xi, z=x!)$ and $W_{P_6}^{\forall}(\xi, z=x!)$ are true for $Q(\xi,y_1,y_1',y_2)$ being the predicate $y_2 \cdot y_1! = \xi! \cdot y_1'!$.

To construct $W_{P_7}^{\exists}(x, z=x!)$ and $W_{P_7}^{\forall}(x, z=x!)$, associate the predicate variable $Q(y,y',z)$ with the function variable $F(y,y')$. Then $W_{P_7}^{\exists}(x, z=x!)$ is:

$$\exists Q\{\forall z[Q(x,0,z) \supset z=x!] \wedge \forall y \forall y' [\text{IF } y=y' \text{ THEN } Q(y,y',1) \text{ ELSE } \forall t[Q(y-1,y',t) \supset \\ Q(y,y',y\cdot t)] \vee \forall t[Q(y,y'+1,t) \supset Q(y,y',(y'+1)\cdot t)]]\} .$$

$W_{P_7}^{\forall}(x, z=x!)$ is similar with the '$\vee$' connective replaced by '$\wedge$'.

The reader can verify easily that for every non-negative integer ξ, both formulas $W_{P_7}^{\exists}(\xi, z=x!)$ and $W_{P_7}^{\forall}(\xi, z=x!)$ are true for $Q(y,y',z)$ being the predicate $z\cdot y'! = y!$.

Acknowledgments

I am indebted to Edward Ashcroft and Stephen Ness for many stimulating discussions and also for their critical reading of the manuscript and subsequent helpful suggestions.

TOWARDS AUTOMATIC PROGRAM SYNTHESIS[1]

by

Zohar Manna and Richard J. Waldinger

1. INTRODUCTION

It is often easier to describe what a computation does than it is to define it explicitly. That is, we may be able to write down the relation between the input and the output variables easily, even when it is difficult to construct a program to satisfy that relation. A program synthesizer is a system that takes such a relational description and tries to produce a program that is guaranteed to satisfy the relationship and, therefore, does not require debugging or verification.

On a more limited scale, we can envision an automatic debugging system that corrects programs written by humans instead of merely verifying them. We can further imagine clever compilers and optimizers that understand the operation of the programs they manipulate and that can transform them intelligently.

Some program synthesizers have already been written, including the Heuristic Compiler (Simon (1963)), DEDUCOM (Slagle(1965)), QA3 (Green (1969a), (1969b)), and PROW (Waldinger and Lee (1969) and Waldinger (1969)). The last three of these systems use a theorem-proving approach: in order to construct a program satisfying a certain input-output relation, the system proves a theorem induced by this relation and extracts the program directly from the proof. All three used the resolution principle of Robinson (1965). However, these systems have been fairly limited; for

[1]The research reported herein was sponsored in part by the Air Force Systems Command, USAF, Department of Defense, through the Air Force Cambridge Research Laboratories, Office of Aerospace Research, under contract number F19628-70-C-0246; and by the Advanced Research Projects Agency of the Department of Defense and the National Aeronautics and Space Administration under contract number NAS 12-2221 (at Stanford Research Institute); and by the Advanced Research Projects Agency of the Office of the Secretary of Defense, ARPA Order Number SD-183 (at Stanford University).

example, they either have been completely unable to produce programs with loops, or they introduced loops by underhanded methods.

When a theorem-proving approach is used in program synthesis, the introduction of loops into the extracted program is closely related to the use of the principle of mathematical induction in the corresponding proof. The induction principle presented special problems to the earlier program-synthesis systems, problems which limited their ability to produce loop programs. These problems are discussed in this paper. We propose to use a variety of different versions of the induction rule, each of which applies to a particular data structure, and each of which induces a different form in the extracted program. The data structures treated are the natural numbers, lists, and trees.

We do not rely on any specific mechanical theorem-proving techniques here, both because we do not wish to restrict our class of readers to those familiar with, say, the resolution principle, and because we believe the approach to be more general and not dependent on one particular theorem-proving method. We give a large number of examples of programs, with the corresponding theorems and proofs used in their synthesis. The proofs we give are informal and in the style of a mathematics textbook. Some of them have been achieved by such systems as PROW and QA3; others we believe to be beyond the powers of existing automatic theorem provers, but none are unreasonably difficult, and we hope that the designers of theorem-proving systems will accept them as a challenge.

Section 2 gives the flavor of the approach illustrated by three examples. In that section, we do not prove the induced theorems, and we present the constructed programs without describing the extraction process. In section 3, we demonstrate the extraction process with complete examples of the synthesis of two programs without loops. We choose loop-free programs for these examples so as to postpone discussion of the principle of mathematical induction.

The heart of the paper is contained in section 4, with the presentation of the induction principles and their corresponding iterative or recursive program forms. One of the examples in this section gives details of the proof and program extraction process. Section 5 demonstrates a more general rule, the complete induction principle.

Section 6 suggests applying program-synthesis techniques to translate recursive programs into iterative programs and presents two examples in which a striking gain in efficiency was achieved. Finally, in section 7 we suggest further research in this field.

2. GENERAL DISCUSSION

We define the problem of automatic program synthesis as follows: given an <u>input predicate</u> $\varphi(\bar{x})$ and an <u>output predicate</u> $\psi(\bar{x},\bar{z})$, construct a program computing a partial function $\bar{z}=f(\bar{x})$ such that if $\bar{x}$ is an <u>input vector</u> satisfying $\varphi(\bar{x})$, then $f(\bar{x})$ is defined and $\psi(\bar{x},f(\bar{x}))$ is true. In short, the predicates $\varphi(\bar{x})$ and $\psi(\bar{x},\bar{z})$ provide the specifications for the program to be written.

In order to construct such a program, we prove the theorem

$$(\forall\bar{x})[\varphi(\bar{x})\supset(\exists\bar{z})\psi(\bar{x},\bar{z})] .$$

The desired program is then implicit in the proof that the <u>output vector</u> $\bar{z}$ exists. The theorem prover must be restricted to show the existence of $\bar{z}$ constructively, so that the appropriate program can be extracted from the proof automatically.

Frequently, $\varphi(\bar{x})$ is identically true; i.e., we are interested in the performance of the program for every input $\bar{x}$. Then the theorem to be proved is simply

$$(\forall\bar{x})[T\supset(\exists\bar{z})\psi(\bar{x},\bar{z})] ,$$

or equivalently,

$$(\forall\bar{x})(\exists\bar{z})\psi(\bar{x},\bar{z}) .$$

In such cases we shall neglect to mention the input predicate.

Let us first illustrate the flavor of this idea with three examples:

(i) The construction of an iterative program to compute the quotient and the remainder of two natural numbers.

(ii) The translation of a LISP recursive program for reversing the top-level elements of a list into an equivalent LISP iterative program.

(iii) The construction of a recursive program for finding the maximum among the terminal nodes in a binary tree with integer terminals.

In each case we give the specifications for the program, the induced theorem, and the automatically synthesized program, without introducing the proofs of the theorems or the extraction of the programs from the proofs. Such details will be given in the examples of our later sections.

In our examples we express our input and output predicates in a modified predicate calculus language. However, this is not essential to the method; any language for describing relations may be used.

Example 1: Construction of an iterative division program.

We wish to construct an iterative program to compute the integer quotient and the remainder of two natural numbers, x_1 and x_2, where $x_2 \neq 0$. The program should set the output variable z_1 to be the quotient of x_1 divided by x_2, and the output variable z_2 to be the corresponding remainder.

Thus, $\bar{x} = x_1, x_2$ and $\bar{z} = z_1, z_2$. Since we are not interested in the program's performance for $x_2 = 0$, our input predicate is

$$\varphi(\bar{x}):\ x_2 \neq 0 .$$

The output predicate is

$$\psi(\bar{x},\bar{z}):\ (x_1 = z_1 \cdot x_2 + z_2) \wedge (z_2 < x_2) .$$

The theorem induced is then

$$(\forall x_1)(\forall x_2)\{x_2 \neq 0 \supset (\exists z_1)(\exists z_2)[(x_1 = z_1 \cdot x_2 + z_2) \wedge (z_2 < x_2)]\} .$$

The program synthesizer proves the theorem, and a program such as that illustrated in Figure 1 is extracted from the proof.[1]

We have assumed that certain symbols, including the "minus" operator and the "less than" predicate, for instance, exist in our programming language; therefore, these operators are said to be primitive. However, if the use of the "minus" operator or the "less than" predicate is not permitted in the language (i.e., if they are non-primitive), the program is illegal.

---

[1] Statements in which n-tuples of terms are assigned to n-tuples of variables represent simultaneous replacements. For example, $(y_1, y_2) \leftarrow (y_1+1, y_2-x_2)$ means that y_1 is replaced by y_1+1 and y_2 by y_2-x_2 simultaneously.

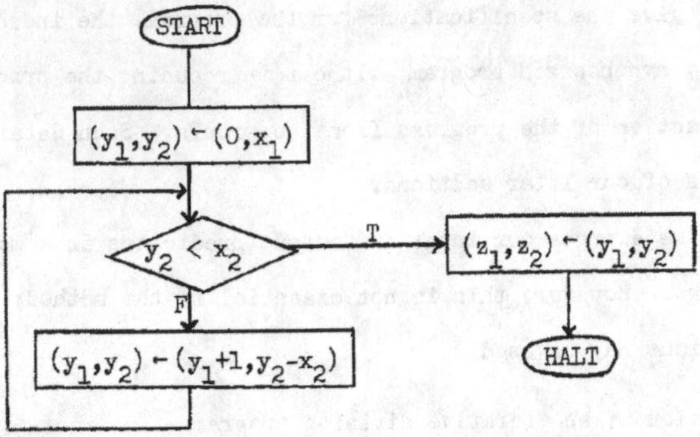

Figure 1. A division program

This suggests that the user must always specify a list of primitive operators, predicates, and constants that the derived program may use. If, for example, we allow our system to use the constant "0" , the "successor" and "predecessor" operators, and the "equality" predicate, but not the "minus" operator or the "less than" operator, the program illustrated in Figure 2 might be constructed.

Henceforth, we shall assume all commonly used symbols are primitive unless we make explicit mention to the contrary.

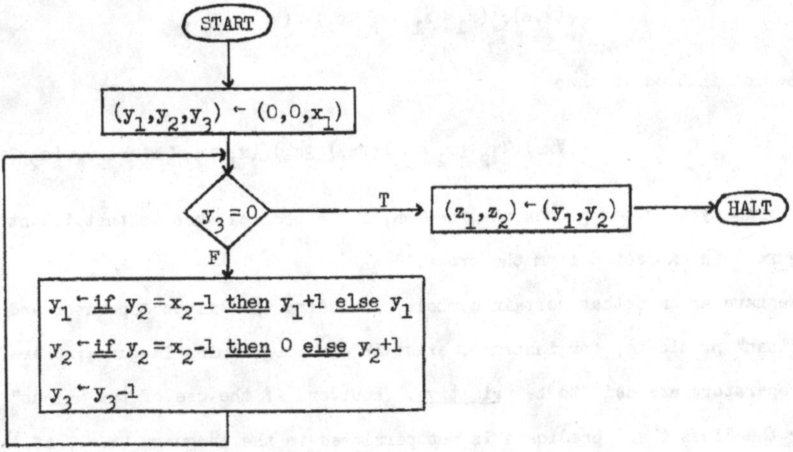

Figure 2. Another division program

Example 2: Translation of a recursive reverse program into an iterative program.

We wish to translate a LISP recursive program for reversing the top-level elements of a list into an equivalent LISP iterative program. For example, if x is the list (a b (c d) e) , then its reverse is (e (c d) b a) .

Here, $\bar{x}=x$, $\bar{z}=z$, and since we want the program to work on all lists, $\varphi(\bar{x})$ is T . The output predicate will be

$$\psi(\bar{x},\bar{z}): \; z=\underline{reverse}(x) \; ,$$

where reverse is defined by the recursive program (see McCarthy (1962))

$$\underline{reverse}(y) \Leftarrow \underline{if}\ \underline{Null}(y)\ \underline{then}\ \text{NIL}\ \underline{else}\ \underline{append}(\underline{reverse}(\underline{cdr}(y)),\underline{list}(\underline{car}(y))) \; .$$

The function $\underline{append}(y_1,y_2)$ concatenates the two lists y_1 and y_2 . For example, if y_1 is the list (a b (c d)) and y_2 is the list (e) , then $\underline{append}(y_1,y_2)$ is the list (a b (c d) e) .

Thus the theorem to be proved is

$$(\forall x)(\exists z)[z=\underline{reverse}(x)] \; .$$

The above theorem has a trivial proof, taking z to be reverse(x) itself. Therefore our program synthesizer might construct the following unsatisfactory program:

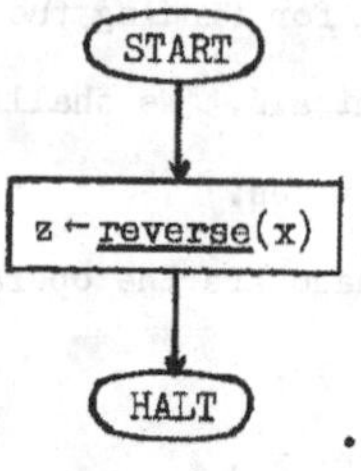

This introduces the problem of primitivity again. The reverse function should not be considered as a primitive in the programming language in this specific task, because we clearly do not want reverse to occur in our iterative program. Henceforth we shall assume that the name of the program to be constructed is never primitive.

If we allow our system to use the constant NIL, the operators car, cdr, and cons, and the predicate Null as primitives, the program illustrated in Figure 3 might be constructed.

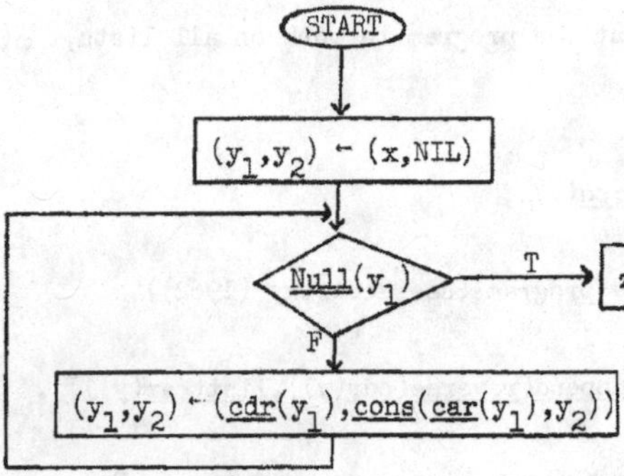

Figure 3. The reverse program

Note that the computation of the derived iterative program consumes less time and space than the computation of the given recursive program. This is not only because of the stacking mechanism necessary in general to implement recursive calls, but also because the repeated use of the append function during execution of the recursive program introduces redundancy in the computation.

Example 3: Construction of a recursive maxtree program.

We wish to construct a recursive program for finding the maximum among the terminal nodes in a binary tree with integer terminals. We shall introduce a special language called TREE for manipulating binary trees.

The primitives allowed in our TREE language are the operators

(a) left(y) : the left subtree of the tree y ,

(b) right(y) : the right subtree of the tree y ,

and the predicate

(c) Atom(y) : is the tree y a single integer?

For example, if y is the binary tree

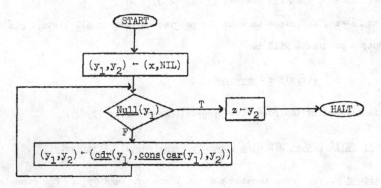

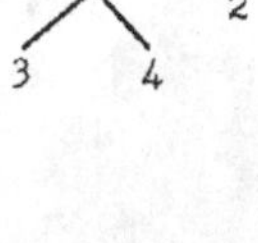

,

then $\underline{left}(y)$ is 

, $\underline{right}(y)$ is 2 , $\underline{Atom}(y)$ is F , and $\underline{Atom}(\underline{right}(y))$ is T .

Let $\bar{x} = x$, $\bar{z} = z$, $\varphi(\bar{x})$ be T , and the output predicate be

$$\psi(\bar{x},\bar{z}): \underline{Terminal}(z,x) \wedge (\forall u)[\underline{Terminal}(u,x) \supset u \leq z] ,$$

where $\underline{Terminal}(y_1,y_2)$ means that the integer y_1 occurs as a terminal in the tree y_2 . The output predicate says that the integer z is a terminal of the tree x not less than any other terminal node of x .

Thus, the theorem to be proved is

$$(\forall x)(\exists z)\{\underline{Terminal}(z,x) \wedge (\forall u)[\underline{Terminal}(u,x) \supset u \leq z]\} .$$

If we allow the $\underline{max}$ operator over the integers,

$$\underline{max}(y_1,y_2): \text{the maximum of the integers } y_1 \text{ and } y_2 ,$$

to be used as a primitive, the recursive program produced might be

$$\begin{cases} z = \underline{maxtree}(x) \ \underline{where} \\ \underline{maxtree}(y) \Leftarrow \underline{if}\ \underline{Atom}(y)\ \underline{then}\ y \\ \qquad\qquad \underline{else}\ \underline{max}(\underline{maxtree}(\underline{left}(y)),\underline{maxtree}(\underline{right}(y))). \end{cases}$$

If we do not allow the $\underline{max}$ operator to be used as a primitive but allow the predicate "less than or equal to", the program produced might be

$$\begin{cases} z = \underline{maxtree}(x) \ \underline{where} \\ \underline{maxtree}(y) \Leftarrow \underline{if}\ \underline{Atom}(y)\ \underline{then}\ y \\ \qquad\qquad \underline{else}\ \underline{if}\ \underline{maxtree}(\underline{left}(y)) \leq \underline{maxtree}(\underline{right}(y)) \\ \qquad\qquad\qquad \underline{then}\ \underline{maxtree}(\underline{right}(y)) \\ \qquad\qquad\qquad \underline{else}\ \underline{maxtree}(\underline{left}(y)) . \end{cases}$$

Note that although the symbol $\underline{maxtree}$, the name of the program, is not primitive, it may be used as a dummy function name in the recursive definitions. Any other function name could have been used instead.

We feel that at this point we should clarify the role of the input predicate. Compare the following program writing tasks: in the first, the input predicate is $\varphi(\bar{x})$ and the output predicate is $\psi(\bar{x},\bar{z})$, while in the second, the input predicate is $\varphi'(\bar{x})$: T and the output predicate is $\psi'(\bar{x},\bar{z})$: $\varphi(\bar{x}) \supset \psi(\bar{x},\bar{z})$. In the first task we do not care how the synthesized program behaves if the input $\bar{x}$ does not satisfy $\varphi(\bar{x})$. In the second case we insist that the program terminates even if $\bar{x}$ does not satisfy $\varphi(\bar{x})$, but we still do not care what the value of the output is.

The theorems induced are

$$(\forall\bar{x})[\varphi(\bar{x}) \supset (\exists\bar{z})\psi(\bar{x},\bar{z})]$$

and

$$(\forall\bar{x})(\exists\bar{z})[\varphi(\bar{x}) \supset \psi(\bar{x},\bar{z})],$$

respectively. Surprisingly enough, these theorems are logically equivalent, even though they represent distinct tasks. This suggests that the program extractor must make use of the input predicate in the process of synthesizing the program.

Suppose, for instance, that in constructing our iterative division program (cf., Example 1), we had given the system the input predicate

$$\varphi(\bar{x}): T$$

and the output predicate

$$\psi'(\bar{x},\bar{z}): x_2 \neq 0 \supset (x_1 = z_1 \cdot x_2 + z_2) \wedge (z_2 < x_2) .$$

The theorem induced in this case would be

$$(\forall x_1)(\forall x_2)(\exists z_1)(\exists z_2)[x_2 \neq 0 \supset (x_1 = z_1 \cdot x_2 + z_2) \wedge (z_2 < x_2)] ,$$

which is logically equivalent to the theorem

$$(\forall x_1)(\forall x_2)\{x_2 \neq 0 \supset (\exists z_1)(\exists z_2)[(x_1 = z_1 \cdot x_2 + z_2) \wedge (z_2 < x_2)]\} .$$

However, the program extracted from the first theorem (Figure 4) halts for every natural number input, whereas the program extracted from the second theorem (Figure 1) does not halt when x_2 is 0.

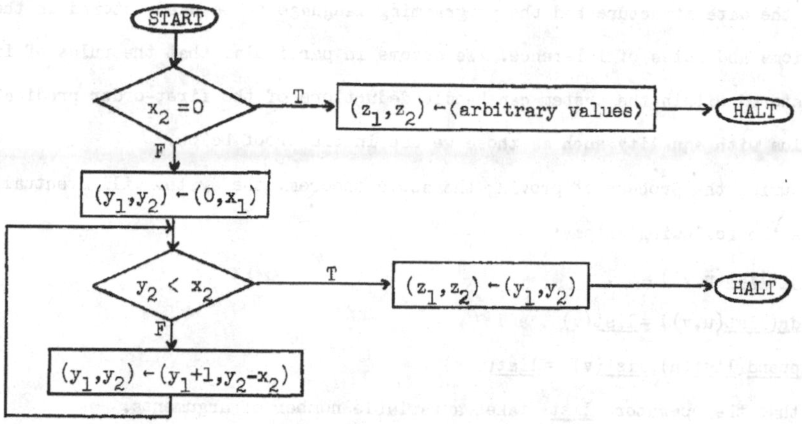

Figure 4. Another division program

3. CONSTRUCTION OF LOOP-FREE PROGRAMS

We would like first to illustrate with two examples the extraction of a program from a proof. The programs we will construct are especially simple, since they have no loops. The program extraction process in this case may be roughly described as follows: substitutions into the output variables in the proof results in assignment statements in iterative programs and operator composition in recursive programs. Case analysis arguments in the proof result in conditional branching in both iterative and recursive programs.

Example 4: Reversing a two-element list.

We wish to construct a LISP program that takes as input a list of two elements, and produces as output the same list with the elements reversed.

Thus, the output predicate is

$$\psi(x,z):\ (\forall u_1)(\forall u_2)[x = \underline{list}(u_1,u_2) \supset z = \underline{list}(u_2,u_1)]\ ,$$

and the theorem to be proved is

$$(\forall x)(\exists z)\{(\forall u_1)(\forall u_2)[x = \underline{list}(u_1,u_2) \supset z = \underline{list}(u_2,u_1)]\}\ .$$

We assume that any system used to prove this theorem has a large supply of facts about the data structure and the programming language to be used, stored in the form of axioms and rules of inference. We assume in particular that the rules of inference stored within the system can handle deductions of the first-order predicate calculus with equality such as those we use in the proof below.

During the process of proving the above theorem, the system will eventually choose the following axioms:

1. car(list(u,v)) = u ,
2. cdr(list(u,v)) = list(v) , and
3. append(list(u),list(v)) = list(u,v) .

Note that the operator list takes a variable number of arguments.

The proof will proceed as follows: Suppose x = list(u_1,u_2) for some arbitrary u_1 and u_2 . Then by Axioms 1 and 2, respectively,

4. car(x) = u_1 , and
5. cdr(x) = list(u_2) .

From 4 we have

6. list(car(x)) = list(u_1) .

Then, combining 5 and 6 using Axiom 3, we obtain

7. append(cdr(x),list(car(x))) = list(u_2,u_1) .

Letting z be append(cdr(x),list(car(x))) , we obtain

8. z = list(u_2,u_1) ,

which is the desired conclusion.

Now, in order to extract the program, we keep track of the substitutions made for z during the proof. In the above proof, we have replaced z by append(cdr(x),list(car(x))); therefore, the desired program is simply

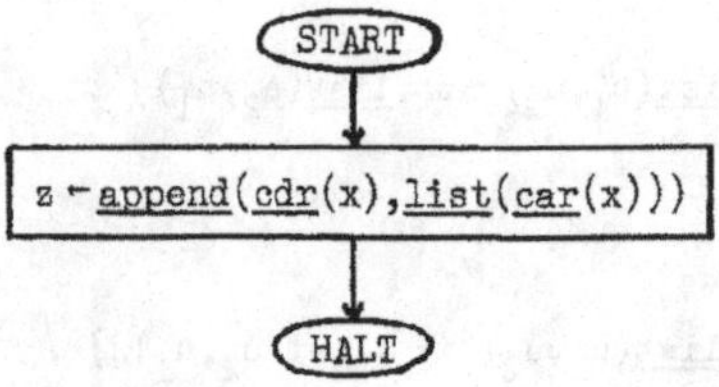

<u>Example 5</u>: The <u>max</u> of two numbers.

The program constructed in this example contains a conditional branch but no loops. We wish to find the maximum of two given integers. Thus, the output predicate is

$$\psi(x_1,x_2,z): (z=x_1 \vee z=x_2) \wedge z\geq x_1 \wedge z\geq x_2 ,$$

and the corresponding theorem is

$$(\forall x_1)(\forall x_2)(\exists z)[(z=x_1 \vee z=x_2) \wedge z\geq x_1 \wedge z\geq x_2] .$$

The proof proceeds by case analysis; it may appear poorly motivated, but it is well within the capacity of existing theorem-proving programs. Translating the theorem into disjunctive normal form, we have

$$(\forall x_1)(\forall x_2)(\exists z)[(z=x_1 \wedge z\geq x_1 \wedge z\geq x_2) \vee (z=x_2 \wedge z\geq x_1 \wedge z\geq x_2)] .$$

If we assume $(u=v)\supset(u\geq v)$ as an axiom, we can simplify the above formula to

$$(\forall x_1)(\forall x_2)(\exists z)[(z=x_1 \wedge z\geq x_2) \vee (z=x_2 \wedge z\geq x_1)] .$$

Now suppose $x_1\geq x_2$; then if we let z be x_1 , the first disjunct is satisfied. On the other hand, suppose $x_1<x_2$; then if we let z be x_2 , the second disjunct is satisfied.

Since the substitution we made for z depends on whether or not x_1 was greater than or equal to x_2 , the program extracted from the proof of the theorem is

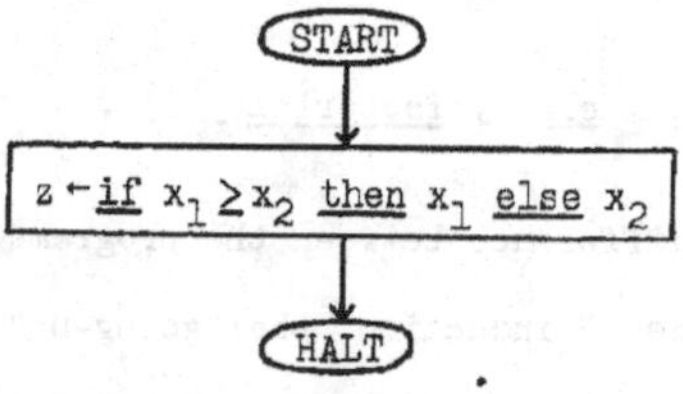

The reader who is unsatisfied with our seat-of-the-pants description of the program extraction process may examine any of the more rigorous accounts in the literature (e.g., Green (1969a,b), Waldinger and Lee (1969), Waldinger (1969), and Luckham and Nilsson (1970)).

The above programs are clearly of limited interest, since neither contains a loop. In order to construct a program with loops, application of some version of the principle of mathematical induction is necessary. Therefore, in the next section we digress into a discussion of the induction principle.

4. THE INDUCTION PRINCIPLE

The induction principle is most commonly associated with proving theorems about the natural numbers, but analogues of it apply to other data structures, such as lists, trees, and strings. Furthermore, for each data structure there are many equivalent forms of the principle. Mathematicians use whichever version is most convenient. Similarly, the theorem prover chooses an appropriate induction principle from a given supply during the course of the proof. This choice directly determines the form of the program to be constructed, since each induction rule has an associated program form stored with it. Therefore, if we want to restrict the form of the extracted program, we must limit the set of available induction principles accordingly.

4.1. Natural numbers

We shall discuss four versions of the induction principle for the natural numbers; two will be appropriate for writing recursive programs and two for writing iterative programs. In each class, one rule will be called a "going-up" principle and the other a "going-down" principle. We will illustrate each of these with a different version of the factorial program. The output predicate is

$$\psi(x,z):\ z=\underline{factorial}(x)\ ,$$

where

$$\underline{factorial}(y) \Leftarrow \underline{if}\ y=0\ \underline{then}\ 1\ \underline{else}\ y\cdot\underline{factorial}(y-1)\ .$$

This example will illustrate clearly the difference between the programs generated by using "going-up" induction and "going-down" induction: the "going-up" programs compute $x!$ in the order $1, 1\cdot 2, 1\cdot 2\cdot 3, \ldots$, while the "going-down" programs compute $x, x\cdot(x-1), x\cdot(x-1)\cdot(x-2), \ldots$.

The proofs required for the synthesis of the programs use two axioms induced by the above definition:

$$\underline{factorial}(0) = 1$$

and

$$u > 0 \supset [\underline{factorial}(u) = u \cdot \underline{factorial}(u-1)] .$$

We will not include those proofs, but merely will give the programs extracted, in order to illustrate the relationship between the form of the induction principle used in the proof and the form of the constructed program.

(a) Iterative going-up induction

The reader is probably familiar with the most common version of the induction principle over the natural numbers,

$$\frac{\begin{array}{c}\mathcal{A}(0)\\ (\forall y_1)[\mathcal{A}(y_1) \supset \mathcal{A}(y_1+1)]\end{array}}{(\forall x)\mathcal{A}(x)} .$$

Intuitively, this means that if a property $\mathcal{A}$ holds for 0 and if whenever it holds for y_1 it holds for y_1+1, then it holds for every natural number x. We call this version iterative going-up induction.

For our purpose, we use a special form of the principle in which $\mathcal{A}(y_1)$ is $(\exists y_2)\mathcal{B}(y_1,y_2)$, where $\mathcal{B}$ still represents an unspecified property. The induction principle now becomes

$$\frac{\begin{array}{c}(\exists y_2)\mathcal{B}(0,y_2)\\ (\forall y_1)[(\exists y_2)\mathcal{B}(y_1,y_2) \supset (\exists y_2)\mathcal{B}(y_1+1,y_2)]\end{array}}{(\forall x)(\exists y_2)\mathcal{B}(x,y_2)} .$$

The program form associated with this rule is illustrated in Figure 5. If the theorem to be proved happens to be of the form

$$(\forall x)(\exists z)\mathcal{B}(x,z) ,$$

and if going-up induction is applied, the program extractor then knows that the program must be of the form illustrated in Figure 5.

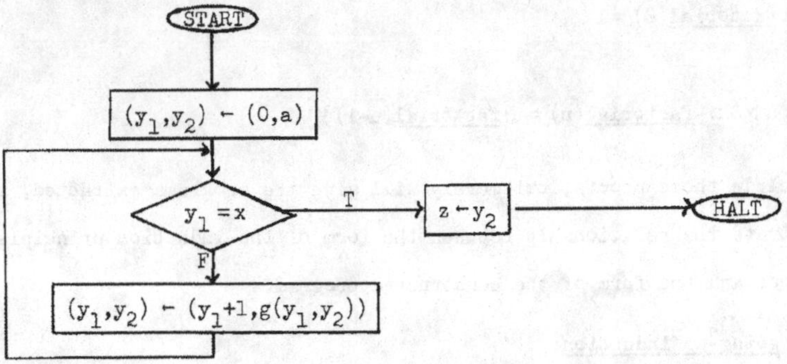

Figure 5. Iterative "going-up" program form

The constant a and the function $g(y_1,y_2)$ are unspecified in the above form. The task of the program constructor is now to write subroutines to compute a and g in such a way that the program of Figure 5 will satisfy the desired relation. This is done as follows.

The theorem to be proved is of the form $(\forall x)(\exists z)\mathcal{B}(x,z)$. This is precisely the form of the consequent of the induction principle. Therefore, if we can prove the two antecedents, then we are done. This suggests that we attempt to prove the two lemmas:

(A) $(\exists y_2)\mathcal{B}(0,y_2)$,

and

(B) $(\forall y_1)[(\exists y_2)\mathcal{B}(y_1,y_2) \supset (\exists y_2)\mathcal{B}(y_1+1,y_2)]$,

or equivalently, translating into prenex normal form,

(B') $(\forall y_1)(\forall y_2)(\exists y_2^*)[\mathcal{B}(y_1,y_2) \supset \mathcal{B}(y_1+1,y_2^*)]$.

The proof of Lemma (A) generates a subroutine with no variables that yields a value for y_2 satisfying $\mathcal{B}(0,y_2)$. This is the desired definition of the constant a ; hence

(1) $\mathcal{B}(0,a)$

is true.

The proof of Lemma (B') generates another subroutine which yields a value of y_2^* in terms of y_1 and y_2. This subroutine provides a definition of $g(y_1,y_2)$ satisfying

(2) $$\mathcal{B}(y_1,y_2) \supset \mathcal{B}(y_1+1,g(y_1,y_2))$$

for all y_1 and y_2.

The proof of the lemmas concludes the proof of the theorem $(\forall x)(\exists z)\mathcal{B}(x,z)$. We have now completely specified a program that computes a function $z=f(x)$ satisfying $\mathcal{B}(x,f(x))$ for all values of x.

For the suspicious reader, we are ready to verify the above assertion. Consider the iterative "going-up" program form labeled as in Figure 6. We will use Floyd's approach (1967a) and show that whenever control passes through arc α, $\mathcal{B}(y_1,y_2)$ is true for the current values of y_1 and y_2. Furthermore, whenever control passes through arc β, $\mathcal{B}(x,z)$ is true for the initial value of x and the final value of z.

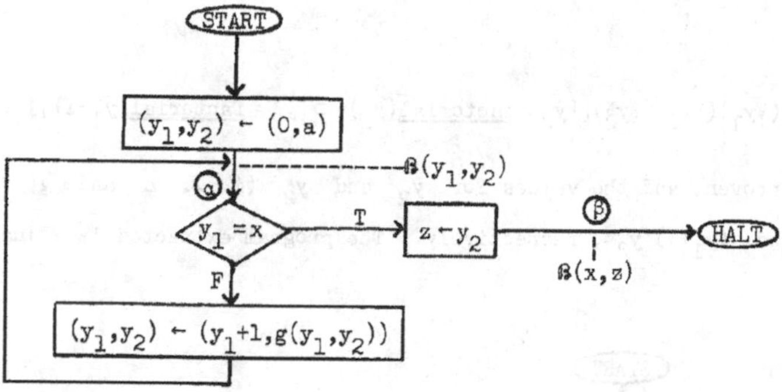

Figure 6. Labeled iterative "going-up" program form

Beginning at the START node, we set y_1 to 0 and y_2 to a, and so when we pass through arc α, $\mathcal{B}(y_1,y_2)$ (i.e., $\mathcal{B}(0,a)$) is true by (1).

Now suppose that at some point in the execution, control is passing through arc α and currently $\mathcal{B}(y_1,y_2)$ is true and $y_1 \neq x$. Then, by (2)

(3) $$\mathcal{B}(y_1+1,g(y_1,y_2))$$

is true. Traveling around the loop, we simultaneously set y_1 to y_1+1 and y_2 to $g(y_1,y_2)$ and reach arc α again. That $\mathcal{B}(y_1,y_2)$ is satisfied at this time follows directly from (3) and our assignments to y_1 and y_2.

Clearly, we must at some time reach arc α with $y_1=x$, since x is a natural number. Then we set z to y_2 and pass to arc β. Since $\mathcal{B}(y_1,y_2)$ was true at arc α and $y_1=x$, $\mathcal{B}(x,z)$ is true at arc β. This concludes the proof that the program constructed has the desired properties.

<u>Example 6</u>: Iterative "going-up" <u>factorial</u> program.

We wish to construct an iterative "going-up" program for computing the <u>factorial</u> function. The theorem to be proved is

$$(\forall x)(\exists z)[z=\underline{factorial}(x)]\ .$$

Applying the iterative going-up induction principle (with $\mathcal{B}(y_1,y_2)$ being $y_2 = \underline{factorial}(y_1)$), we are presented with the two lemmas

(A) $$(\exists y_2)[y_2=\underline{factorial}(0)]$$

and

(B') $$(\forall y_1)(\forall y_2)(\exists y_2^*)\{[y_2=\underline{factorial}(y_1)]\supset[y_2^*=\underline{factorial}(y_1+1)]\}\ .$$

The lemmas are proven, and the values for y_2 and y_2^* (i.e., a and $g(y_1,y_2)$) found are 1 and $(y_1+1)\cdot y_2$, respectively. The program extracted is illustrated in Figure 7.

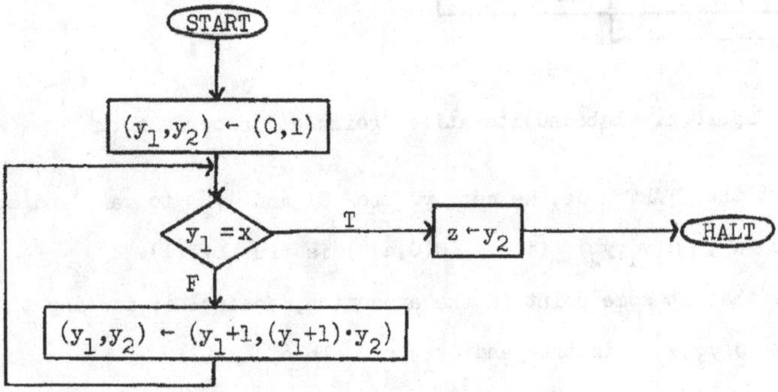

Figure 7. Iterative "going-up" <u>factorial</u> program

Note that for simplicity we have assumed in the above discussion that the program to be constructed has only one input variable x and one output variable z. This restriction may be waived by a straightforward generalization of the induction principle given above as illustrated in Examples 13 and 15.

(b) Recursive going-up induction

We present another going-up induction principle that leads to a different program form. The principle

$$\frac{\begin{array}{l}\mathcal{A}(0)\\ (\forall y_1)[y_1 \neq 0 \wedge \mathcal{A}(y_1-1) \supset \mathcal{A}(y_1)]\end{array}}{(\forall x)\mathcal{A}(x)}$$

is logically equivalent to the first version but leads to the construction of a recursive program of the form

$$\begin{cases} z = f(x) \ \underline{\text{where}} \\ f(y) \Leftarrow \underline{\text{if}}\ y=0\ \underline{\text{then}}\ a\ \underline{\text{else}}\ g(y,f(y-1)) \end{cases} .$$

We call this version recursive going-up induction. Note that the f is a dummy function symbol that need not be declared primitive.

We have omitted the details concerning the derivation of this program, but they are quite similar to those involved in section (a) above.

(c) Iterative going-down induction

Another form of the induction principle is the iterative going-down form

$$\frac{\begin{array}{l}(\exists y_1)\mathcal{A}(y_1)\\ (\forall y_1)[y_1 \neq 0 \wedge \mathcal{A}(y_1) \supset \mathcal{A}(y_1-1)]\end{array}}{\mathcal{A}(0)} .$$

The reader may verify that this rule is equivalent to the recursive going-up induction, replacing $\mathcal{A}$ by $\sim\mathcal{A}$ and twice using the fact that $p \wedge q \supset r$ is logically equivalent to $\sim r \wedge q \supset \sim p$.

In this case we use a special form of the principle in which $\mathcal{A}(y_1)$ is of the form $(\exists y_2)\mathcal{B}(x,y_1,y_2)$, where x is a free variable. The induction principle now becomes

$$(\exists y_1)(\exists y_2)\mathcal{B}(x,y_1,y_2)$$

$$\frac{(\forall y_1)[y_1 \neq 0 \wedge (\exists y_2)\mathcal{B}(x,y_1,y_2) \supset (\exists y_2)\mathcal{B}(x,y_1-1,y_2)]}{(\exists y_2)\mathcal{B}(x,0,y_2)} \quad .$$

Suppose now that the theorem to be proved is of the form

$$(\forall x)(\exists z)\mathcal{B}(x,0,z) \ .$$

The theorem may be deduced from the conclusion of the above induction principle. If the iterative going-down induction principle is used, the program to be extracted is automatically of the form illustrated in Figure 8.

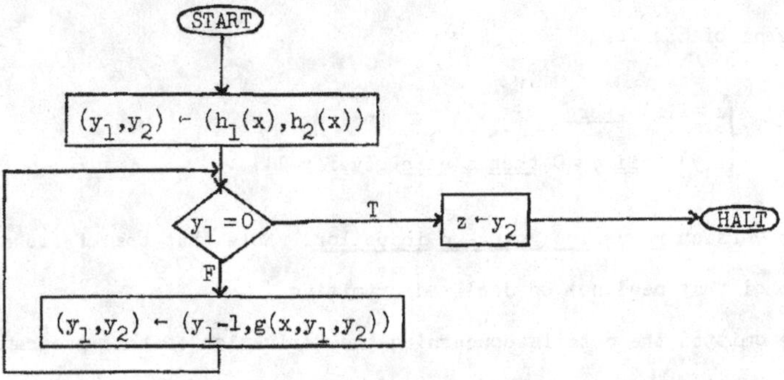

Figure 8. Iterative "going-down" program form

Thus, all that remains is to construct subroutines to compute the functions $h_1(x)$, $h_2(x)$, and $g(x,y_1,y_2)$.

The antecedents of the induction principle give us the two lemmas to be proved

(A) $\qquad (\exists y_1)(\exists y_2)\mathcal{B}(x,y_1,y_2)$

and

(B) $\qquad (\forall y_1)[y_1 \neq 0 \wedge (\exists y_2)\mathcal{B}(x,y_1,y_2) \supset (\exists y_2)\mathcal{B}(x,y_1-1,y_2)]$,

or, equivalently,

(B') $$(\forall y_1)(\forall y_2)(\exists y_2^*)[y_1 \neq 0 \wedge \mathcal{B}(x,y_1,y_2) \supset \mathcal{B}(x,y_1-1,y_2^*)] .$$

The proof of Lemma (A) yields subroutines to compute y_1 and y_2 in terms of x, which define the desired functions $h_1(x)$ and $h_2(x)$, respectively. The proof of Lemma (B') yields a single subroutine to compute y_2^* in terms of x, y_1, and y_2, thus defining the desired function $g(x,y_1,y_2)$. The program is then completely specified, and its correctness and termination can be demonstrated using Floyd's approach, as was done before.

Note that iterative going-down induction is of value only if the constant 0 occurs in the theorem to be proved. Otherwise, the theorem prover must manipulate the theorem to introduce 0.

Example 7: Iterative "going-down" factorial program.

We wish to construct an iterative "going-down" program for computing the factorial function. The theorem to be proved is again

$$(\forall x)(\exists z)[z = \underline{factorial}(x)] .$$

The theorem contains no occurrence of the constant 0. Thus, the theorem prover tries to introduce 0, using the first part of the definition of the factorial function (i.e., $\underline{factorial}(0) = 1$) and its supply of axioms ($u \cdot 1 = u$, in particular), deriving as a subgoal

$$(\exists z)[\underline{factorial}(0) \cdot z = \underline{factorial}(x)] ,$$

where x is a free variable. This theorem is in the form of the consequent of the iterative going-down induction, i.e., $(\exists y_2)\mathcal{B}(x,0,y_2)$; hence, the theorem prover chooses the induction hypothesis $\mathcal{B}(x,y_1,y_2)$ to be $\underline{factorial}(y_1) \cdot y_2 = \underline{factorial}(x)$.

The lemmas proposed are

(A) $$(\exists y_1)(\exists y_2)[\underline{factorial}(y_1) \cdot y_2 = \underline{factorial}(x)] ,$$

and

(B') $$(\forall y_1)(\forall y_2)(\exists y_2^*)\{y_1 \neq 0 \wedge [\underline{factorial}(y_1) \cdot y_2 = \underline{factorial}(x)] \supset [\underline{factorial}(y_1-1) \cdot y_2^* = \underline{factorial}(x)]\} .$$

The values obtained for y_1, y_2, and y_2^* (i.e., $h_1(x)$, $h_2(x)$, and $g(x,y_1,y_2)$), respectively, are x, 1, and $y_1 \cdot y_2$. The program constructed is illustrated in Figure 9.

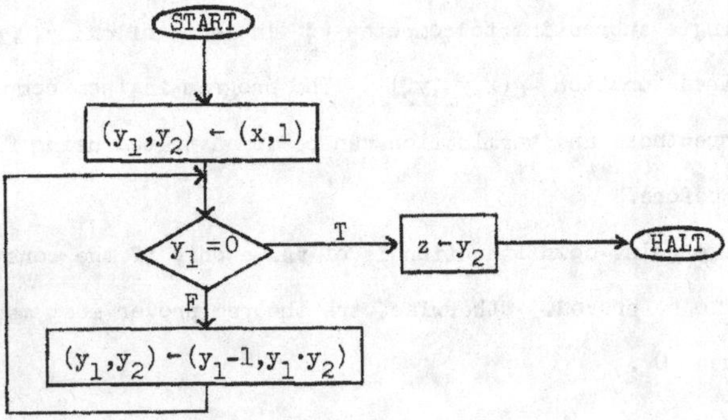

Figure 9. Iterative "going-down" <u>factorial</u> program

(d) <u>Recursive going-down induction</u>

The recursive going-up induction was very similar to the iterative going-up induction. In the same way, the <u>recursive going-down</u> induction.

$$\frac{\begin{array}{c}(\exists y_1)\mathcal{A}(y_1)\\ (\forall y_1)[\mathcal{A}(y_1+1) \supset \mathcal{A}(y_1)]\end{array}}{\mathcal{A}(0)} ,$$

is very similar to the iterative going-down induction. The form we are most interested in is

$$\frac{\begin{array}{c}(\exists y_1)(\exists y_2)\mathcal{B}(x,y_1,y_2)\\ (\forall y_1)[(\exists y_2)\mathcal{B}(x,y_1+1,y_2) \supset (\exists y_2)\mathcal{B}(x,y_1,y_2)]\end{array}}{(\exists y_2)\mathcal{B}(x,0,y_2)} ,$$

where x is a free variable.

If the rule is used in generating a program, the two appropriate lemmas allow us to construct $h_1(x)$, $h_2(x)$, and $g(x,y_1,y_2)$ as before, and the program extracted is

$$\begin{cases} z = f(x) = f'(x,0) \ \ \underline{\text{where}} \\ f'(x,y) \Leftarrow \underline{\text{if}}\ y = h_1(x)\ \underline{\text{then}}\ h_2(x)\ \underline{\text{else}}\ g(x,y,f'(x,y+1)) . \end{cases}$$

Example 8: Recursive "going-down" factorial program.

The program we wish to construct this time is a recursive "going-down" program to compute the factorial function. Again the theorem

$$(\forall x)(\exists z)[z = \underline{factorial}(x)]$$

is transformed into

$$(\exists z)[\underline{factorial}(0)\cdot z = \underline{factorial}(x)] \ .$$

We continue as before, and the program generated is

$$\begin{cases} z = f(x) = f'(x,0) \ \underline{where} \\ f'(x,y) \Leftarrow \underline{if}\ y = x\ \underline{then}\ 1\ \underline{else}\ (y+1)\cdot f'(x,y+1) \ . \end{cases}$$

4.2. Lists

Our treatments of lists and natural numbers are in some ways analogous, since the constant NIL and the function cdr in LISP play the same role as the constant 0 and the "predecessor" function, respectively, in number theory. The induction principles of both data structures are closely related, but since there is no exact analogue in LIST to the "sucessor" function in number theory, there are no iterative going-up and recursive going-down list induction principles. Hence, we shall only deal with two induction rules in this section: recursive (going-up) and iterative (going-down) list inductions. In the discussion in this section, we shall omit details, since they are similar to the details in the previous section.

We shall illustrate the use of both induction rules by constructing two programs for sorting a given list of integers. The output predicate is

$$\psi(x,z):\ z = \underline{sort}(x) \ ,$$

where

$$(\forall y)(\forall z)\{[z = \underline{sort}(y)] \equiv \underline{if}\ \underline{Null}(y)\ \underline{then}\ \underline{Null}(z)$$
$$\underline{else}\ (\forall u)[\underline{Member}(u,y) \supset z = \underline{merge}(u,\underline{sort}(\underline{delete}(u,y)))]\} \ .$$

Here,

<u>Member</u>(u,y) means that the integer u is a member of the list y ,

<u>delete</u>(u,y) is the list obtained by deleting the integer u from the list y , and

<u>merge</u>(u,v) , where v is a sorted list that does not contain the integer u , is the list obtained by placing u in its place on the list v , so that the ordering is preserved.

The theorem to be proved is

$$(\forall x)(\exists z)[z = \underline{sort}(x)] .$$

(a) <u>Recursive list induction</u>

The recursive (going-up) list induction principle is

$$\frac{\begin{array}{l}\mathcal{A}(\mathrm{NIL}) \\ (\forall y_1)[\sim \underline{\mathrm{Null}}(y_1) \wedge \mathcal{A}(\underline{\mathrm{cdr}}(y_1)) \supset \mathcal{A}(y_1)]\end{array}}{(\forall x)\mathcal{A}(x)} .$$

The program-synthesis form of the rule is

$$\frac{\begin{array}{l}(\exists y_2)\mathcal{B}(\mathrm{NIL},y_2) \\ (\forall y_1)[\sim \underline{\mathrm{Null}}(y_1) \wedge (\exists y_2)\mathcal{B}(\underline{\mathrm{cdr}}(y_1),y_2) \supset (\exists y_2)\mathcal{B}(y_1,y_2)]\end{array}}{(\forall x)(\exists y_2)\mathcal{B}(x,y_2)} .$$

The corresponding program form generated is

$$\begin{cases} z = f(x) \ \underline{\text{where}} \\ f(y) \Leftarrow \underline{\text{if}}\ \underline{\mathrm{Null}}(y)\ \underline{\text{then}}\ a\ \underline{\text{else}}\ g(y,f(\underline{\mathrm{cdr}}(y))) . \end{cases}$$

<u>Example 9</u>: Recursive <u>sort</u> program.

The <u>sort</u> program obtained using the recursive list induction principle is

$$\begin{cases} z = \underline{\mathrm{sort}}(x) \ \underline{\text{where}} \\ \underline{\mathrm{sort}}(y) \Leftarrow \underline{\text{if}}\ \underline{\mathrm{Null}}(y)\ \underline{\text{then}}\ \mathrm{NIL}\ \underline{\text{else}}\ \underline{\mathrm{merge}}(\underline{\mathrm{car}}(y),\underline{\mathrm{sort}}(\underline{\mathrm{cdr}}(y))) . \end{cases}$$

(b) <u>Iterative list induction</u>

The reader can undoubtedly guess that the iterative (going-down) list induction principle is

$$\frac{(\exists y_1)\mathcal{A}(y_1) \qquad (\forall y_1)[\sim \underline{\text{Null}}(y_1) \wedge \mathcal{A}(y_1) \supset \mathcal{A}(\underline{\text{cdr}}(y_1))]}{\mathcal{A}(\text{NIL})} .$$

We are especially interested in the form

$$\frac{(\exists y_1)(\exists y_2)\mathcal{B}(x,y_1,y_2) \qquad (\forall y_1)[\sim \underline{\text{Null}}(y_1) \wedge (\exists y_2)\mathcal{B}(x,y_1,y_2) \supset (\exists y_2)\mathcal{B}(x,\underline{\text{cdr}}(y_1),y_2)]}{(\exists y_2)\mathcal{B}(x,\text{NIL},y_2)} ,$$

where x is a free variable.

The corresponding program form generated is illustrated in Figure 10; it employs the construction known among LISP programmers as the "cdr loop".

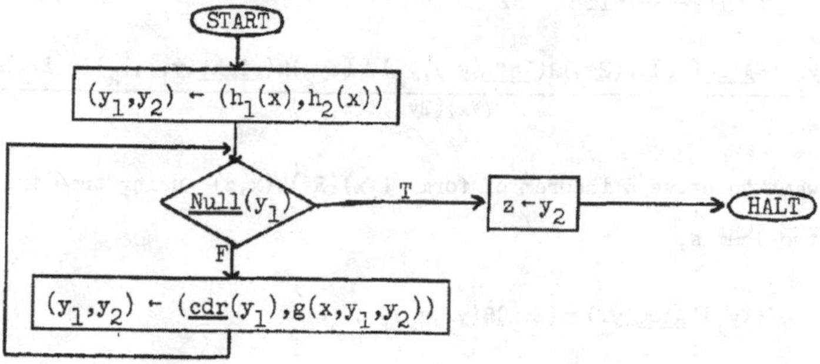

Figure 10. Iterative list program form

Example 10: Iterative sort program.

Using the iterative list induction, we can extract the sort program of Figure 11.

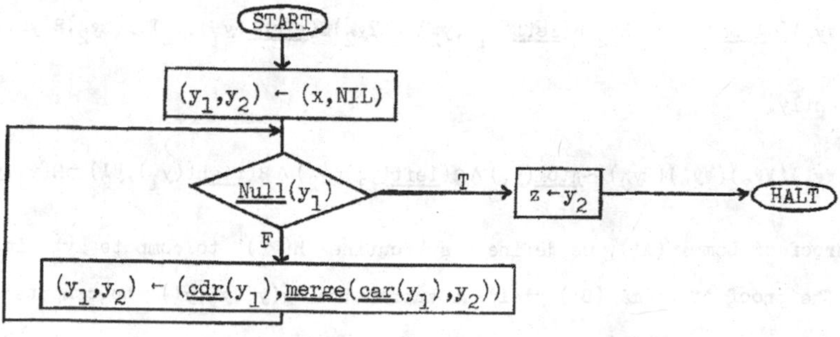

Figure 11. Iterative sort program

4.3. Trees

There is no simple induction rule for tree structures which gives rise to an iterative program form, because such a program would have to use a complex mechanism to keep track of its place in the tree. However, there is a simple recursive tree induction rule:

$$\frac{\begin{array}{c}(\forall y_1)[\underline{Atom}(y_1) \supset \mathcal{A}(y_1)] \\ (\forall y_1)[\sim\underline{Atom}(y_1) \wedge \mathcal{A}(\underline{left}(y_1)) \wedge \mathcal{A}(\underline{right}(y_1)) \supset \mathcal{A}(y_1)]\end{array}}{(\forall x)\mathcal{A}(x)} .$$

In the automatic program synthesizer, we are chiefly interested in the following form:

$$\frac{\begin{array}{c}(\forall y_1)[\underline{Atom}(y_1) \supset (\exists y_2)\mathcal{B}(y_1,y_2)] \\ (\forall y_1)[\sim\underline{Atom}(y_1) \wedge (\exists y_2)\mathcal{B}(\underline{left}(y_1),y_2) \wedge (\exists y_2)\mathcal{B}(\underline{right}(y_1),y_2) \supset (\exists y_2)\mathcal{B}(y_1,y_2)]\end{array}}{(\forall x)(\exists y_2)\mathcal{B}(x,y_2)} .$$

If we want to prove a theorem of form $(\forall x)(\exists z)\mathcal{B}(x,z)$ using tree induction, we must prove two lemmas,

(A) $\qquad (\forall y_1)[\underline{Atom}(y_1) \supset (\exists y_2)\mathcal{B}(y_1,y_2)]$,

or equivalently,

(A') $\qquad (\forall y_1)(\exists y_2)[\underline{Atom}(y_1) \supset \mathcal{B}(y_1,y_2)]$,

and

(B) $\qquad (\forall y_1)[\sim\underline{Atom}(y_1) \wedge (\exists y_2)\mathcal{B}(\underline{left}(y_1),y_2) \wedge (\exists y_2)\mathcal{B}(\underline{right}(y_1),y_2) \supset (\exists y_2)\mathcal{B}(y_1,y_2)]$

or equivalently,

(B') $\qquad (\forall y_1)(\forall y_2)(\forall y_2')(\exists y_2^*)[\sim\underline{Atom}(y_1) \wedge \mathcal{B}(\underline{left}(y_1),y_2) \wedge \mathcal{B}(\underline{right}(y_1),y_2') \supset \mathcal{B}(y_1,y_2^*)]$.

From the proof of Lemma (A'), we define a subroutine $h(y_1)$ to compute y_2 in terms of y_1. The proof of Lemma (B') yields a subroutine $g(y_1,y_2,y_2')$ to compute y_2^* in terms of y_1, y_2, and y_2'.

The corresponding program form is

$$\begin{cases} z = f(x) \ \underline{\text{where}} \\ f(y_1) \Leftarrow \underline{\text{if}}\ \underline{\text{Atom}}(y_1)\ \underline{\text{then}}\ h(y_1)\ \underline{\text{else}}\ g(y_1, f(\underline{\text{left}}(y_1)), f(\underline{\text{right}}(y_1))) \ . \end{cases}$$

Note that this program form employs two recursive calls.

Example 11: Recursive maxtree program (see Example 3).

We wish to give the synthesis of a TREE recursive program for finding the maximum among the terminal nodes in a binary tree with integer terminals. This will be the first detailed example of the construction of a program containing loops.

The theorem to be proved is

$$(\forall x)(\exists z)[z = \underline{\text{maxtree}}(x)] \ ,$$

where

(1) $$[z = \underline{\text{maxtree}}(x)] \equiv [\underline{\text{Terminal}}(z,x) \wedge (\forall u)[\underline{\text{Terminal}}(u,x) \supset u \leq z]] \ ,$$

and

(2) $$\underline{\text{Terminal}}(u,v) \Leftarrow \underline{\text{if}}\ \underline{\text{Atom}}(v)\ \underline{\text{then}}\ u = v \ \underline{\text{else}}\ \underline{\text{Terminal}}(u, \underline{\text{left}}(v)) \vee \underline{\text{Terminal}}(u, \underline{\text{right}}(v)) \ .$$

We assume that maxtree itself is not primitive.

The theorem is of the form $(\forall x)(\exists z)\mathcal{B}(x,z)$, where $\mathcal{B}(x,z)$ is $z = \underline{\text{maxtree}}(x)$. Taking z to be y_2 , this is precisely the conclusion of the tree induction principle. Therefore, we apply the induction with

$$\mathcal{B}(y_1,y_2):\ y_2 = \underline{\text{maxtree}}(y_1) \ .$$

Hence, it suffices to prove the following two lemmas:

(A') $$(\forall y_1)(\exists y_2)[\underline{\text{Atom}}(y_1) \supset y_2 = \underline{\text{maxtree}}(y_1)] \ ,$$

and

(B') $$(\forall y_1)(\forall y_2)(\forall y_2')(\exists y_2^*)[\sim\underline{\text{Atom}}(y_1) \wedge y_2 = \underline{\text{maxtree}}(\underline{\text{left}}(y_1)) \wedge y_2' = \underline{\text{maxtree}}(\underline{\text{right}}(y_1)) \supset y_2^* = \underline{\text{maxtree}}(y_1)] \ .$$

The proof of the lemmas will rely on the definition of <u>maxtree</u> (Axiom (1)) and the following two axioms induced by the recursive definition (2) of the <u>Terminal</u> predicate:

(2a) $\underline{Atom}(v) \supset [\underline{Terminal}(u,v) \equiv (u=v)]$,

and

(2b) $\sim\underline{Atom}(v) \supset [\underline{Terminal}(u,v) \equiv [\underline{Terminal}(u,\underline{left}(v)) \vee \underline{Terminal}(u,\underline{right}(v))]]$.

First we prove Lemma (A'). By Axiom (1), it follows that we want to prove

$$(\forall y_1)(\exists y_2)\{\underline{Atom}(y_1) \supset [\underline{Terminal}(y_2,y_1) \wedge (\forall u)[\underline{Terminal}(u,y_1) \supset u \leq y_2]]\} ,$$

or equivalently, using Axiom (2a) (with u being y_2 and v being y_1),

$$(\forall y_1)(\exists y_2)\{\underline{Atom}(y_1) \supset [y_2 = y_1 \wedge (\forall u)[u = y_1 \supset u \leq y_2]]\} .$$

It clearly suffices to take y_2 to be y_1 to complete the proof of the lemma. Therefore, the subroutine for $h(y_1)$ that we derive from the proof of this lemma is simply $h(y_1) = y_1$.

The proof of Lemma (B') is a bit more complicated. Let us assume that y_1 is a tree such that $\sim\underline{Atom}(y_1)$ and let $y_2 = \underline{maxtree}(\underline{left}(y_1))$ and $y_2' = \underline{maxtree}(\underline{right}(y_1))$. We want to find a y_2^* (in terms of y_1 , y_2 , and y_2') for which $y_2^* = \underline{maxtree}(y_1)$. This means by Axiom (1) that we want y_2^* such that

$$\underline{Terminal}(y_2^*,y_1) \wedge (\forall u)[\underline{Terminal}(u,y_1) \supset u \leq y_2^*] .$$

This implies, by Axiom (2b) and our assumption $\sim\underline{Atom}(y_1)$, that we have to find a y_2^* satisfying the following three conditions:

(i) $\underline{Terminal}(y_2^*,\underline{left}(y_1)) \vee \underline{Terminal}(y_2^*,\underline{right}(y_1))$,

(ii) $(\forall u)[\underline{Terminal}(u,\underline{left}(y_1)) \supset u \leq y_2^*]$,

and

(iii) $(\forall u)[\underline{Terminal}(u,\underline{right}(y_1)) \supset u \leq y_2^*]$.

It was assumed that $y_2 = \underline{maxtree}(\underline{left}(y_1))$ and $y_2' = \underline{maxtree}(\underline{right}(y_1))$. Thus, using Axiom (1), condition (i) implies that $y_2^* \le y_2 \vee y_2^* \le y_2'$, condition (ii) implies that $y_2 \le y_2^*$, and condition (iii) implies that $y_2' \le y_2^*$. This suggests that we take y_2^* to be $\underline{max}(y_2, y_2')$, which indeed satisfies the three conditions. Therefore, the subroutine for $g(y_1, y_2, y_2')$ that we extract from the proof of this lemma is $g(y_1, y_2, y_2') = \underline{max}(y_2, y_2')$.

The complete program derived from the proof is

$$\begin{cases} z = \underline{maxtree}(x) \ \underline{where} \\ \underline{maxtree}(y_1) \Leftarrow \underline{if}\ \underline{Atom}(y_1)\ \underline{then}\ y_1 \\ \qquad\qquad \underline{else}\ \underline{max}(\underline{maxtree}(\underline{left}(y_1)), \underline{maxtree}(\underline{right}(y_1))) . \end{cases}$$

5. COMPLETE INDUCTION

The so-called complete induction principle is of the form

$$\frac{(\forall y_1)\{(\forall u)[u < y_1 \supset \mathcal{Q}(u)] \supset \mathcal{Q}(y_1)\}}{(\forall x)\mathcal{Q}(x)} .$$

Intuitively, this means that if a property $\mathcal{Q}$ holds for a natural number y_1 whenever it holds for every natural number u less than y_1, then it holds for every natural number x.

Although this rule is, in fact, logically equivalent to the earlier number-theoretic induction rules (see, for example, Mendelson (1964)), we shall see that it leads to a more general program form than the previous rules, and therefore it is more powerful for program-synthetic purposes. However, it puts more of a burden on the theorem prover because less of the program structure is fixed in advance and more is specified during the proof process.

We are most interested in the version of this rule in which $\mathcal{Q}(y_1)$ has the form $(\exists y_2)\mathcal{B}(y_1, y_2)$, i.e.,

$$\frac{(\forall y_1)\{(\forall u)[u < y_1 \supset (\exists y_2)\mathcal{B}(u, y_2)] \supset (\exists y_2)\mathcal{B}(y_1, y_2)\}}{(\forall x)(\exists y_2)\mathcal{B}(x, y_2)} .$$

Thus, in order to prove a theorem of the form

$$(\forall x)(\exists z)\mathcal{B}(x,z) ,$$

it suffices to prove a single lemma of the form

$$\text{(A')} \qquad (\forall y_1)(\exists u)(\forall y_2)(\exists y_2^*)\{[u<y_1 \supset \mathcal{B}(u,y_2)] \supset \mathcal{B}(y_1,y_2^*)\} .$$

From a proof of the lemma, we extract one subroutine for computing the value of u in terms of y_1 (called $h(y_1)$), and another for computing the value of y_2^* in terms of y_1 and y_2 (called $g(y_1,y_2)$). These functions satisfy the relation

$$\text{(1)} \qquad [h(y_1)<y_1 \supset \mathcal{B}(h(y_1),y_2)] \supset \mathcal{B}(y_1,g(y_1,y_2))$$

for every y_1 and y_2 .

The program form associated with the complete induction rule is then the recursive program form

$$\text{(2)} \qquad \begin{cases} z=f(x) \ \underline{\text{where}} \\ f(y) \Leftarrow g(y,f(h(y))) \ . \end{cases}$$

This form requires some justification.

Assume that the function f satisfies the output predicate

$$\text{(3)} \qquad \mathcal{B}(u,f(u))$$

for all $u<x$. We will try to show

$$\mathcal{B}(x,f(x)) .$$

First, suppose $h(x)<x$. Then by the hypothesis (3), $\mathcal{B}(h(x),f(h(x)))$. Therefore, from (1) (taking y_1 to be x and y_2 to be $f(h(x))$), we obtain $\mathcal{B}(x,g(x,f(h(x))))$, i.e., by (2) $\mathcal{B}(x,f(x))$.

Now suppose $h(x)\geq x$. Then taking y_1 to be x , the antecedent of (1) is true vacuously, and we conclude $\mathcal{B}(x,g(x,f(h(x))))$, i.e., by (2), $\mathcal{B}(x,f(x))$.

<u>Example 12</u>: The recursive quotient program (see Example 1).

We want to construct a recursive program to find (the integer part of) the quotient of two natural numbers x_1 and x_2 , given that $x_2\neq 0$. Our output predicate is therefore

$$\psi(x_1,x_2,z):\ (\exists r)[x_1=z\cdot x_2+r\wedge r<x_2]\ .$$

The theorem is then

$$(\forall x_1)(\forall x_2)\{x_2\neq 0\supset(\exists z)(\exists r)[x_1=z\cdot x_2+r\wedge r<x_2]\}\ .$$

Assume now that x_2 is a fixed positive integer. Then we wish to prove

$$(\forall x_1)(\exists z)(\exists r)[x_1=z\cdot x_2+r\wedge r<x_2]\ .$$

The theorem is now in the same form as the conclusion of the complete induction rule, taking x to be x_1, y_2 to be z, and

$$\beta(y_1,y_2):\ (\exists r)[y_1=y_2\cdot x_2+r\wedge r<x_2]\ .$$

Therefore, the single lemma to be proved is

$$(A')\qquad (\forall y_1)(\exists u)(\forall y_2)(\exists y_2^*)\{[u<y_1\supset(\exists r)[u=y_2\cdot x_2+r\wedge r<x_2]]\\ \supset(\exists r)[y_1=y_2^*\cdot x_2+r\wedge r<x_2]\}\ ,$$

or equivalently,

$$(\forall y_1)(\exists u)(\forall y_2)(\exists y_2^*)(\forall r)(\exists r^*)\{[u<y_1\supset[u=y_2\cdot x_2+r\wedge r<x_2]]\\ \supset[y_1=y_2^*\cdot x_2+r^*\wedge r^*<x_2]\}\ .$$

If $y_1<x_2$, we satisfy the conclusion of the lemma by taking y_2^* to be 0 (and r^* to be y_1). If, on the other hand, $y_1\geq x_2$, we take u to be y_1-x_2, y_2^* to be y_2+1 (and r^* to be r); then the conclusion follows using an appropriate set of axioms for arithmetic. The program derived is then

$$\begin{cases} z=\underline{div}(x_1,x_2)\ \underline{where}\\ \underline{div}(y_1,x_2)\Leftarrow\underline{if}\ y_1<x_2\ \underline{then}\ 0\ \underline{else}\ \underline{div}(y_1-x_2,x_2)+1\ .\end{cases}$$

Although the program we constructed has two input variables, we were able to use the single-variable induction principle in its synthesis by treating the second input x_2 as a free variable. Typically when constructing programs with more than one input variable, we shall have to use a suitably generalized induction rule.

The next example will use two input variables, and we will not be able to treat either of them as a free variable. Therefore, we take this opportunity to demonstrate how to generalize the complete induction principle to construct programs with two inputs.

The form of complete induction was

$$\frac{(\forall y_1)\{(\forall u)[u<y_1 \supset \mathcal{A}(u)] \supset \mathcal{A}(y_1)\}}{(\forall x)\mathcal{A}(x)} .$$

For the two-input-variable case, we take $\mathcal{A}(y_1)$ to be $(\forall y_2)(\exists y_3)\mathcal{B}(y_1,y_2,y_3)$, obtaining the version

$$\frac{(\forall y_1)\{(\forall u)[u<y_1 \supset (\forall y_2)(\exists y_3)\mathcal{B}(u,y_2,y_3)] \supset (\forall y_2)(\exists y_3)\mathcal{B}(y_1,y_2,y_3)\}}{(\forall x_1)(\forall x_2)(\exists y_3)\mathcal{B}(x_1,x_2,y_3)} .$$

Suppose we want to prove a theorem of the form

$$(\forall x_1)(\forall x_2)(\exists z)\mathcal{B}(x_1,x_2,z) .$$

This is the same as the consequent of the complete induction rule. Thus, it suffices to prove the antecedent as a lemma:

(A) $$(\forall y_1)\{(\forall u)[u<y_1 \supset (\forall y_2)(\exists y_3)\mathcal{B}(u,y_2,y_3)] \supset (\forall y_2)(\exists y_3)\mathcal{B}(y_1,y_2,y_3)\} ,$$

or equivalently,

(A') $$(\forall y_1)(\forall y_2)(\exists u)(\exists y_2^*)(\forall y_3)(\exists y_3^*)\{[u<y_1 \supset \mathcal{B}(u,y_2^*,y_3)] \supset \mathcal{B}(y_1,y_2,y_3^*)\} .$$

From the proof of this lemma, we extract three subroutines $h_1(y_1,y_2)$, $h_2(y_1,y_2)$, and $g(y_1,y_2,y_3)$ corresponding to u , y_2^* , and y_3^* , respectively. The program extracted from the proof of the theorem will be of the form

$$\begin{cases} z = f(x_1,x_2) \text{ \underline{where}} \\ f(y_1,y_2) \Leftarrow g(y_1,y_2,f(h_1(y_1,y_2),h_2(y_1,y_2))) . \end{cases}$$

<u>Example 13</u>: The greatest common divisor program.

The program to be constructed must find the greatest common divisor (<u>gcd</u>) of two positive integers x_1 and x_2 . For simplicity, we ignore the possibility of

x_1 or x_2 being 0 , and the theorem to be proved is then

$$(\forall x_1)(\forall x_2)(\exists z)[z = \underline{gcd}(x_1,x_2)] ,$$

where

$$[z = \underline{gcd}(x_1,x_2)] \equiv \{z|x_1 \wedge z|x_2 \wedge \forall u[u|x_1 \wedge u|x_2 \supset u \leq z]\} .$$

Here, $u|v$ means " u divides v evenly". Recall that the function gcd should not be considered to be primitive.

The theorem is in the same form as the conclusion of the complete induction rule (for two input variables), taking y_3 to be z , and

$$\mathfrak{B}(y_1,y_2,y_3): y_3 = \underline{gcd}(y_1,y_2) .$$

Therefore, we must prove the following lemma:

(A') $$(\forall y_1)(\forall y_2)(\exists u)(\exists y_2^*)(\forall y_3)(\exists y_3^*)\{[u < y_1 \supset y_3 = \underline{gcd}(u,y_2^*)] \supset y_3^* = \underline{gcd}(y_1,y_2)\} .$$

This is one of the proofs we consider to be a challenge for existing theorem-proving systems. We suggest taking

u to be $\underline{rem}(y_2,y_1)$,

y_2^* to be y_1 , and

y_3^* to be $\underline{if}\ \underline{rem}(y_2,y_1) = 0\ \underline{then}\ y_1\ \underline{else}\ y_3$,

where $\underline{rem}(y_2,y_1)$ is the remainder when y_2 is divided by y_1 . Therefore,

$h_1(y_1,y_2)$ is $\underline{rem}(y_2,y_1)$,

$h_2(y_1,y_2)$ is y_1 , and

$g(y_1,y_2,y_3)$ is $\underline{if}\ \underline{rem}(y_2,y_1) = 0\ \underline{then}\ y_1\ \underline{else}\ y_3$.

The complete gcd program extracted is, therefore,

$$\begin{cases} z = \underline{gcd}(x_1,x_2)\ \underline{where} \\ \underline{gcd}(y_1,y_2) \Leftarrow \underline{if}\ \underline{rem}(y_2,y_1) = 0\ \underline{then}\ y_1\ \underline{else}\ \underline{gcd}(\underline{rem}(y_2,y_1),y_1) . \end{cases}$$

6. TRANSLATION FROM RECURSION TO ITERATION

Iterative programs and recursive programs compute the same class of functions (namely, the partial recursive functions). However, recursive programs are commonly far more inefficient in time and space than the corresponding iterative programs. Although it is straightforward to transform an iterative program into an equivalent recursive program, the reverse transformation presents difficulties. (See, for example, McCarthy (1963 a), Strong (1970), and Paterson and Hewitt(1970).)

LISP and ALGOL compilers, for example, translate recursive programs into iterative programs that use stacks, without changing the essence of the computation. Using program-synthetic techniques, it is sometimes possible to perform the transformation in such a way that the resulting iterative program performs the computation in a fundamentally better way than the original recursive program. Although we have no mechanism to ensure this improvement in general, we shall see how this occurs in the two examples presented in this section, the first concerning the reverse function and the second the Fibonacci sequence.

Example 14: The reverse function (see Example 2).

We are given a recursive reverse program

$$\begin{cases} z = \underline{reverse}(x) \ \underline{where} \\ \underline{reverse}(y) \Leftarrow \underline{if} \ \underline{Null}(y) \ \underline{then} \ NIL \\ \qquad\qquad\qquad \underline{else} \ \underline{append}(\underline{reverse}(\underline{cdr}(y)), \underline{list}(\underline{car}(y))) \ . \end{cases}$$

As we mentioned earlier, this definition is quite inefficient, since it involves repeated computation of the append function, which in itself requires a relatively complex computation.

The theorem to be proved is

$$(\forall x)(\exists z)[z = \underline{reverse}(x)] \ .$$

Recall that the reverse function is not considered to be primitive. For efficiency, we also omit the append function from the list of primitives.

Since we want to write an iterative program, we must use the iterative list induction rule:

$$\frac{\begin{array}{l}(\exists y_1)(\exists y_2)\mathbb{B}(x,y_1,y_2)\\ (\forall y_1)[\sim\underline{Null}(y_1) \land (\exists y_2)\mathbb{B}(x,y_1,y_2) \supset (\exists y_2)\mathbb{B}(x,\underline{cdr}(y_1),y_2)]\end{array}}{(\exists y_2)\mathbb{B}(x,NIL,y_2)} \ .$$

Aside from this rule, we have two axioms that result directly from the given definition of the reverse function:

(1a) $\underline{reverse}(NIL) = NIL$;

(1b) $\sim \underline{Null}(y) \supset [\underline{reverse}(y) = \underline{append}(\underline{reverse}(\underline{cdr}(y)), \underline{list}(\underline{car}(y)))]$.

Furthermore, the system will use the following axioms chosen from its supply during the course of the proof:

(2a) $\underline{append}(NIL,u) = u$;

(2b) $\underline{append}(u,NIL) = u$;

(3) $\underline{append}(u,\underline{append}(v,w)) = \underline{append}(\underline{append}(u,v),w)$; and

(4) $\underline{append}(\underline{list}(u),v) = \underline{cons}(u,v)$.

The theorem to be proved, $(\forall x)(\exists z)[z = \underline{reverse}(x)]$, is not in the correct form to apply the iterative induction, because NIL does not occur in it. However, by Axiom (2a) and the definition of reverse (Axiom (1a)), our theorem prover will translate the theorem into the following satisfactory form:

$$(\forall x)(\exists z)[\underline{append}(\underline{reverse}(NIL),z) = \underline{reverse}(x)] .$$

Therefore, we can apply the iterative list induction rule with

$$\mathcal{B}(x,y_1,y_2): \underline{append}(\underline{reverse}(y_1),y_2) = \underline{reverse}(x) ,$$

and the two lemmas to be proved are

(A) $(\exists y_1)(\exists y_2)[\underline{append}(\underline{reverse}(y_1),y_2) = \underline{reverse}(x)]$,

and

(B') $(\forall y_1)(\forall y_2)(\exists y_2^*)[\sim\underline{Null}(y_1) \wedge \underline{append}(\underline{reverse}(y_1),y_2) = \underline{reverse}(x)$
$\supset \underline{append}(\underline{reverse}(\underline{cdr}(y_1)),y_2^*) = \underline{reverse}(x)]$.

Using Axiom (2b), the system chooses $\underline{y_1 \text{ to be } x}$ and $\underline{y_2 \text{ to be NIL}}$, concluding the proof of Lemma (A).

To prove Lemma (B'), the system assumes $\sim\underline{Null}(y_1)$ and $\underline{append}(\underline{reverse}(y_1),y_2) = \underline{reverse}(x)$. Using the definition of $\underline{reverse}$ (Axiom (1b)), and the assumption that $\sim\underline{Null}(y_1)$, it derives

$$\underline{reverse}(y_1) = \underline{append}(\underline{reverse}(\underline{cdr}(y_1)),\underline{list}(\underline{car}(y_1))) .$$

Substituting in the hypothesis, it deduces

$$\underline{append}(\underline{append}(\underline{reverse}(\underline{cdr}(y_1)),\underline{list}(\underline{car}(y_1))),y_2) = \underline{reverse}(x) .$$

Using the associative rule for $\underline{append}$ (Axiom (3)), it obtains

$$\underline{append}(\underline{reverse}(\underline{cdr}(y_1)),\underline{append}(\underline{list}(\underline{car}(y_1)),y_2)) = \underline{reverse}(x) .$$

Then from Axiom (4), it derives

$$\underline{append}(\underline{reverse}(\underline{cdr}(y_1)),\underline{cons}(\underline{car}(y_1),y_2)) = \underline{reverse}(x) .$$

Comparing this formula with the desired conclusion, the system takes y_2^* to be $\underline{cons}(\underline{car}(y_1),y_2)$, concluding the proof.

Such a proof is well within the capabilities of existing theorem provers. In fact, the above proof of Lemma (B') has actually been found (see Brice and Derksen (1970)) using the QA3 theorem-proving system (Green and Raphael (1968)) with Morris's E-resolution (1969).

Since in the proof of Lemma (A) y_1 and y_2 were replaced by x and NIL, respectively, and in the proof of Lemma (B') y_2^* was replaced by $\underline{cons}(\underline{car}(y_1),y_2)$, the iterative program illustrated in Figure 12 will be constructed. Note that this program is far more efficient than its recursive counterpart.

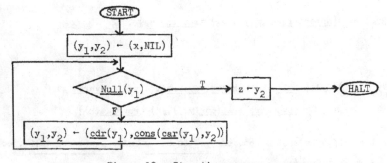

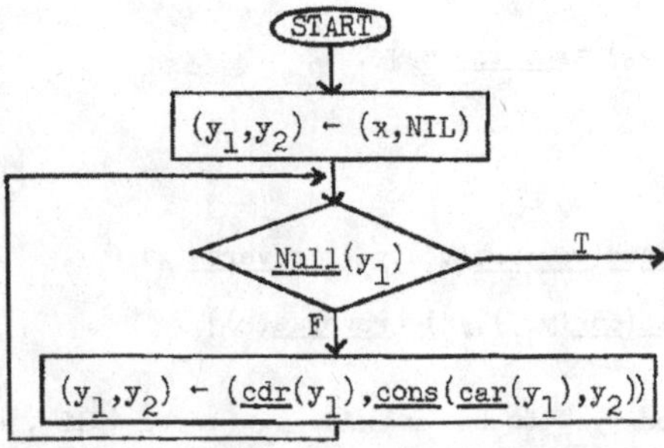

Figure 12. Iterative reverse program

Example 15: The Fibonacci sequence.

The advantage of iteration over recursion is particularly apparent in the computation of the Fibonacci series

$$1, 1, 2, 3, 5, 8, 13, 21, 34, 55, \ldots ,$$

each of whose terms (after the second) is the sum of the preceding two. Given a natural number x , the value of the xth Fibonacci number is most simply defined by the recursive program

$$\begin{cases} z = \underline{fibonacci}(x) \ \underline{where} \\ \underline{fibonacci}(y) \Leftarrow \underline{if} \ (y=0 \vee y=1) \ \underline{then} \ 1 \\ \qquad\qquad\qquad\qquad \underline{else} \ \underline{fibonacci}(y-1) + \underline{fibonacci}(y-2) \ . \end{cases}$$

In practice, this program is grossly inefficient, involving many repetitions of the same computation. We would like to use our approach to translate this program into an efficient iterative program with no redundant computation.

The theorem to be proved is simply

$$(\forall x)(\exists z)[z = \underline{fibonacci}(x)] \ .$$

The recursive definition of the fibonacci function implies the axioms

(1a) $\quad (u=0 \vee u=1) \supset \underline{fibonacci}(u) = 1 ,$

and

(1b) $\quad (u \geq 2) \supset \underline{fibonacci}(u) = \underline{fibonacci}(u-1) + \underline{fibonacci}(u-2) ,$

or equivalently,

$$\underline{fibonacci}(u'+2) = \underline{fibonacci}(u'+1) + \underline{fibonacci}(u') \ .$$

Axiom (1a) suggests to the theorem prover that the case $(x=0 \vee x=1)$ be treated separately; in this event, we take z to be 1 , and the output relation is satisfied.

It remains to prove

$$(\forall x)\{x \geq 2 \supset (\exists z)[z = \underline{\text{fibonacci}}(x)]\},$$

or equivalently (using Axiom (1b)),

$$(\forall x)\{x \geq 2 \supset (\exists z)[z = \underline{\text{fibonacci}}(x-1) + \underline{\text{fibonacci}}(x-2)]\}.$$

Thus, <u>taking x' to be $x-2$</u>, we have

$$(\forall x')(\exists z)[z = \underline{\text{fibonacci}}(x'+1) + \underline{\text{fibonacci}}(x')].$$

Since the "plus" operator is primitive, <u>taking z to be $z_1 + z_2$</u>, it suffices to prove

$$(\forall x')(\exists z_1)(\exists z_2)[z_1 = \underline{\text{fibonacci}}(x'+1) \wedge z_2 = \underline{\text{fibonacci}}(x')].$$

Note that we now have two output variables z_1 and z_2 rather than one. However, the proof procedure is precisely analogous to the single variable case; the iterative going-up induction principle used is

$$\frac{(\exists y_2)(\exists y_3)\mathcal{B}(0,y_2,y_3) \qquad (\forall y_1)[(\exists y_2)(\exists y_3)\mathcal{B}(y_1,y_2,y_3) \supset (\exists y_2)(\exists y_3)\mathcal{B}(y_1+1,y_2,y_3)]}{(\forall x)(\exists y_2)(\exists y_3)\mathcal{B}(x,y_2,y_3)},$$

with

$$\mathcal{B}(y_1,y_2,y_3):\ y_2 = \underline{\text{fibonacci}}(y_1+1) \wedge y_3 = \underline{\text{fibonacci}}(y_1).$$

<u>Taking z_1 to be y_2 and z_2 to be y_3</u> (i.e., z is $y_2 + y_3$), the conclusion of this induction rule is identical to the modified theorem. Thus, the two lemmas to be proved are

(A) $$(\exists y_2)(\exists y_3)[y_2 = \underline{\text{fibonacci}}(1) \wedge y_3 = \underline{\text{fibonacci}}(0)]$$

and

(B') $$(\forall y_1)(\forall y_2)(\forall y_3)(\exists y_2^*)(\exists y_3^*)\{[y_2 = \underline{\text{fibonacci}}(y_1+1) \wedge y_3 = \underline{\text{fibonacci}}(y_1)] \supset [y_2^* = \underline{\text{fibonacci}}(y_1+2) \wedge y_3^* = \underline{\text{fibonacci}}(y_1+1)]\}.$$

Lemma (A) is proved using Axiom (1a) <u>taking y_2 and y_3 both to be 1</u>.

To prove Lemma (B') we assume $y_2 = \underline{\text{fibonacci}}(y_1 + 1)$ and $y_3 = \underline{\text{fibonacci}}(y_1)$. Then taking $y_2^* = y_2 + y_3$ as suggested by Axiom (1b), and $y_3^* = y_2$ as suggested by the hypothesis, we have completed the proof of Lemma (B').

The program extractor combines all the replacements and substitutions made in the proof to form the program of Figure 13, which exhibits none of the crude inefficiencies of the original recursive program. The reader may observe how closely the operations in the program mirror the steps of the proof.

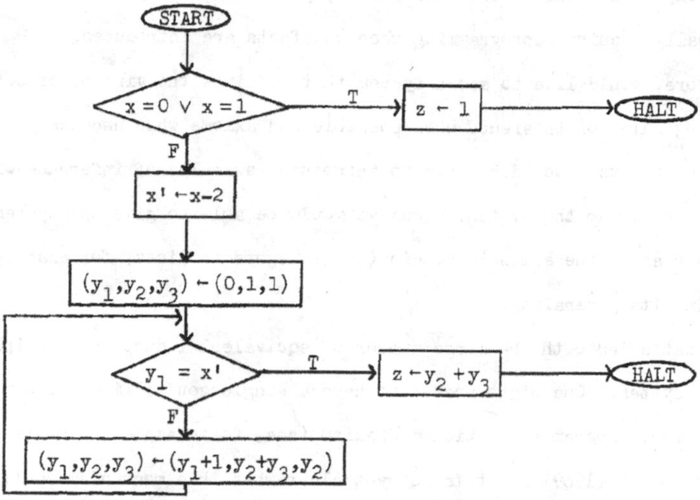

Figure 13. Iterative fibonacci program

7. FUTURE RESEARCH

Clearly, the results reported in this note represent but a step in the direction of automatic program synthesis. Our chief goal was not to present a complete work, but rather to stimulate other people to examine these problems.

(a) Suggested theorem-proving research

The foundation of our approach, and its chief weakness, lie in the theorem prover. We have mentioned that many of our proofs are probably beyond the state of the art of mechanical theorem proving, although none of them are terribly difficult. We, therefore, can use our experience to pinpoint some weaknesses in the current methods and to suggest some directions for theorem-proving research.

Any theorem-proving system stores its knowledge either in the form of axioms (which are simply assertions) or rules of inference (which are methods for transforming assertions). A system that relies mainly on axioms is very general; new facts may be introduced without modifying the system, because new axioms may be added long after the system is written. However, without restrictive strategies about how each axiom is to be used, such systems tend to thrash and flounder. On the other hand, systems such as King's (1969) (see also King and Floyd (1970)) which rely on rules of inference applying to a specific semantic domain, proceed with a great sense of direction but usually require reprogramming when new facts are introduced.

We, therefore, would like to see a system that combines the virtues of both approaches, using rules of inference when possible and axioms when necessary. We further hope that the user would be able to introduce new rules of inference without being forced to reprogram the system. Thus we would be able to give the system special knowledge about the semantic domain (the integers or lists, for example) without affecting its generality.

We are dissatisfied with the large number of equivalent induction principles required by our system. One might prefer to have a single general induction rule with a more powerful program extraction mechanism (see, for example, Burstall (1969), Park (1970), and Scott (1969)). It is not yet clear what this mechanism would be, and we are not sure that the machine implementation of such a rule in a theorem-proving system would be feasible.

Finally, it occurred to us during the preparation of this paper that partial function logic (see McCarthy (1963 b)) would be a more appropriate vehicle for program synthesis, because in this language we may discuss partial functions, whereas in the usual predicate calculus all operations and predicates are assumed to be total. We believe the techniques we have already outlined above apply to partial function logic as well. Some work has already been done by Hayes (1969) towards the machine implementation of this logic. Taking this remark in conjunction with a paper by Manna and McCarthy (1970) suggests that partial function logic may be the most natural language for program analysis and synthesis.

(b) Language and representation

In our discussion, we have used a modified predicate calculus in specifying the program to be constructed. This suggests that predicate calculus could be used as a higher-level programming language, where the compiler would be a program synthesis system, extracting a program in a lower-level language.

On the other hand, we are not bound to the use of predicate calculus as our source language. Programmers might find such a language lacking in readability and conciseness. However, any language that might be developed for expressing input and output relations would be satisfactory so long as the system could translate it into the language its theorem prover understood.

Of course, there are cases in which it is as easy to write the program itself as to write input and output relations describing it. However, this is more likely to be the case with trivial examples than with complex realistic programs.

(c) Interactive program synthesis

We have not considered the possibility that the synthesizer might interact with the user in constructing its programs. However, an interactive approach might lead immediately to a more practical system. For example, if the theorem prover were interactive, the power of the program synthesizer would be greatly increased. Alternatively, we might interact by allowing the user to suggest program segments to the synthesizer, allowing the system to incorporate them into the program.

(d) Program modification

We have not approached the problem of constructing efficient programs in any systematic way. We have contented ourselves with the construction of correct programs, and have seldom been very critical of the programming quality exhibited. Although in section 6 we illustrated that we can write more efficient programs by avoiding recursion and declaring inefficient subroutines non-primitive, more general work in this direction is clearly needed.

Once we have developed a method for controlling the efficiency of the extracted program, we not only can produce better programs with the purely synthetic approach, but also can use our techniques to write better compilers and program optimizers, which transform programs written by human beings. We take such a program (or a portion thereof) and transform it into its representation in predicate calculus (see Ashcroft (1970), Burstall (1970), Manna (1969a), and Manna and Pnueli (1970)), which is then taken as the specification of a new, more efficient reconstruction.

Another way program-synthetic techniques may be used in the improvement of an already existing program is in the construction of an automatic debugging system. Current program verification methods give us a way to detect and locate errors in a program; we then can use the program-synthetic approach to replace the incorrect segment without affecting the remainder of the program.

Acknowledgments

We would like to thank Claude Brice and Jan Derksen for programming in connection with this work. We are indebted to Johns Rulifson for discussions leading to the suggestion of examples presented in the paper. We are also grateful to Edward Ashcroft, Terry Davis, Jan Derksen, Stephen Ness, and Nils Nilsson for critical reading of the manuscript and subsequent helpful suggestions. Nils Nilsson obtained simplifications in several of our examples.

THE LATTICE OF FLOW DIAGRAMS

by

Dana Scott

TABLE OF CONTENTS

§0. Introduction
§1. Flow Diagrams
§2. Constructing Lattices
§3. Completing the Lattice
§4. The Algebra of Diagrams
§5. Loops and Other Infinite Diagrams
§6. The Semantics of Flow Diagrams
§7. The Meaning of a While Loop
§8. Equivalence of Diagrams
§9. Conclusion

0. INTRODUCTION. This paper represents an initial chapter in a development of a mathematical theory of computation based on lattice theory and especially on the use of continuous functions defined on complete lattices. For a general orientation, the reader may consult Scott (1970).

Let D be a complete lattice. We use the symbols:

$$\sqsubseteq, \bot, \top, \sqcup, \sqcap, \bigsqcup, \bigsqcap$$

to denote respectively the *partial ordering*, the *least* element, the *greatest* element, the *join* of two elements, the *meet* of two elements, the *join* of a set of elements, and the *meet* of the set of elements. The definitions and mathematical properties of these notions can be found in many places, for example Birkhoff (1967). Our notation is a bit altered from the standard notation to avoid confusion with the differently employed notations of set theory and logic.

The main reason for attempting to use lattices systematically throughout the discussion relates to the following well-known result of Tarski:

<u>THE FIXED-POINT THEOREM</u>. *Let* $f:D \to D$ *be a monotonic function defined on the complete lattice* D *and taking values also in* D. *Then* f *has a minimal fixed point* $p = f(p)$ *and in fact*

$$p = \bigsqcap \{x \in D : f(x) \sqsubseteq x\} .$$

For references and a proof see Birkhoff (1970), p. 115, and Bekić (1970). A function is called *monotonic* if whenever $x, y \in D$ and $x \sqsubseteq y$, then $f(x) \sqsubseteq f(y)$. Clearly, from the definition of p the element is $\sqsubseteq$ all the fixed points of f (if any). The only trick is to use the monotonic property of f to prove that p is indeed a fixed point. In the case of continuous functions we can be rather more specific.

Continuous functions preserve limits. It turns out that in complete lattices the most useful notion of limit is that of forming the join of a directed subset. A subset $X \subseteq D$ is called *directed* if every finite subset of X has an upper bound (in the sense of $\sqsubseteq$) belonging to X. This applies to the empty subset, so X must be non-empty. This also applies to any pair $x, y \in X$, so there must exist an element $z \in X$ with $x \sqcup y \sqsubseteq z$. An obvious example of a directed set is a *chain*:

$$X = \{x_0, x_1, \ldots, x_n, \ldots\}$$

where

$$x_0 \sqsubseteq x_1 \sqsubseteq \ldots \sqsubseteq x_n \sqsubseteq \ldots .$$

The *limit* of the directed set is the element $\bigsqcup X$. In the case of a chain (or any sequence for that matter) we write the limit (join) as:

$$\bigsqcup_{n=0}^{\infty} x_n .$$

A function $f:D \to D$ is called *continuous* if whenever $X \subseteq D$ is directed, then

$$f(\bigsqcup X) = \bigsqcup\{f(x) : x \in X\} .$$

It is easy to show that continuous functions are monotonic. Note, too, that the definition also applies to functions $f:D \to D'$ between two different lattices; in which case we read the right-hand side of the above equation as the join-operation in the second lattice D'. In the case of continuous $f:D \to D$, the fixed point turns out to be:

$$p = \bigsqcup_{n=0}^{\infty} f^n(\bot)$$

where $f^0(x) = x$ and $f^{n+1}(x) = f(f^n(x))$.

This all seems very abstract, but there is a large variety of quite useful complete lattices, and the fixed-point theorem is exactly the right way in which to introduce functions defined by *recursion*. This has been known for a long time, but the novelty of the present study centers around the *choice* of the lattices to which this idea may be applied. In particular, we are going to show that the familiar flow diagrams can be embedded in a useful way in an interesting complete lattice, and *then* that the semantics of flow diagrams can be obtained from a continuous function defined with the aid of fixed points. Of course, this is only one small application of the method, but it should be instructive.

1. FLOW DIAGRAMS. Intuitively, a flow diagram looks very roughly like Figure 1. There is a distinguished *entry point* into which the input information "flows" and an *exit point* out of which the result or output will (hopefully) come. The main question, then, is what goes on inside the "black box". Now, the box may represent a *primitive operation* which we do not analyze further, or the box may be compounded from other diagrams.

A trivial example of compounding may be the combination of *no* diagrams whatsoever. The result is the "straight arrow" of Figure 2.

Figure 1

A FLOW DIAGRAM

Figure 2

THE IDENTITY

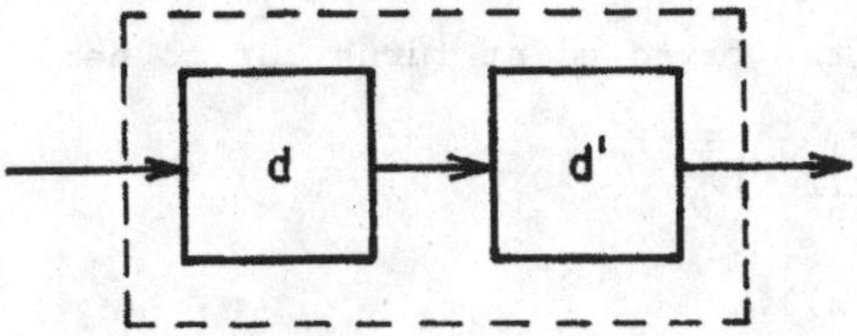

Figure 3

A PRODUCT

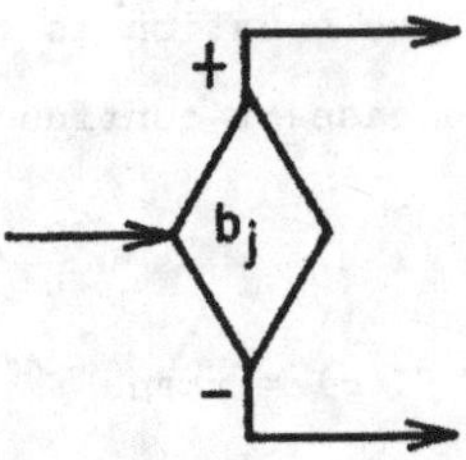

Figure 4

A SWITCH

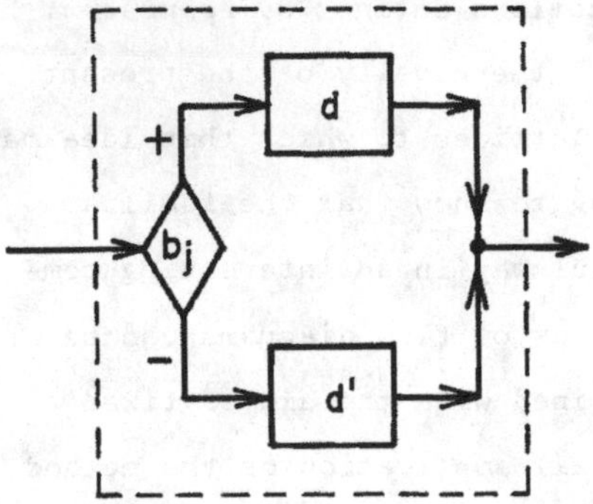

Figure 5

A SUM

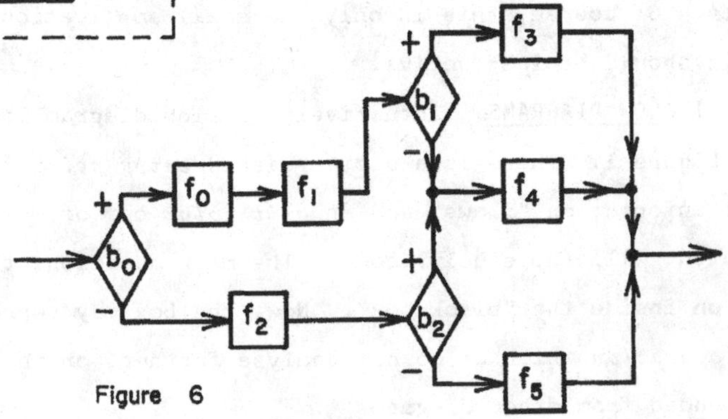

Figure 6

A LARGE DIAGRAM

The information flowing along such a channel exits untransformed; and so that diagram represents the *identity function*. A non-trivial compound is shown in Figure 3. In this combination, called a *product*, the output of the first box is fed directly into the input of the second with the obvious result.

With products alone not much useful could be done. As information flows, it must be tested and switched into proper channels according to the outcomes of the tests. For these switches we shall assume for simplicity in this paper that a *fixed* stock of primitive ones are given. This is not a serious restriction, and the method can just as well be applied when various forms of compounding of switches are allowed. We shall assume, by the way, that information flowing through a switch, though tested, exits untransformed. In diagrams a switch is represented as in Figure 4. In case the result of the test is positive, the information flows out of the top; if negative, from the bottom. A switch by itself is not a flow diagram because it has two exits. If these "wires" are attached to the inputs of the two boxes, and then if the outputs of the two boxes are brought together, we have a proper flow diagram. It is shown in Figure 5. We call this construction a *sum* (of the two boxes) for short, but it is also called a *conditional* because the outcome is conditional on the test.

Sums and products are the basic compounding operations for flow diagrams; iterating them leads to large diagrams such as the one shown in Figure 6. Here, the primitive boxes and switches have been labeled for reference and to distinguish them. The attentive reader will notice that we have cheated in the diagram in that the (-) and (+) leads from b_1 and b_2 have been brought together. The reason for doing this was to avoid duplicating box f_4. Strictly speaking, such shortcuts are *not* allowed: all repetitions must be written out. The diagrams will thus have a "tree" structure with switches at the branch points and with strings of boxes (any number including zero) along the branches. (We draw these trees sideways.) At the "top" of the tree all the leads are brought together for the output.

What is wrong in Figure 6? That is to say, what is lacking? Obviously, the answer is that there are no *loops*; all good flow diagrams permit feedback around loops. The proper way to allow looping is discussed in the next section; first, we must connect flow diagrams in the intuitive sense with the mathematical theory of lattices.

Some notation will help. We have already used the notation

$$b_0, b_1, b_2, \ldots ,$$
$$f_0, f_1, f_2, \ldots$$

for the switches and boxes, respectively. (The "b" recalls *Boolean* or *binary*; while the "f" is used because the boxes represent *functions* on information.) For the identity (or "dummy") diagram we may use the notation I. Suppose d and d' are two diagrams, then the *product* is denoted by:

$$(d;d')$$

where the order is the same left-right order as in Figure 3. The *sum* is written:

$$(b_j \rightarrow d,d')$$

which is the familiar "conditional expression" used here in an adapted form for diagrams. The diagram of Figure 6 may now be written as:

$$(b_0 \rightarrow (f_0;(f_1;(b_1 \rightarrow f_3,f_4))),(f_2;(b_2 \rightarrow f_4,f_5))) .$$

This expression has many too many parentheses, but we shall have to discuss problems of equivalence before we can eliminate any. In any case, it is clear that instead of diagrams we may talk of *expressions* generated from the f_i and I by repeated applications of the various sum and product operations. The expressions may get long, but it is a bit more obvious what we are talking about.

The totality of all expressions obtained in the way described above is a natural and well-determined whole, but just the same, we are going to embed it in a much larger complete lattice by a method similar to the expansion of the rationals to the reals. The first step is to introduce a sense of *approximation*, and the second step is to introduce *limits*. In our particular case, a very convenient way to achieve the desired goal is to introduce *approximate* (or: *partial*) expressions which interact with the "perfect" expressions we already know in useful ways not directly analogous to the common notion of approximation in the reals. (There is an exactly parellel way to treat reals, however.) Existing between approximate expressions is a partial ordering relation $\sqsubseteq$ which provides the required sense of approximation of one expression by another. We now turn to the details of setting up this relation.

If the relationship

$$d \sqsubseteq d'$$

between partial diagrams is to mean that d *approximates* d', then it seems very likely that in a large number of cases d can approximate many *different* d'. In particular, we may as well also assume the existence of the *worst* (or most incomplete) diagram $\perp$ which approximates *everything*; that is,

$$\perp \sqsubseteq d'$$

will hold for all d'. In pictures we may draw $\perp$ as a "vague" box whose contents are undetermined. Now, these incomplete boxes may occur as *parts* of other diagrams, as has been indicated in Figure 7. The expression for Figure 7 of course would be written as:

$$(b_0 \rightarrow (f_0;\perp),(f_1;(b_1 \rightarrow f_2,I))) \;.$$

If we are going to allow incomplete parts of diagrams, then we must also allow ourselves the option of *filling in* the missing parts.

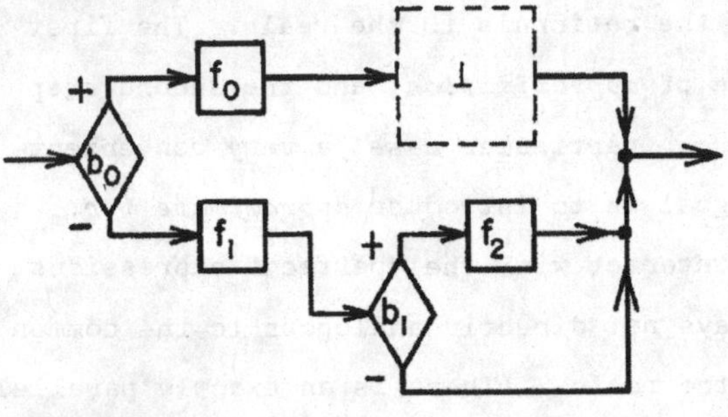

Figure 7
AN INCOMPLETE DIAGRAM

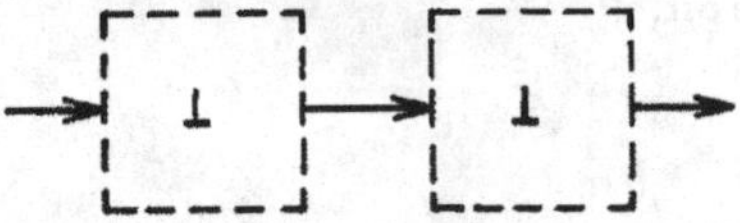

Figure 8
A PRODUCT OF
TWO INCOMPLETES

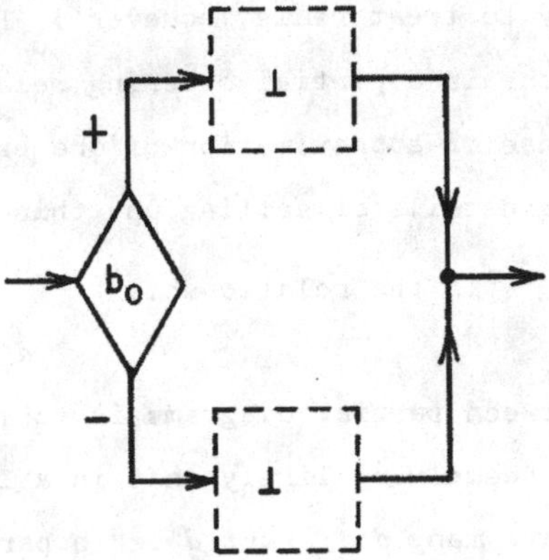

Figure 9
A SUM OF TWO
INCOMPLETES

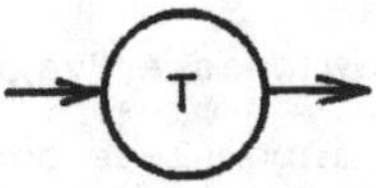

Figure 10
THE OVERDETERMINED
DIAGRAM

Thus, if d is incomplete, then a more precise reading of the relationship

$$d \sqsubseteq d'$$

is that d' is like d except that some of the parts left vague in d have been filled in. That reading is quite correct for the relationship $\bot \sqsubseteq d'$ that must always hold. In compound cases we can assure the desired results by assuming that sum and product formations are *monotonic* in the following precise sense:

$$\begin{aligned}&\textit{if } d_0 \sqsubseteq d_1 \textit{ and } d_0' \sqsubseteq d_1', \textit{ then}\\ &(d_0;d_0') \sqsubseteq (d_1;d_1') \textit{ and}\\ &(b_j \to d_0,d_0') \sqsubseteq (b_j \to d_1,d_1') \ .\end{aligned}$$

Besides this, the relation $\sqsubseteq$ must be assumed to be *reflexive*, *transitive*, and *antisymmetric* ($\sqsubseteq$ is a partial ordering).

As an illustration we could fill in the box of Figure 7 and prove by the above assumptions that:

$$(b_0 \to (f_0;\bot),(f_1;(b_1 \to f_2,I))) \sqsubseteq (b_0 \to (f_0;(f_1;f_0)),(f_1;(b_1 \to f_2,I)))$$

In working out these relationships it seems reasonable to assume in addition that:

$$(\bot;\bot) = \bot$$

but *not* to assume that:

$$(b_j \to \bot,\bot) = \bot \ ,$$

as may be appreciated from the pictures in Figures 8 and 9.

For the sake of mathematical symmetry (and to avoid making exceptions in certain definitions) we also introduce an exceptional diagram denoted by $\top$ about which we assume:

$$d \sqsubseteq \top$$

for all d. We can think of $\bot$ as being the *underdetermined* diagram, and $\top$ as being *overdetermined*. The diagram $\top$ is something like a short circuit -- we will make its "meaning" quite precise in the section on semantics. We assume that

$$(\top;\top) = \top \ ,$$

but *not* that

$$(b_j \to \top,\top) = \top \ ,$$

again for reasons that will be semantically motivated. Other equations that might seem reasonable (say, $(d;\top) = \top$) are postponed to the discussion of equivalence.

Taking stock of where we are now, we can say that we begin with certain "atomic" symbols (representing elementary diagrams); namely:

$$\bot, f_0, f_1, \ldots, f_n, \ldots, I, \top \ .$$

Then, we form all combinations generated from these using:

$$(d;d') \text{ and } (b_j \to d,d').$$

These expressions are partially ordered by a relation $\sqsubseteq$ about which we demand first that

$$\bot \sqsubseteq d \sqsubseteq \top$$

for all d; and then which we subject to the reflexive, transitive, and monotonic laws (the so-generated relation will automatically be antisymmetric).

This is the "symbolic" method which is quite reasonable and is well motivated by the pictures. We could even pursue it further and make the totality of expressions into a lattice in the following way. The join and meet operations must satisfy these laws:

$$\begin{aligned} d \sqcup d' &= d' \sqcup d & \qquad d \sqcap d' &= d' \sqcap d \\ d \sqcup d &= d & d \sqcap d &= d \\ d \sqcup \bot &= d & d \sqcap \bot &= \bot \\ d \sqcup \top &= \top & d \sqcap \top &= d \ . \end{aligned}$$

In addition for the atomic expressions other than $\bot$ and $\top$ we stipulate:

$$\begin{aligned} f_i \sqcup f_j &= \top & \qquad f_i \sqcap f_j &= \bot \\ f_i \sqcup I &= \top & f_i \sqcap I &= \bot \ , \end{aligned}$$

where $i \neq j$. For the case of products we have:

$$(\bot;\bot) = \bot \qquad (\top;\top) = \top ,$$

and in the following assume that the pair d,d' is not either of the exceptional pairs $\bot,\bot$ or $\top,\top$:

$$f_i \sqcup (d;d') = \top \qquad f_i \sqcap (d;d') = \bot$$
$$(b_j \to d_0,d_0') \sqcup (d;d') = \top \qquad (b_j \to d_0,d_0') \sqcap (d;d') = \bot$$
$$f_i \sqcup (b_j \to d_0,d_0') = \top \qquad f_i \sqcap (b_j \to d_0,d_0') = \bot$$
$$I \sqcup (d;d') = \top \qquad I \sqcap (d;d') = \bot$$
$$I \sqcup (b_j \to d_0,d_0') = \top \qquad I \sqcap (b_j \to d_0,d_0') = \bot$$

where d_0,d_0' is arbitrary. Moreover, for any two pairs d_0,d_0' and d_1,d_1' we assume:

$$(d_0;d_0') \sqcup (d_1;d_1') = (d_0 \sqcup d_1;d_0' \sqcup d_1')$$
$$(d_0;d_0') \sqcap (d_1;d_1') = (d_0 \sqcap d_1;d_0' \sqcap d_1')$$
$$(b_j \to d_0,d_0') \sqcup (b_j \to d_1,d_1') = (b_j \to d_0 \sqcup d_1,d_0' \sqcup d_1')$$
$$(b_j \to d_0,d_0') \sqcap (b_j \to d_1,d_1') = (b_j \to d_0 \sqcap d_1,d_0' \sqcap d_1') .$$

Finally, it *might* seem reasonable to assume:

$$(b_j \to d_0,d_0') \sqcup (b_k \to d_1,d_1') = \top$$
$$(b_j \to d_0,d_0') \sqcap (b_k \to d_1,d_1') = \bot$$

when $j \neq k$; but we postpone this decision.

This large number of rules allows us to compute joins and meets for any two expressions (in a recursive way running from the longer to the shorter expression), and it could be shown that in this manner the expressions do indeed form a lattice with the $\sqsubseteq$ relation as the partial ordering. The proof would be long and boring, however, as is always the case with symbolic methods. The reasons one must exercise care in this approach are in the main these two: one must be sure that all cases are covered, and one must be certain that different orders in carrying out symbolic operations do not lead to inconsistent results. Now, it would be quite possible to do all this for our construction of the lattice of diagrams, but it is quite unnecessary because a better method is available.

The idea of the better approach is to work with structures that are *known* to be lattices from the very start; hence we shall never have to check the lattice laws except in some trivial cases. Next, some operations on structures are carried out which are known to transform lattices into lattices (in our case this will correspond to the formation of compound expressions). Finally, (and this is the main virtue of the approach) the extension to a *complete* lattice may be described in a neat way. The lattice of expressions to the extent to which it has been apprehended up to this point is *not* complete; and the adjunction of limits requires a certain amount of care: the structural approach will make the exercise of this care more or less automatic. It must be stressed, however, that after the desired structures are created as lattices a certain amount of argument is required to see that the structures conform to our intuitive ideas about expressions. Though necessary, this will not be difficult, as we demonstrate in the next section.

2. <u>CONSTRUCTING LATTICES</u>. The initial part of the lattice we are trying to construct corresponds to the atomic symbols $f_0, f_1, \ldots$ and I. Since these symbols play slightly different roles, we separate I from the others. Now, all we really know about the f_i is that they are pairwise distinct; hence it will be sufficient to represent them by elements of a lattice illustrated in Figure 11. In such pictures of lattices the partial ordering is represented by the ascending lines; the weaker (smaller) elements are below and the stronger (larger) elements are above. (By the way, a lattice is *not* a flow diagram; the two kinds of pictures should not be confused. We are trying to make flow diagrams *elements* of a lattice.) What the

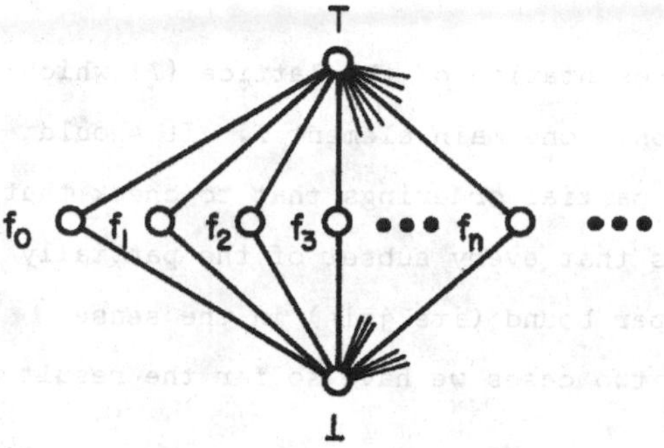

Figure 11

THE LATTICE F

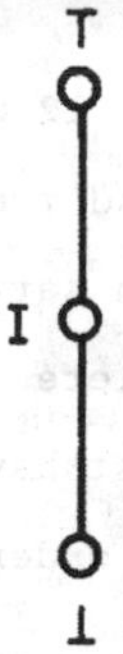

Figure 12

THE LATTICE {I}

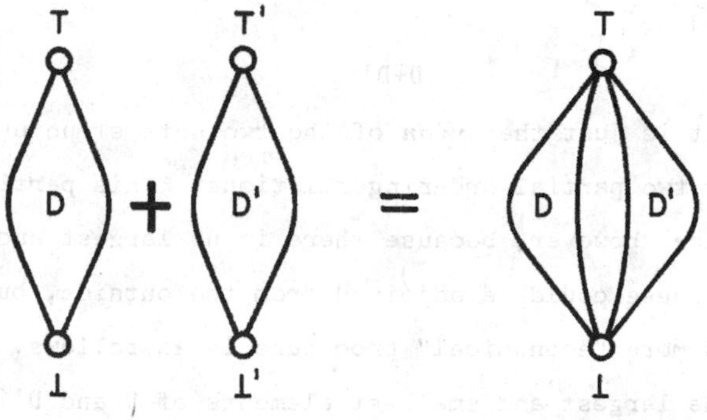

Figure 13

THE SUM OF TWO LATTICES

picture of the lattice F in Figure 11 shows is that the only partial ordering relations allowed are:

$$\bot \sqsubseteq f_i \sqsubseteq \top$$

for all i.

In Figure 12 we have a representation of the lattice $\{I\}$ which beside the $\bot$ and $\top$ elements has only one main element I. It should be mentioned in setting up these partial orderings that to check that they form complete lattices means that every subset of the partially ordered set must have a least upper bound (its join) in the sense of the partial ordering. In the two cases we have so far the result is obvious.

Suppose now that D and D' are two given complete lattices with partial orderings $\sqsubseteq$ and $\sqsubseteq'$, respectively. Inasmuch as it is only structure that is important, we may assume as sets of elements that D and D' are *disjoint*. We wish to combine D and D' together in one unified lattice: it will be called the sum of the two lattices and will be denoted by

$$D+D' .$$

Essentially, it is just the *union* of the two sets structured by the "union" of the two partial ordering relations. This partial ordering is not a lattice, however, because there is no largest and no smallest element. These could be adjoined from the outside, but a more convenient and more "economical" procedure is as follows. Let $\top,\top'$ and $\bot,\bot'$ be the largest and smallest elements of D and D', respectively. We have been regarding them as distinct (*all* the elements of D were to be distinct from the elements of D'), but now just these two pairs will be made equal. That is, we shall decree for D+D' that $\top=\top'$ and $\bot=\bot'$; though all the other elements are kept separate. The resulting partial ordering is easily seen to be a complete lattice. The process of forming this sum of lattices is illustrated in Figure 13.

The initial lattice of atomic expressions (diagrams) we wish to consider, then, is the lattice:

$$F+\{I\} .$$

It will be noted that the notion of sum just introduced could easily be extended to infinitely many factors. Thus, if we considered lattices $\{f_i\}$ that structurally were isomorphic to $\{I\}$ (but with different elements), then the lattice F could be defined by:

$$F = \{f_0\}+\{f_1\}+\ldots+\{f_n\}+\ldots .$$

Though they are not by themselves atomic expressions, the symbols b_i will also be thought of as elements of a lattice B defined by:

$$B = \{b_0\}+\{b_1\}+\ldots+\{b_n\}+\ldots .$$

The lattices F, B, and $F+\{I\}$ are all isomorphic as lattices but are different because they have different elements. These, however, are very trivial lattices, and we need much more complicated structures.

Suppose D and D' are lattices whose elements represent "diagrams" we wish to consider. If we want to form products of diagrams, then according to the intuitive discussion in the last section, the partial ordering on products should be defined so that

$$(d_0;d_0') \sqsubseteq (d_1;d_1') \textit{ if and only if } d_0 \sqsubseteq d_1 \textit{ and } d_0' \sqsubseteq' d_1'$$

for all d_0, $d_1 \in$ D and all d_0', $d_1' \in$ D'. Abstractly, we usually write $< d_0,d_0' >$ as an *ordered pair* in place of $(d_0;d_0')$, and then write:

$$D{\times}D'$$

for the set of all ordered pairs $< d,d' >$ with $d \in$ D and $d' \in$ D'. The above biconditional defines a partial ordering on D×D' called the (cartesian) product ordering, and, as is well known, the result is again a complete lattice. The largest and smallest elements of D×D' are the pairs $< \top,\top' >$ and $< \bot,\bot' >$, respectively.

Let the lattice

$$D_0 = F+\{I\}$$

be the lattice of atomic expressions. Then the lattice

$$D_0+(D_0\times D_0)$$

could be regarded as the lattice which in addition to the atomic expressions has compound expressions which can be thought of as products of two atomic expressions. In fact, there is no compelling reason to use the abstract notation $\langle d,d' \rangle$: we can use the more suggestive $(d;d')$ remembering that lattice-theoretically this is just an ordered pair. Notice in this regard that by our definitions of sums and products of lattices we have the equations

$$(\bot;\bot) = \bot \qquad (\top;\top) = \top$$

automatically.

What about diagrams? Well, even though we wrote

$$(b_j \to d,d') ,$$

abstractly all we have is an ordered triple

$$\langle b_j,d,d' \rangle .$$

This is just an element of the lattice

$$B\times D\times D .$$

(if the reader wants to be especially pedantic he can take $B\times D\times D$ to be $B\times(D\times D)$ and $\langle b_j,d,d' \rangle = \langle b_j,\langle d,d' \rangle\rangle$, or he can introduce an independent notion of ordered triple. Structurally, all approaches give isomorphic lattices.) Hence, the next lattice we wish to consider would be

$$D_1 = D_0+(D_0\times D_0)+(B\times D_0\times D_0)$$

Again, there is no reason to use the abstract notation so that $(b_j \to d,d')$ can just as well stand for an ordered triple. Notice that we have in this way introduced some elements not considered as diagrams before:

$$(\top \to d,d') \text{ and } (\bot \to d,d') ;$$

but we shall find that it is easy to interpret them semantically, so that this extra generality costs us no special effort. If we like, we can also use the more suggestive notation for the lattices themselves and write:

$$D_1 = D_0+(D_0;D_0)+(B \rightarrow D_0,D_0) ,$$

but for the time being it may be better to retain the abstract notation to emphasize the fact that we know all these structures as lattices.

Clearly, D_1 contains as elements only very short diagrams. To obtain the larger diagrams we must proceed recursively, iterating our compounding of expressions. Abstractly, this means forming ever more complex lattices:

$$D_{n+1} = D_0+(D_n \times D_n)+(B \times D_n \times D_n) .$$

The way we are construing the elements of these lattices, D_0 is a *subset* of each D_n:

$$D_0 \subseteq D_n ;$$

and, in fact, D_0 is a *sublattice*. This means that partial ordering on D_0 is the *restriction* of the intended partial ordering on D_n (restricted to the subset). *And* besides the join of any subset of D_0 formed within the lattice D_0 is exactly the same as the join formed within D_n. (This last is very important to remember.) The same goes for meets, but this fact is not so important.

Consider that

$$D_0 \subseteq D_1 ,$$

and that this implies that

$$D_0 \times D_0 \subseteq D_1 \times D_1$$

both as a subset *and* as a sublattice. Similarly, we have:

$$B \times D_0 \times D_0 \subseteq B \times D_1 \times D_1 .$$

It then follows that

$$D_0+(D_0 \times D_0)+(B \times D_0 \times D_0) \subseteq D_0+(D_1 \times D_1)+(B \times D_1 \times D_1)$$

both as a subset and as a sublattice. By definition we have:

$$D_1 \subseteq D_2 ,$$

and continuing in this way, we prove:

$$D_n \subseteq D_{n+1} .$$

Therefore,

$$D_n \subseteq D_m$$

whenever $n < m$.

What we have just done is to take advantage of general properties of the sum and product constructions on lattices as regards sublattices. These general properties about the comparisons of the partial orderings and the joins and meets are very simple to prove abstractly, and the reader is urged to work out the details for himself including the assertions of the last paragraph. As a result of these considerations it will be seen that the *union set*

$$\bigcup_{n=0}^{\infty} D_n$$

has a coherent partial ordering. Is this a lattice? It is not a complete lattice (we shall see why, later). On the other hand, many joins and meets do exist; in particular, the join of every *finite* subset exists in the union. (The reason is that any finite subset is wholly contained in one of the D_n.) So, the union of the lattices is a finitely complete lattice (a kind of structure that is ordinarily called just a lattice).

What are the elements of this union lattice? They are exactly all the finite combinations we desired generated from the atomic diagrams by means of the two modes of composition. Furthermore, the abstract lattice structure obtained in this way provides perfectly all the laws of computation we listed in the last section. Thus, the abstract approach gives us a structure which we know is a (finitely complete) lattice on the basis of simple, general principles. Then, by reference to the construction, the laws of computation are worked out. Having worked them out in this case, we can see by inspection of cases that we have all we need because there are only a limited number of types of elements formed in an iterative fashion. The next step is to complete the lattice and then to figure out what is obtained.

3. COMPLETING THE LATTICE. Every lattice can be completed (as in Birkhoff (1967), p. 126), but we shall want to complete the lattice of flow diagrams in a special way that allows us to apprehend the nature of the limit elements very clearly. In particular, the notion of approximation will be made quite precise.

Roughly speaking, the elements of the lattice D_n are diagrams of "length" at most n. More exactly, they can be obtained from the generators by nesting the two modes of composition to a level of at most n. This suggests that the elements of D_{n+1} might be approximable by elements of D_n. Consider D_0 and D_1. If $d \in D_1$, then it may belong to $D_0 \subseteq D_1$ or it may not. If $d \in D_0$, then it is its own best approximation. If $d \notin D_0$, then since the elements of D_0 are not compounds (except in a trivial sense) the best we can do in D_0 is to approximate d by $\perp$. In other words, we have defined a mapping

$$\psi_0 : D_1 \to D_0 ,$$

where for $d \in D_1$ we have:

$$\psi_0(d) = \begin{cases} d & \textit{if } d \in D_0 ; \\ \perp & \textit{if not}. \end{cases}$$

As can easily be established this mapping is *continuous* (in fact, a more general theorem relating to sum formation of lattices is provable), and this is important as all the mapping we employ ought to be continuous.

Now, consider D_{n+2} and D_{n+1}. We wish to define

$$\psi_{n+1} : D_{n+2} \to D_{n+1} .$$

For $d \in D_{n+2}$, the element $\psi_{n+1}(d)$ will be the best approximation to d by an element in D_{n+1}. Recall that

$$D_{n+1} = D_0+(D_n \times D_n)+(B \times D_n \times D_n)$$

and

$$D_{n+2} = D_0+(D_{n+1} \times D_{n+1})+(B \times D_{n+1} \times D_{n+1}) .$$

Inductively, we may assume that we have already defined the mapping

$$\psi_n : D_{n+1} \to D_n .$$

Clearly, what is called for is this definition:

$$\psi_{n+1}(d) = \begin{cases} d & \textit{if } d \in D_0 ; \\ (\psi_n(d');\psi_n(d'')) & \textit{if } d = (d';d'') ; \\ (b \to \psi_n(d');\psi_n(d'')) & \textit{if } d = (b \to d',d'') . \end{cases}$$

Now, these three cases are strictly speaking *not* mutually exclusive, but on the only possibilities of overlap we find agreement because $\psi_n(\top) = \top$ and $\psi_n(\bot) = \bot$. By a proof that need not detain us here, we show that ψ_{n+1} is continuous. Note also that we may prove inductively for all n that for $d \in D_{n+1}$ we have:

$$d \in D_n \textit{ if and only if } \psi_n(d) = d .$$

The mapping $\psi_n : D_{n+1} \to D_n$ is easily illustrated. In Figure 14 two diagrams are given: the first belongs to D_6 and the second is the result of applying ψ_5 to the first. It will be noted that drawn diagrams are slightly ambiguous; this ambiguity is removed when one choses an expression for the diagram. In this example we chose to associate to the right and to interpret a long arrow without boxes as a single occurrence of I and not as a product of several I's. The upper diagram is complete; while what we might call its *projection* from D_6 into D_5 is necessarily incomplete. Clearly, we can recapture the upper figure by removing the vagueness in one position of the lower figure. This is the way approximation works. It is a very simple idea.

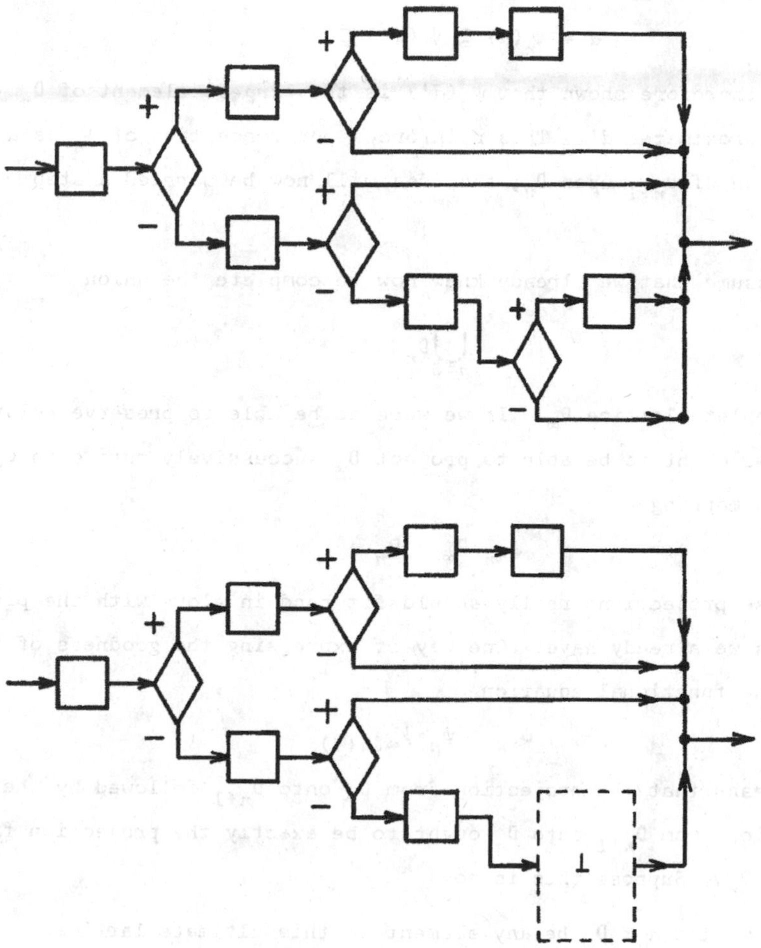

Figure 14

PROJECTING A DIAGRAM

Suppose $d' \in D_{n+1}$ and $d \in D_n$ and $d \sqsubseteq d'$. Now ψ_n is not only continuous, but also monotonic. Thus,

$$d = \psi_n(d) \sqsubseteq \psi_n(d') \sqsubseteq d' .$$

We have therefore shown that $\psi_n(d')$ is the *largest* element of D_n which approximates d'. This reinforces our conception of ψ_n as a projection of D_{n+1} upon D_n; the idea will now be carried a step further.

Assume that we already knew how to complete the union

$$\bigcup_{n=0}^{\infty} D_n$$

to a complete lattice D_∞. If we were to be able to preserve relationships, we ought to be able to project D_∞ successively onto each D_n, say by a mapping

$$\psi_{\infty n} : D_\infty \to D_n .$$

But these projections really should fit hand in glove with the projections we already have. One way of expressing the goodness of fit is by the functional equation

$$\psi_{\infty n} = \psi_n \circ \psi_{\infty(n+1)}$$

which means that the projection from D_∞ onto D_{n+1} followed by the projection from D_{n+1} onto D_n ought to be exactly the projection from D_∞ onto D_n. Suppose this is so.

Now, let $d \in D_\infty$ be any element of this ultimate lattice. Define a sequence of elements in the known lattices by the equation:

$$d_n = \psi_{\infty n}(d)$$

for all n. By what we have conjectured

$$\psi_n(d_{n+1}) = d_n$$

holds for all n, and so

$$d_0 \sqsubseteq d_1 \sqsubseteq \ldots \sqsubseteq d_n \sqsubseteq d_{n+1} \sqsubseteq \ldots .$$

How does the limit element d fit into the picture? Easy. We claim

$$d = \bigsqcup_{n=0}^{\infty} d_n .$$

Since $\psi_{\infty n}$ is a projection, we at least have $d_n \sqsubseteq d$ for all n; thus, the limit of the d_n must also approximate d. But why the equality? Well, since D_∞ is to be the completion of the union, each element of D_∞ is determined as the directed join (limit) of all elements of the union $\sqsubseteq$ it. (All elements of D_∞ must be approximable as closely as we please by elements from the union lattice.) If d' belongs to the union and $d' \sqsubseteq d$, then since $d' \in D_n$ for some n we must have $d' \sqsubseteq d_n$. Hence, the equality.

We have seen that each element $d \in D_\infty$ determines a sequence

$$< d_n >_{n=0}^{\infty}$$

such that

$$\psi_n(d_{n+1}) = d_n$$

holds for all n. Furthermore, distinct elements of D_∞ **determine** distinct sequences. (Because each d is determined as the limit of its corresponding sequence.) Suppose conversely that such a sequence of elements $d_n \in D_n$ is given and we *define* $d \in D_\infty$ by the equation

$$d = \bigsqcup_{n=0}^{\infty} d_n .$$

We are going to prove that for all n:

$$d_n = \psi_{\infty n}(d) .$$

In the first place, since these projections are continuous we have:

$$\psi_{\infty n}(d) = \bigsqcup_{m=0}^{\infty} \psi_{\infty n}(d_m) .$$

For $m \leq n$, since $d_m \in D_n$, it follows that

$$\psi_{\infty n}(d_m) = d_m .$$

For $m \geq n$, we are going to prove that

$$\psi_{\infty n}(d_m) = d_n .$$

This is true for $m = n$. We argue by induction on the quantity $(m-n)$. Having just checked it for the value 0, suppose the value is positive and that we know the result for the previous value. Thus, $m > n$. We use the equation relating the various projections and compute:

$$\begin{aligned}\psi_{\infty n}(d_m) &= \psi_n(\psi_{\infty(n+1)}(d_m)) \\ &= \psi_n(d_{n+1}) \\ &= d_n .\end{aligned}$$

Then since the required equation is proved we see:

$$\begin{aligned}\psi_{\infty n}(d) &= d_0 \sqcup d_1 \sqcup \ldots \sqcup d_n \sqcup d_n \sqcup d_n \sqcup \ldots \\ &= d_n .\end{aligned}$$

That is to say, in the infinite join all the terms after $n = m$ are d_n but the previous ones are $\sqsubseteq d_n$ anyway.

In other words, we have shown that there is a *one-one correspondence* between the elements of D_∞ and the sequences $< d_n >_{n=0}^{\infty}$ which satisfy the equations

$$\psi_n(d_{n+1}) = d_n .$$

Mathematically, this is very satisfactory because it means that instead of *assuming* that we know D_∞ we can *construct* it as actually *being* the set of these sequences. In this way, our intuitive ideas are shown to be mathematically consistent. This construction is particularly pleasant because the partial ordering on D_∞ has this easy sequential definition:

$d \sqsubseteq d'$ if and only if $d_n \sqsubseteq d'_n$ for all n.

The other lattice operations also have easy definitions based on the sequences. But for the moment the details need not detain us. All we really need to know is that the desired lattice D_∞ does indeed exist and that the projections behave nicely. In fact, it can be proved quite generally that each $\psi_{\infty n}$ is not only continuous but also is *additive* in the sense that

$$\psi_{\infty n}(\bigsqcup X) = \bigsqcup\{\psi_{\infty n}(x) : x \in X\}$$

for *all* $X \subseteq D_\infty$. Hence, we can obtain a reasonably clear picture of the lattice structure of D_∞. But D_∞ has "algebraic" structure as well, and we now turn to its examination.

4. THE ALGEBRA OF DIAGRAMS. Because the D_n were constructed in a special way, the complete lattice D_∞ is much more than just a lattice. Since we want to interpret the elements of D_∞ as diagrams, we replace the more abstract notation of the previous section by our earlier algebraic notation. Thus, by construction, if d, $d' \in D_n$ and if $b \in B$, then both

$$(d;d') \text{ and } (b \to d,d')$$

are elements of D_{n+1}. What if d, $d' \in D_\infty$? Will these algebraic combinations of elements make sense?

In order to answer this interesting question, we shall employ for elements $d \in D_\infty$ this abbreviated notation for projection:

$$d_n = \psi_n(d)$$

Remember that we regard each $D_n \subseteq D_\infty$ and so d_n is the *largest element of* D_n *which approximates* d. If $d \in D_{n+1}$, then

$$d_n = \psi_n(d)$$

also.

Using these convenient subscripts, we may then define for d, $d' \in D_\infty$:

$$(d;d') = \bigsqcup_{n=0}^{\infty}(d_n;d_n') \quad , \quad \text{and} \quad (b \to d,d') = \bigsqcup_{n=0}^{\infty}(b \to d_n,d_n') .$$

The idea is that the new elements of D_∞ will have the following projections:

$$(d;d')_{n+1} = (d_n;d_n') \quad , \quad \text{and} \quad (b \to d,d')_{n+1} = (b \to d_n,d_n') .$$

(The projections onto D_0 behave differently in view of the special nature of ψ_0 as defined in Section 3.) It can be shown that these operations $(d;d')$ and $(b \to d,d')$ defined on D_∞ are not only continuous but additive. (This answers the question at the end of Section 2.) Hence, D_∞ is a lattice enriched with algebraic operations (called *products* and *sums* and *not* to be confused with products and sums of *whole lattices*.)

Let $(D_\infty;D_\infty)$ be the totality of all elements $(d;d')$ with $d, d' \in D_\infty$. This is a sublattice of D_∞. Similarly, $(B \to D_\infty,D_\infty)$ is a sublattice of D_∞. In view of the construction of D_{n+1} from D_n we can show that in fact:

$$D_\infty = D_0+(D_\infty;D_\infty)+(B \to D_\infty,D_\infty) .$$

Because if $d \in D_\infty$ and if $d \notin D_0$, then we can find elements $d'_n, d''_n \in D_n$ such that either for all n:

$$d_{n+1} = (d'_n;d''_n)$$

or there is some $b \in B$ such that for all n:

$$d_{n+1} = (b \to d'_n,d''_n) .$$

Setting

$$d' = \bigsqcup_{n=0}^{\infty} d'_n \text{ and } d'' = \bigsqcup_{n=0}^{\infty} d''_n$$

we find that *either*

$$d = (d';d'') \text{ or } d = (b \to d',d'').$$

(One must also check that the $\top$ and $\bot$ elements match.) Since there can obviously not be any partial ordering relationships holding between the three different types of elements, we thus see why D_∞ decomposes into the sum of three of its sublattices.

Inasmuch as D_∞ is our ultimate lattice of *expressions* for diagrams, it will look neater if we call it E from now on. Having obtained the algebra and the decomposition, we shall find very little need to refer back to the projections. Thus, we can write:

$$E = F+\{I\}+(E;E)+(B \rightarrow E;E) ,$$

an equation which can be read very smoothly in words:

Every expression is either a function symbol,

or is the identity symbol,

or is the product of two expressions,

or is the sum of two expressions.

These words are very suggestive but in a way are a bit vague. We show next how to specify additional structure on E that will turn the above sentence into a mathematical equation.

To carry out this last program we need to use a very important lattice: the lattice T of *truth values*. It is illustrated in Figure 15. Aside from the ubiquitous $\bot$ and $\top$ it has two special elements 0 (false) and 1 (true). Defined on this lattice are the Boolean operations $\wedge$ (and), $\vee$ (or), $\neg$ (not) given by the tables of Figure 16. For our present purposes, these operations are not too important however, and we discuss them no further. What is much more important is the *conditional*.

Given an arbitrary lattice D, the conditional is a function

$$\supset : T\times D\times D \rightarrow D$$

such that

$$\supset(t,d,d') = \begin{cases} d \sqcup d' & if\ t = \top ; \\ d & if\ t = 1 ; \\ d' & if\ t = 0 ; \\ \bot & if\ d = \bot . \end{cases}$$

The reason for the choice of this definition is to make $\supset$ an *additive* function on T×D×D. Intuitively, we can read $\supset(t,d,d')$ as telling us to test t. If the result is 1 (true), we take d as the value of the conditional. If the result is 0 (false) we take d'. If the result of the test is *underdetermined*, so is the value of the conditional. If the result is *overdetermined*, we take the join of the

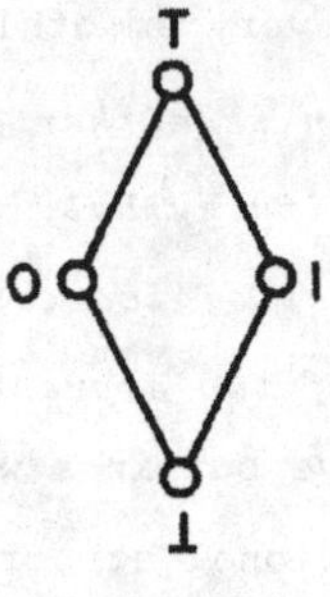

Figure 15

THE LATTICE **T**

∧	⊥	0	1	T
⊥	⊥	0	⊥	0
0	0	0	0	0
1	⊥	0	1	T
T	0	0	T	T

∨	⊥	0	1	T
⊥	⊥	⊥	1	1
0	⊥	0	1	T
1	1	1	1	1
T	1	T	1	T

¬	
⊥	⊥
0	1
1	0
T	T

Figure 16

THE BOOLEAN OPERATIONS

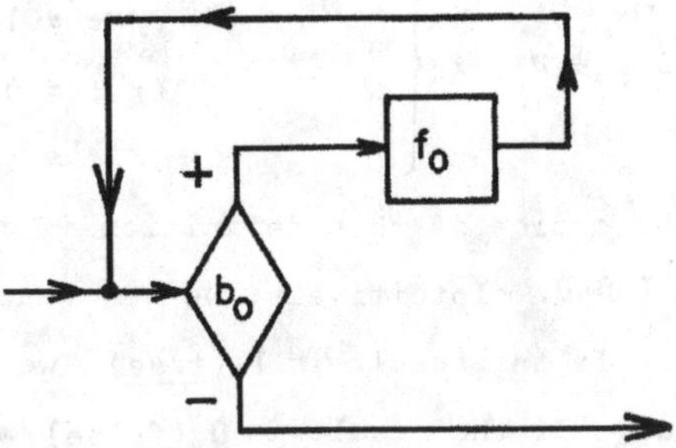

Figure 17

A SIMPLE LOOP

values we would have obtained in the self-determined cases. This last is conventional, but it seems to be the most convenient convention. It will be easier to read if we write

$$(t \supset d,d') = \supset(t,d,d') ,$$

and say in words:

$$\text{if } t \text{ then } d \text{ else } d' .$$

It is common to write → in place of ⊃, but we have chosen the latter to avoid confusion with the conditional *expression* in E.

Returning now to our lattice E there are four fundamental functions:

$$\mathsf{func}:\mathsf{E} \to \mathsf{T} ,$$

$$\mathsf{idty}:\mathsf{E} \to \mathsf{T} ,$$

$$\mathsf{prod}:\mathsf{E} \to \mathsf{T} ,$$

$$\mathsf{sum}:\mathsf{E} \to \mathsf{T} .$$

All of these functions map ⊤ to ⊤ and ⊥ to ⊥. For elements with $d \neq \top$ and $d \neq \bot$ we have

$$\mathsf{func}(d) = \begin{cases} 1 & \textit{if } d \in \mathsf{F} ; \\ 0 & \textit{if } d \notin \mathsf{F} . \end{cases}$$

$$\mathsf{idty}(d) = \begin{cases} 1 & \textit{if } d \in \{I\} ; \\ 0 & \textit{if } d \notin \{I\} . \end{cases}$$

$$\mathsf{prod}(d) = \begin{cases} 1 & \textit{if } d \in (\mathsf{E};\mathsf{E}) ; \\ 0 & \textit{if } d \notin (\mathsf{E};\mathsf{E}) . \end{cases}$$

$$\mathsf{sum}(d) = \begin{cases} 1 & \textit{if } d \in (\mathsf{B} \to \mathsf{E},\mathsf{E}) ; \\ 0 & \textit{if } d \notin (\mathsf{B} \to \mathsf{E},\mathsf{E}) . \end{cases}$$

These functions are all continuous (even:additive). They are the functions that correspond to the decomposition of E into four kinds of expressions.

Besides these there are five other fundamental functions:

$$\mathsf{first}:E \to E$$

$$\mathsf{secnd}:E \to E$$

$$\mathsf{left}:E \to E$$

$$\mathsf{right}:E \to B$$

$$\mathsf{bool}:E \to B$$

In case $d \in (E;E)$, we have:

$$d = (\mathsf{first}(d);\mathsf{secnd}(d)) \ ;$$

otherwise:

$$\mathsf{first}(d) = \mathsf{secnd}(d) = \bot.$$

In case $d \in (B \to E,E)$, we have:

$$d = (\mathsf{bool}(d) \to \mathsf{left}(d),\mathsf{right}(d)) \ ;$$

otherwise:

$$\mathsf{left}(d) = \mathsf{right}(d) = \bot \text{ (in E)} \ ,$$

and

$$\mathsf{bool}(d) = \bot \text{ (in B)} \ .$$

These functions are all continuous.

These nine functions together with the notions of products and sums of elements of E give a complete analysis of the structure of E. In fact, we can now rewrite the informal statement mentioned previously as the following equation which holds for all $d \in E$:

$$\begin{aligned} d = (&\mathsf{func}(d) \supset d \ , \\ &(\mathsf{idty}(d) \supset I \ , \\ &(\mathsf{prod}(d) \supset (\mathsf{first}(d);\mathsf{secnd}(d)) \ , \\ &(\mathsf{sum}(d) \supset (\mathsf{bool}(d) \to \mathsf{left}(d),\mathsf{right}(d)),\bot)))) \end{aligned}$$

Another way to say what the result of our construction is would be this: the lattice E replaces the usual notions of syntax. This lattice is constructed "synthetically", but what we have just verified is the basic equation of "analytic" syntax. All we really need to know about E is that it is a complete lattice that decomposes into a sum of its algebraic parts. These algebraic parts are either generators or products and sums. The complete analysis of an element (down *one* level) is provided by the above equation which shows that the algebraic terms out of which an element is formed are uniquely determined as *continuous* functions of the element itself.

Except for stressing the lattice-theoretic completeness and the continuity of certain functions, this sounds just like ordinary syntax. The parallel was intended. But our syntax is *not* ordinary; it is an essential generalization of the ordinary notions as we now show.

5. LOOPS AND OTHER INFINITE DIAGRAMS. In Figure 17 we have the most well-known construction of a flow diagram which allows the information to flow in circles: the so-called while-loop. It represents, as everyone knows, one of the very basic ideas in programming languages. Intuitively, the notion is one of the simplest: information enters and is tested (by b_0). If the test is positive, the information is transformed (by f_0) and is channeled back to the test in preparation for recirculation around the loop. While tests turn out positive, the circulation continues. Eventually, the cumulative effects of the repeated transformations will produce a negative test result (*if* the procedure is to allow output), and then the information exits.

None of our finite diagrams in E (that is, diagrams in any of the D_n lattices) has this form. It might then appear that we had overlooked something. But we did not, and that was the point of making E complete. To appreciate this, ask whether the diagram involving a loop in Figure 17 is not an abbreviation for a more ordinary diagram. There are many shortcuts one can take in the drawing of diagrams to avoid tiresome repetitions; we have noted several previously. Loops may just be an extreme case of abbreviation. Indeed,

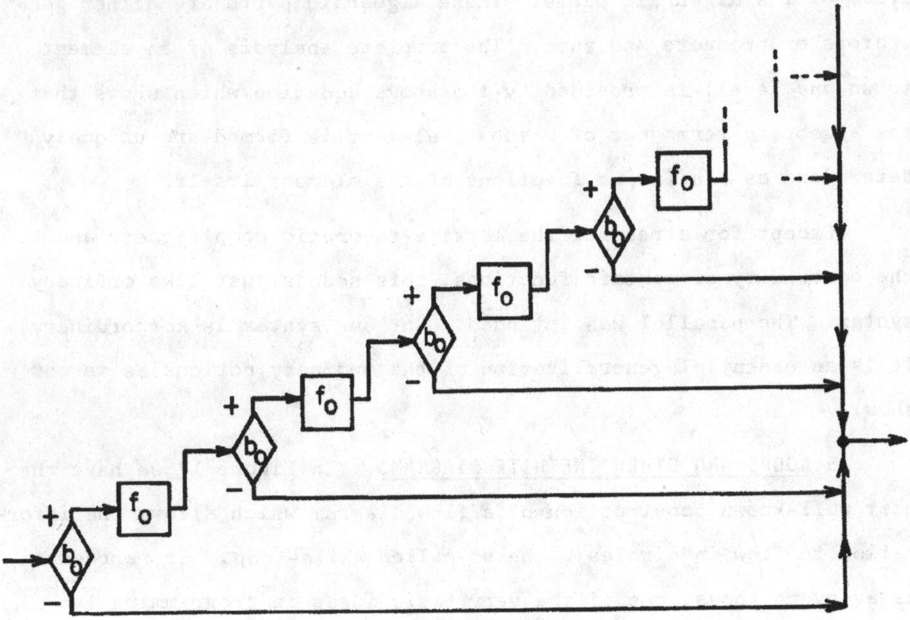

Figure 18

THE INFINITE VERSION OF THE LOOP

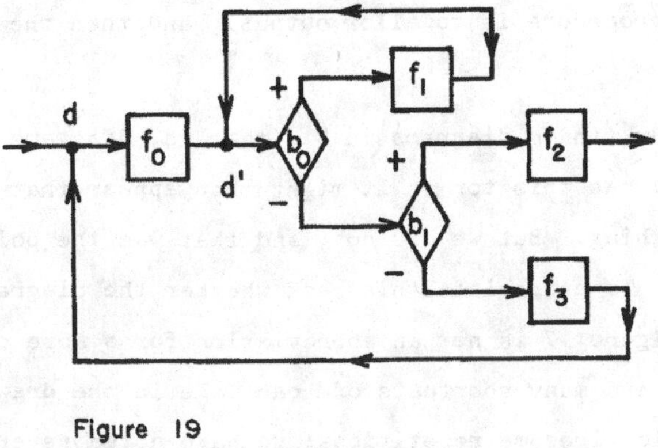

Figure 19

A DOUBLE LOOP

instead of bending the channel back around to the front of the diagram, we could write the test again. And, after the next transformation, we could write it out again. And again. And again, and again, and ... . The beginning of the *infinite* diagram that will thereby be produced is shown in Figure 18. Obviously, the infinite diagram will produce the same results as the loop. (Actually, this assertion requires proof.)

Does what we have just said make any sense? Some symbolization will help to see that it does. We have symbols for the test b_0 and the transformation f_0. Let the diagram we seek be called d. Look again at Figure 18. After the first test and transformation the diagram repeats itself. This simple pattern can easily be expressed in symbols thus:

$$d = (b_0 \rightarrow (f_0;d),I) \ .$$

In other words, we have a test with an exit on negative. If positive, on the other hand, we compound f_0 with the same procedure immediately following. Therefore, the diagram contains *itself* as a part.

That is all very pretty, but does this diagram d really exist in E? To see that it does, recall that all our algebraic operations are *continuous* on E. Consider the function $\Phi : E \rightarrow E$ defined by the equation:

$$\Phi(x) = (b_0 \rightarrow (f_0;x),I) \ .$$

The function Φ is evidently a continuous mapping of diagrams. *Every continuous function on a complete lattice into itself has a fixed point.* In this case, we of course want d to be the *least* fixed point:

$$d = \Phi(d) \ ,$$

because the diagram should have no other quality aside from the endless repetition. The infinite diagram d *does* exist. (It cannot be finite, as is obvious.) We can now see why we did not introduce loops in the beginning: their existence follows from completeness and continuity. In any case, they are very special and only one among many diverse types of infinite diagrams.

Figure 19 shows a slightly more complex example with a double loop. We shall not attempt to draw the infinite picture, since that exercise is unnecessary. The figure with loops is explicit enough to allow us to pass directly to the correct symbolization. To accomplish this, label the two re-entry points d and d'. Following the flow of the diagram we can write these equations:

$$d = (f_0;d')$$

and

$$d' = (b_0 \to (f_1;d'),(b_1 \to f_2,(f_3;d)))$$

Substituting the first equation in the second we find:

$$d' = (b_0 \to (f_1;d'),(b_1 \to f_2,(f_3;(f_0;d')))) .$$

Now, the "polynomial"

$$\Psi(x) = (b_0 \to (f_1;x),(b_1 \to f_2,(f_3;(f_0;x))))$$

is a bit more complex than the previous $\Phi(x)$, but just the same it is continuous and has its least fixed point d'. Thus, d', and therefore d, does exist in E.

Sometimes, the simple elimination procedure we have just illustrated does not work. A case in point is shown in Figure 20. The loops (whose entry points are marked d and d') are so nested in one another that each fully involves the other. (By now, an attempt at drawing the infinite diagram is quite hopeless.) The symbolization is easy, however:

$$d = (b_0 \to f_0,(b_1 \to (f_1;d),(f_2;d')))$$

and

$$d' = (b_2 \to f_3,(b_3 \to (f_4;d),(f_5;d'))).$$

In this situation any substitution of either equation in the other leaves us with an expression still containing *both* letters d and d'.

That is to say, the two diagrams called d and d' have to be constructed *simultaneously*. Is this possible? It is. Consider the fact that E×E is also a complete lattice. Introduce the function

$$\Theta: E\times E \rightarrow E\times E$$

defined as follows:

$$\Theta(\langle x,y\rangle) = \langle(b_0 \rightarrow f_0,(b_1 \rightarrow (f_1;x),(f_2;y))),(b_2' \rightarrow f_3,(b_3 \rightarrow (f_4;x),(f_5;y)))\rangle$$

Now, this function Θ is continuous and has a least fixed point:

$$\langle d,d'\rangle = \Theta(\langle d,d'\rangle) ,$$

and this pair is exactly the pair of diagrams we wanted.

This method can now be seen to be flexible and of wide applicability. For example, if using our algebra on E, we write down any system of polynomials in several variables:

$$\Pi_0(x_0,x_1,x_2,\ldots),\Pi_1(x_0,x_1,x_2,\ldots),\Pi_2(x_0,x_1,x_2,\ldots),\ldots ,$$

then on a suitable product lattice:

$$E\times E\times E\times\ldots$$

we can solve for fixed points:

$$\begin{aligned} d_0 &= \Pi_0(d_0,d_1,d_2,\ldots) \\ d_1 &= \Pi_1(d_0,d_1,d_2,\ldots) \\ d_2 &= \Pi_2(d_0,d_1,d_2,\ldots) \\ &\ldots . \end{aligned}$$

Diagrams constructed in this way may be called *algebraic* elements of E. The finite diagrams in the union of the D_n may be called *rational*. This classification does not by far exhaust the elements of E: there are besides a continuum number of *transcendental* elements. (The reader may construct one from Figure 18 by replacing the sequence of boxes f_0, f_0, f_0, ... by the sequence f_0, f_1, f_2, ... or by some other nonrepeating sequence.) Whether these other elements of E are of any earthly good remains to be seen. They are there, in any case. If you do not care to look at them, you need not do so. It will be your loss not theirs.

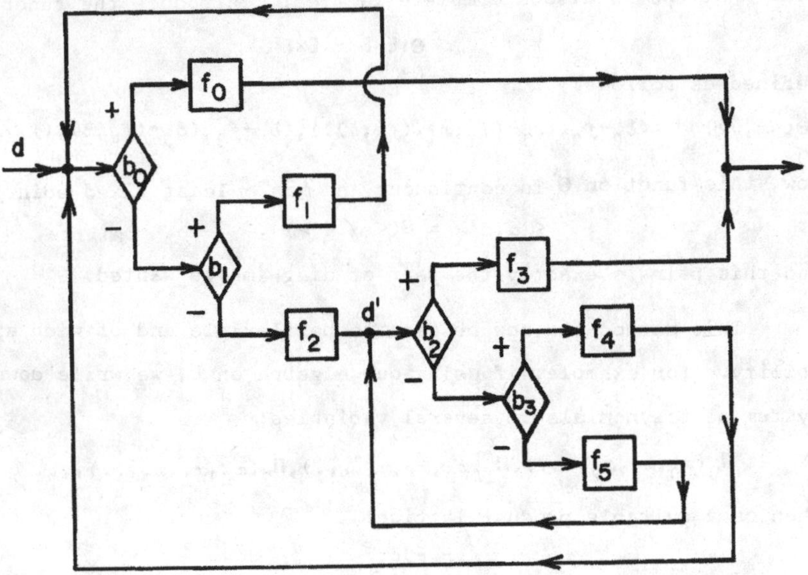

Figure 20

NESTED LOOPS

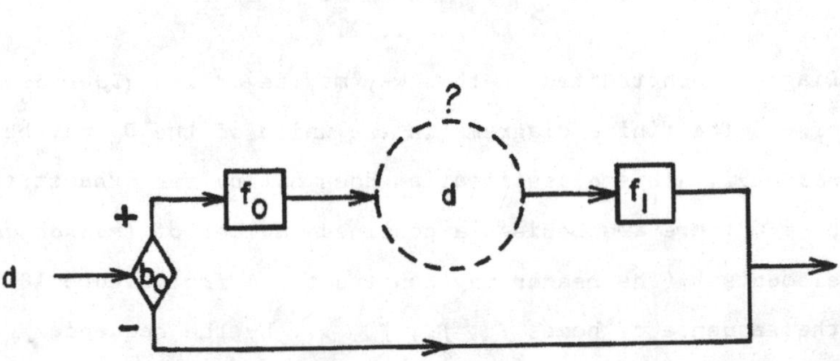

Figure 21

A DIFFICULT DIAGRAM

It is not too easy to draw pictures of some of the algebraic elements of E. Take, for example, this defining equation:

$$d = (b_0 \to (f_0;(d;f_1)),I)$$

A first and an unsatisfactory attempt to draw this as a diagram is shown in Figure 21. The question is what to fill in the middle. We need another copy of d itself; but this involves still another copy of d. And, so on. There seem to be no shortcuts available. Any attempt to introduce loops will not make it clear that in any one tour of the channels the *same* number of f_0 boxes as f_1 boxes must be visited. But this is a failure of the picture language. The algebraic language is unambiguous (hence, better)! Nevertheless, this example does suggest that there is a classification of the *algebraic* elements of E that needs additional thought.

Now that we see something of the scope of E, we can organize the study of its elements with the aid of further notations. For example, the while-loop is so fundamental that it deserves its own notation:

$$(b * d) \ ,$$

which stands for the least fixed point of the function

$$(b \to (d;x),I) \ .$$

It can be easily shown that $*$ is a continuous function on B×E into E. There are many others.

This is the place to clear up a continuing notational confusion. Since, in order to communicate mathematical facts, we need to write formulas involving symbols, we have to be clear about the distinction between a symbol and what it denotes. This distinction becomes particularly critical when we study the theory of syntax, as we have been doing here. So, let us be very pedantic about the nature of the constructions we have been discussing. What are the elements of E actually? Either they are elements of F or of $\{I\}$ or they are pairs or triples of pairs or triples of ... of elements of B and E. *Or* they are limit points of these, which strictly speaking, are infinite ("convergent") sequences of rational elements of E. Alas, E contains no *symbols*, only mathematical constructs.

But defined on E is a whole array of functions and constants:

$$f_0, f_1, \ldots, I, (x;y), (b \rightarrow x,y) ,$$
$$\text{func}(x), \ldots, \text{first}(x), \ldots, b{*}x ,$$
$$etc.$$

Thus, such things as subscripts, capital letters, parentheses, semicolons, arrows, commas, bold-face letters, and stars do not actually occur as parts of any of the elements of our "expression" space E. Rather, E is to be regarded as *a mathematical model of a theory of expressions*. It is only one of many similar models. Or, if you like, E is a model for *a theory of geometric diagrams*, and a quite satisfactory theory at that. The lattice E does not care what applications you care to make of it. E is abstract. E gives you a fixed structure to guide your thoughts. It is the same with the theory of the real numbers and analytic geometry. These structures are "pure": it is up to us to supply the plot and to write exciting stories about them using a careful choice of language (that is, functions, relations, etc.). In the case of E, however, we can ask not only what it *is*, and what its elements *do*, but also what do they *mean*.

6. THE SEMANTICS OF FLOW DIAGRAMS. We have spoken all along of the flow of information through a diagram. It is intuitively clear what is meant, but eventually one must introduce some precise definitions if he ever hopes to get any definite results. In other words, it is now time to present in detail a mathematical model of the concept of *flowing*. Up to this point, everything is static: the elements of E do not move; they do not light up, make noise, or otherwise show signs of life. We have sketched many pictures of elements

of E and *on the paper*, on these diagrams, we can move *our* fingers or shift *our* eyes back and forth. The abstract elements of E remain impassive, however, and must remain so, frozen in the eternal realm of ideas. But they neither expect or want our pity. And, we are free to study them, to talk about them as we do of works of art.

Clearly, the first requirement in the study of the meaning of the artifacts in E is a theory of *information.* Disappointingly enough, in this paper we shall not make a very deep study of this essential notion. We shall take it as axiomatic that *the quanta of information form a lattice called*:

$$S \ .$$

If you prefer, you can also consider the lattice S as being the lattice of *states* (states of "nature"). Where the lattice comes from, we do not say. We shall give some examples, by and by, but shall not be able to discuss lattices in general here. It should be reasonably evident from the success we have had in constructing lattices with useful properties, that this assumption is no loss of generality. Indeed, it can be argued that the requirement is a *gain* of generality.

In order to specify the meanings of the elements E, we must begin with the $f_i \in F$. Here, we have great freedom: their meanings can be determined at will -- within certain limits. The limits are set by this reasoning: as information passes through a box it is transformed. If the box is labeled with the symbol f_i, then the meaning of f_i is this transformation. That is, corresponding to each f_i is a *function*

$$\mathfrak{F}(f_i):S \to S$$

which provides the means of transforming S. Note that the transformation depends only on the label and not on the context of occurrence of a box, because we intend like labeled boxes to perform the same transformation. Since we have gone to the trouble of saying that

S is a complete lattice, we will also require each function $\mathfrak{F}(f_i)$ to be *continuous*.

Think for a moment of the collection of all continuous functions from S into S. If u and v are such, there is a most natural way of defining what it means for u to approximate v:

$$u \sqsubseteq v \textit{ if and only if } u(\sigma) \sqsubseteq v(\sigma) \textit{ for all } \sigma \in S .$$

It can easily be established that the set of continuous functions becomes in this way a complete lattice itself. We denote this lattice by $[S \to S]$. (*Caution*: do not confuse this notation with the earlier $(B \to E, E)$, which is a certain sublattice of E.)

In a highly useful short-hand way we can say that

$$\mathfrak{F}: F \to [S \to S] .$$

We even require $\mathfrak{F}$, as a mapping, to be *continuous*. Thus,

$$\mathfrak{F} \in [F \to [S \to S]] .$$

In this manner, we indicate succinctly what is called the *logical type* of $\mathfrak{F}$ as a mapping. Attention paid to logical types is attention well spent.

The next project is to attach meanings to elements of B. If $b \in B$ it designates some test that may be applied to elements of S. The outcome of a test is a truth value. For us, that means an element of the lattice T. Hence, to have meanings is to have a (continuous) function

$$\mathfrak{B}: B \to [S \to T]$$

Both $[S \to S]$ and $[S \to T]$ have largest elements (both are lattices). In $[S \to S]$ it is the constant function $\top$ (obviously, a continuous function). We should write

$$\top_{[S\to S]} \in [S \to S]$$

where for all $\sigma \in S$:

$$\top_{[S\to S]}(\sigma) = \top_S .$$

But we drop the subscripts and write $\top(\sigma) = \top$. Similarly, for $\bot(\sigma) = \bot$. The same slightly ambiguous notation is used for $[S \to T]$. For simplicity, we require both $\mathfrak{F}$ and $\mathfrak{B}$ to have the property that

$$\mathfrak{F}(\top) = \top , \quad \mathfrak{F}(\bot) = \bot ,$$

$$\mathfrak{B}(\top) = \top , \quad \mathfrak{B}(\bot) = \bot ,$$

where it is left to the reader to determine to which lattices each of the $\top$'s and $\bot$'s belong.

The functions $\mathfrak{F}$ and $\mathfrak{B}$ may be chosen freely within their respective logical types -- but that is all the freedom we have. The meanings of all the other elements of E are uniquely determined relative to this choice of $\mathfrak{F}$ and $\mathfrak{B}$.

To show how this works out, we shall determine a function $\mathfrak{V}$ (again:continuous) such that

$$\mathfrak{V}: E \to [S \to S] .$$

If $d \in E$, then $\mathfrak{V}(d)$ is the "value" of d (given $\mathfrak{F}$ and $\mathfrak{B}$). The intention is that if $\sigma \in S$ is the initial state of the information entering the flow diagram d, then

$$\mathfrak{V}(d)(\sigma)$$

is the final state upon exiting. We thus do not teach you how to swim through the channels of the flow diagram, but content ourselves with telling you what you will look like when you come out as a function of what you looked like when you jumped in. The transformation is, of course, continuous, And, merely knowing this transformation (over all d and all σ) is sufficient for a mathematical theory of *flowing*.

The precise definition of $\mathfrak{V}$ is obtained by simply writing out an equation that corresponds to what you yourself would do in swimming through a diagram. We write it first and then read it:

$$\mathfrak{V}(d)(\sigma) = (\mathrm{func}(d) \supset \mathfrak{F}(d)(\sigma),$$
$$(\mathrm{idty}(d) \supset \sigma,$$
$$(\mathrm{prod}(d) \supset \mathfrak{V}(\mathrm{secnd}(d))(\mathfrak{V}(\mathrm{first}(d))(\sigma)),$$
$$(\mathrm{sum}(d) \supset (\mathfrak{B}(\mathrm{bool}(d))(\sigma) \supset \mathfrak{V}(\mathrm{left}(d))(\sigma), \mathfrak{V}(\mathrm{right}(d)(\sigma)), \bot))))$$

(One small point: we may regard $\mathfrak{F}$ as being of type $\mathfrak{F}: E \to [S \to S]$ because $\mathfrak{F}(d) = \bot$ is a good value if $d \notin F$. Or, we should replace $\mathfrak{F}(d)$ by $\mathfrak{F}(|d|)$ where $|d| = d$ if $d \in F$, and $|d| = \bot \in F$ if $d \notin F$.)

The translation of the above equation runs as follows:

> *To compute the outcome of the passage of σ through d, first ask whether d is a function symbol. If it is, the outcome is* $\mathfrak{F}(d)(\sigma)$. *If it is not, ask whether d is the identity symbol. If it is, then the outcome is σ. If it is not, ask whether d is a product. If it is, find the first and second terms of d. Pass σ through the first term of d obtaining the proper outcome. Take this outcome and pass it through the second term of d. That gives the desired final outcome. If d is not a product, ask whether it is a sum. If it is, find the boolean part of d and test σ by it. Depending on the result of the test, pass σ through either the left or the right branch of d, obtaining the desired outcome. If d is not a sum (this case will not arise), the outcome is* ⊥.

One soon learns to appreciate equations. And, the equations are more precise as well as being more perspicuous -- though sometimes they become so involved as to be unreadable. Note, for example, how our equation for $\mathfrak{V}$ tells us exactly what to do in case

$$\mathfrak{B}(\mathrm{bool}(d))(\sigma) = \bot \ or = \top \ .$$

This would be rather tiresome to put in words. The question we need to ask now, however, is whether this equation really defines ϑ. Obviously, it is not an *explicit* definition because ϑ occurs on both sides of the equation. Hence, we cannot claim straight off that ϑ exists. To prove that it does, some fixed points must be found in some rather sophisticated lattices.

It was not just an idle remark to point out that

$$[S \to S]$$

is a complete lattice. Knowing this, we have by the same token that

$$[E \to [S \to S]]$$

is also complete. And this lattice gives the logical type of ϑ :

$$\vartheta \in [E \to [S \to S]] \ .$$

To find this ϑ, then, as a fixed point, we would need a function

$$\Xi \in [[E \to [S \to S]] \to [E \to [S \to S]]]$$

which is a lattice somewhat removed from everyday experience. But that does not matter: we know all the general definitions.

Here is the specific principle we need. In the following expression the variables $\mathfrak{X}$, d, and σ occur of types [E → [S → S]], E, and S respectively. The expression is:

$(\mathrm{func}(d) \supset \mathfrak{F}(d)(\sigma)$,
$(\mathrm{idty}(d) \supset \sigma$
$(\mathrm{prod}(d) \supset \mathfrak{X}(\mathrm{secnd}(d))(\mathfrak{X}(\mathrm{first}(d))(\sigma))$,
$(\mathrm{sum}(d) \supset (\mathfrak{B}(\mathrm{bool}(d))(\sigma) \supset \mathfrak{X}(\mathrm{left}(d))(\sigma), \mathfrak{X}(\mathrm{right}(d))(\sigma)), \bot))))$

This is a function of three variables. We can prove, just by looking at it, that it is *continuous* in its three variables.

Forget about the exact form of the above expression and imagine any such continuous expression:

$$(\ldots \mathfrak{X} \ldots d \ldots \sigma) \ .$$

Holding $\mathfrak{X}$ and d *fixed* we have a function of σ. The logical type of the value of the expression is also S. Thus, there is a function $\Xi'(\mathfrak{X}, d)$ depending on given $\mathfrak{X}$, d such that

$$\Xi'(\mathfrak{X}, d)(\sigma) = (\ldots \mathfrak{X} \ldots d \ldots \sigma) .$$

The logical type of $\Xi'(\mathfrak{X}, d)$ is $[S \rightarrow S]$. In other words, $\Xi'(\mathfrak{X}, d)$ is an "expression" whose value depends on $\mathfrak{X}$ and d. We can show that $\Xi'(\mathfrak{X}, d)$ is *continuous* in $\mathfrak{X}$ and d. Going around again, there must be a function (a uniquely determined function) $\Xi(\mathfrak{X})$ such that

$$\Xi(\mathfrak{X})(d) = \Xi'(\mathfrak{X}, d) ,$$

so that

$$\Xi(\mathfrak{X}) \in [E \rightarrow [S \rightarrow S]] .$$

But this correspondence is *continuous in* $\mathfrak{X}$. So, really

$$\Xi \in [[E \rightarrow [S \rightarrow S]] \rightarrow [E \rightarrow [S \rightarrow S]]] .$$

All continuous functions have fixed points (when they map a lattice into itself), and so our ϑ is given by

$$\vartheta = \Xi(\vartheta)$$

with the understanding that we take the least such ϑ (as an element of the lattice $[E \rightarrow [S \rightarrow S]]$).

Yes, the argument is abstract, but then it is very general. The easiest thing to do is simply to accept the existence of a continuous (minimal) ϑ and to carry on from there. One need not worry about the lattice theory -- as long as he is sure that all functions that he defines are continuous. Generally, they seem to take care of themselves. Intuitively, the definition of ϑ is nothing more than a recursive definition which gives the meaning of one diagram in terms of "smaller" diagrams. Such definitions are common and are well understood. In the present context, we might only begin to worry when we remember that a portion of an infinite diagram is not really "smaller" (it may even be *equal* to the original). It is this little worry which the method of fixed points lays to rest. Let us examine what happens with the **while**-loop.

7. THE MEANING OF A WHILE-LOOP. Let $b \in B$ and $d \in E$. Recall the definition of

$$(b{*}d) .$$

It is the least element of E satisfying the equation:

$$x = (b \rightarrow (d;x), I) .$$

We see that $(b{*}d) \in (B \rightarrow E,E)$ and

$$\mathbf{bool}\ ((b{*}d)) = b$$

$$\mathbf{left}\ ((b{*}d)) = (d;(b{*}d))$$

$$\mathbf{right}\ ((b{*}d)) = I .$$

Hence, by the definition of $\mathcal{V}$, for $\sigma \in S$ we have:

$$\mathcal{V}((b{*}d))(\sigma) = (\ \mathcal{B}(b)(\sigma) \supset \mathcal{V}((d;(b{*}d)))(\sigma),\sigma)$$

But $(d;(b{*}d)) \in (E;E)$ and

$$\mathbf{first}((d;(b{*}d))) = d$$

$$\mathbf{secnd}((d;(b{*}d))) = (b{*}d) .$$

So we find that:

$$\mathcal{V}((b{*}d))(\sigma) = (\ \mathcal{B}(b)(\sigma) \supset \mathcal{V}((b{*}d))(\ \mathcal{V}(d)(\sigma)),\sigma) .$$

The equation is too hard to read with comfort. Let $w = (b{*}d)$

$$\bar{b} = \mathcal{B}(b) \in [S \rightarrow T] ;$$

and

$$\bar{d} = \mathcal{V}(d) \in [S \rightarrow S] .$$

We may suppose that $\bar{b}$ and $\bar{d}$ are "known" functions. The diagram is the **while**-loop formed from b and d, and the semantical equation above now reads:

$$\mathcal{V}(w)(\sigma) = (\bar{b}(\sigma) \supset \mathcal{V}(w)(\bar{d}(\sigma)),\sigma)$$

The equation is still too fussy, because of all those σ's. Let

$$\bar{I} \in [S \rightarrow S]$$

be such that for all $\sigma \in S$:

$$\overline{I}(\sigma) = \sigma \text{ .}$$

For any two functions u, $v \in [S \to S]$, let

$$u \cdot v \in [S \to S]$$

be such that for all $\sigma \in S$:

$$(u \cdot v)(\sigma) = v(u(\sigma)) \text{ .}$$

(This is functional composition, but note the order.) For any $p \in [S \to S]$, and u, $v \in [S \to S]$, let

$$(p \gg u,v) \in [S \to S]$$

be such that for all $\sigma \in S$:

$$(p \gg u,v)(\sigma) = (p(\sigma) \supset u(\sigma),v(\sigma)) \text{ .}$$

Now, we have enough notation to suppress all the σ's and to write:

$$\mathcal{V}(w) = (\overline{b} \gg (\overline{d} \cdot \mathcal{V}(w)),\overline{I}) \text{ ,}$$

at last an almost readable equation.

It is important to notice that $\cdot$ and $\gg$ are *continuous* functions (of several variables) on the lattices $[S \to T]$ and $[S \to S]$. Indeed, we have a certain parallelism:

E	$[S \to S]$
B	$[S \to T]$
f_i	$\overline{f}_i$
I	$\overline{I}$
b_j	$\overline{b}_j$
$(x;y)$	$(u \cdot v)$
$(b \supset x,y)$	$(p \gg u,v)$

We could even say that $[S \to S]$ is an algebra with *constants* $\overline{f}_i$ and $\overline{I}$, with a *product* $(u \cdot v)$ and with *sums* $(\overline{b}_j \gg u,v)$ (and, if they are of interest, also $(\top \gg u,v)$ and $(\bot \gg u,v)$ where $\top$, $\bot \in [S \to T]$ are the obvious functions). The function $\mathcal{V}: E \to [S \to S]$ then proves to be a *continuous algebraic homomorphism*. It is *not* in general a lattice homomorphism since it is continuous and not join preserving.

It does, however, preserve all products and sums -- as we have illustrated in one case.

Let us make use of this observation. If we let $\Phi:E \to E$ be such that

$$\Phi(x) = (b \to (d;x),I) ,$$

then our while loop $w = (b*d)$ is given by:

$$w = \bigsqcup_{n=0}^{\infty} \Phi^n(\perp)$$

The function Φ is an algebraic operation on E; we shall let $\overline{\Phi}:[S \to S] \to [S \to S]$ be the corresponding operation such that

$$\overline{\Phi}(u) = (\overline{b} \to (\overline{d}\cdot u),\overline{I}) .$$

From what we have said about the continuity and algebraic properties of ϑ, it follows that

$$\vartheta(w) = \bigsqcup_{n=0}^{\infty} \overline{\Phi}^n(\perp) .$$

This proves that $\vartheta(w)$ is the least solution $u \in [S \to S]$ of the equation

$$u = (\overline{b} \gg (\overline{d}\cdot u),\overline{I})$$

Thus, ϑ preserves while; more precisely, there is an operation on $[S \to T]\times[S \to S]$ analogous to the $*$ operation on $B\times E$, and we have shown that ϑ is also a homomorphism with respect to this operation.

Actually, the solution to the equation

$$x = (b \to (d;x),I)$$

is *unique* in E. It is not so in the algebra $[S \to S]$ that the equation

$$u = (\overline{b} \gg (\overline{d}\cdot u),\overline{I})$$

has only one solution. But we have shown that ϑ picks out the least solution. This observation could be applied more generally also.

In any case, we can now state definitely that the quantity

$$\mathcal{V}(w)(\sigma)$$

is computed by the following iterative scheme: first $\bar{b}(\sigma)$ is computed. If the result is 1 (true), then $\bar{d}(\sigma) = \sigma'$ is computed and the whole procedure is started over on

$$\mathcal{V}(w)(\sigma') .$$

If $\bar{b}(\sigma) = 0$, the result is σ at once. (If $\bar{b}(\sigma) = \bot$, the result is $\bot$. If $\bar{b}(\sigma) = \top$, the result is $\mathcal{V}(w)(\sigma') \sqcup \sigma$, which generally is not too interesting.) The minimality of the solution to the equation in $[S \to S]$ means that we get *nothing more* than what is strictly implied by this computation scheme.

This result is hardly surprising; it was not meant to be. What it shows is that our definition is *correct*. Everyone computes a while in the way indicated and the function $\mathcal{V}(w)$ gives us just what was expected: no more, no less. We can say that $\mathcal{V}$ is the semantic function which maps the diagram w to its "value" or "meaning" $\mathcal{V}(w)$. And, we have just shown that the meaning of a while-loop is exactly a function in $[S \to S]$ to be computed in the usual while-manner. The meaning of the diagramatic while is the while-process. No one would want it any other way.

It is to be hoped that the reader can extend this style of argument to other configurations that may interest him.

8. EQUIVALENCE OF DIAGRAMS. Strictly speaking, the semantical interpretation defined and illustrated in the last two sections depends not only on the choice of S but on that of $\mathcal{F}$ and $\mathcal{B}$. Indeed, we should write more fully:

$$\mathcal{V}_S(\mathcal{F})(\mathcal{B})(d)(\sigma) ,$$

and take the logical type of $\mathcal{V}$ to be:

$$\mathcal{V}_S \in [[F \to [S \to S]] \to [[B \to [S \to T]] \to [E \to [S \to S]]]] .$$

Of course, $\mathcal{V}_S(\mathcal{F})(\mathcal{B})$ is continuous in $\mathcal{F}$ and in $\mathcal{B}$.

If we like, we can call the set S the set *states of a machine*. The functions $\mathfrak{F}$ and $\mathfrak{B}$ give the behavior of the "hardware" of a machine. Thus, the lattice

$$[F \rightarrow [S \rightarrow S]] \times [B \rightarrow [S \rightarrow T]]$$

may be called the *lattice of machines* (relative to the given S). This is obviously a very superficial analysis of the nature of machines: we have not discussed the "content" of the states in S, nor have we explained how a function $\mathfrak{F}(f)$ manages to produce its values. Thus, for example, the functions have no estimates of "cost" of execution attached to them (e.g. the time required for computation or the like). The level of detail, however, is that generally common in studies in automata theory (cf. Arbib (1969) as a recent reference), but it is sufficient to draw some distinctions. Certainly, lattices are capable of providing the structure of finite state machines with partial functions, as in Scott (1967),and much more: the uses of *continuous* functions on certain infinite lattices S are more subtle than ordinary employment of point-to-point functions. The demonstration that the present generalization is really fruitful will have to wait for future publications, though. (Cf. also Bekić, and Park (1969))

Whenever one has some semantical construction that assigns "meaning" to "syntactical" objects, it is always possible to introduce a relationship of "synonymity" of expressions. We shall call the relation simply: *equivalence*, and for x, $y \in E$ write:

$$x \equiv y$$

to mean that for *all* S and all $\mathfrak{F}$ and $\mathfrak{B}$ relative to this S we have:

$$\mathfrak{V}_S(\mathfrak{F})(\mathfrak{B})(x) = \mathfrak{V}_S(\mathfrak{F})(\mathfrak{B})(y) .$$

This relationship obviously has all the properties of an equivalence relation, but it will be the "algebraic" properties that will be of more interest. In this connection, there is one algebraic relationship that suggests itself at once. For x, $y \in E$ we write:

$$x \sqsubseteq y$$

to mean that for all S and all $\mathfrak{F}$ and $\mathfrak{B}$ relative to this S we have:

$$\mathcal{V}_S(\mathfrak{F})(\mathfrak{B})(x) \sqsubseteq \mathcal{V}_S(\mathfrak{F})(\mathfrak{B})(y) .$$

These relationships are very strong -- but not as strong as *equality*, as we shall see. The $\approx$ and $\sqsubseteq$ are related; for, as it is easy to show, $\sqsubseteq$ is *reflexive* and *transitive*, and further

$$x \approx y \textit{ if and only if } x \sqsubseteq y \textit{ and } y \sqsubseteq x .$$

But these are only the simplest properties of $\approx$ and $\sqsubseteq$.

For additional properties we must refer back to the exact definition of $\mathcal{V}$ in Section 6. In the first place, the definition of $\mathcal{V}$ was tied very closely to the *algebra* of E involving products and sums of diagrams. The meaning of a product turned out to be *composition* of functions; and that of a sum, a *conditional* "join" of functions. The meanings of $\approx$ and $\sqsubseteq$ are *equality* and *approximation* in the function space $[S \to S]$, respectively. Hence, it follows from the monotonic character of compositions and conditionals that:

$$x \sqsubseteq x' \textit{ and } y \sqsubseteq y' \textit{ implies } (x;y) \sqsubseteq (x';y')$$

$$\textit{and } (b \to x,y) \sqsubseteq (b \to x',y') .$$

In view of the connection between $\approx$ and $\sqsubseteq$ noted in the last paragraph, the corresponding principle with $\sqsubseteq$ replaced by $\approx$ also follows.

A somewhat more abstract way to state the fact just noted can be obtained by passing to equivalence classes. For $x \in E$, we write

$$x/\approx \; = \{x' \in E : x \approx x'\}$$

for the *equivalence class* of x under $\approx$. We also write

$$E/\approx$$

to denote the set of all such equivalence classes, the so-called *quotient algebra.* And the point is that $E/\approx$ *is* an algebra in the sense that products and sums are well defined on the equivalence classes, as we have just seen.

We shall be able to make E seem even more like an algebra, if we write:

$$x \dotplus y = (\top \to x, y)$$

for all $x, y \in E$. Now, in Section 6 we restricted consideration to those $\mathcal{B} \in [B \to [S \to T]]$ such that

$$\mathcal{B}(\top) = \top \ .$$

Thus

$$\mathcal{V}_S(\mathcal{F})(\mathcal{B})(x \dotplus y)(\sigma) = \mathcal{V}_S(\mathcal{F})(\mathcal{B})(x)(\sigma) \sqcup \mathcal{V}_S(\mathcal{F})(\mathcal{B})(y)(\sigma) \ ;$$

that is the meaning of $x \dotplus y$ is the lattice-theoretic *join* (or full sum) of the functions assigned to x and to y. This, of course, seems very special. As a diagram we would draw $x \dotplus y$ as in Figure 21. The intended interpretation is that flow of information is directed through *both* x and y and is "joined" at the output. The sense of "join" being used is that of the join in the lattice S.

Pushing the algebraic analogy a bit further we can write certain conditionals as *scalar products*:

$$b \cdot x = (b \to x, \bot) \ ,$$

and

$$(1-b) \cdot y = (b \to \bot, y) \ .$$

These two compounds are diagramed in Figure 22. The first passes information through b provided b is true; the second, through y provided b is false. Now, our first really "algebraic" result is this equivalence:

$$(b \to x, y) \approx (b \cdot x) \dotplus (1-b) \cdot y \ .$$

That is, up to equivalence, the conditional sum can be "defined" by the full sum with the aid of scalar multiples. A fact that can

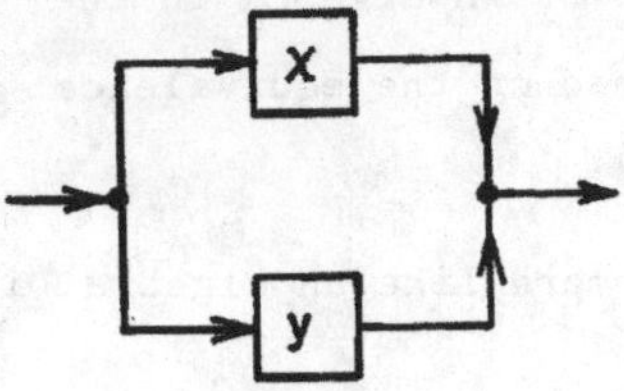

Figure 21
A (FULL) SUM

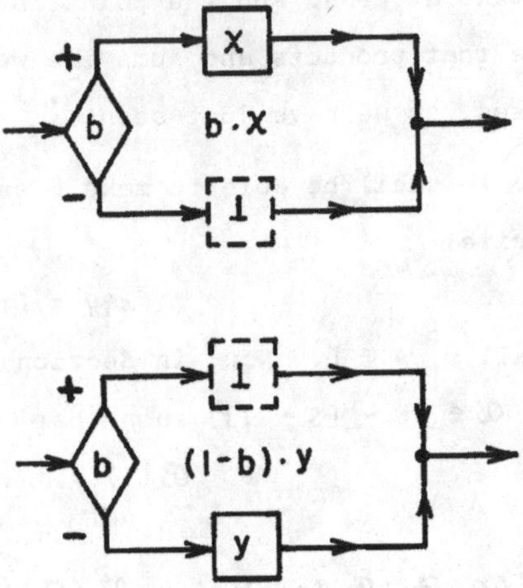

Figure 22
TWO SCALAR PRODUCTS

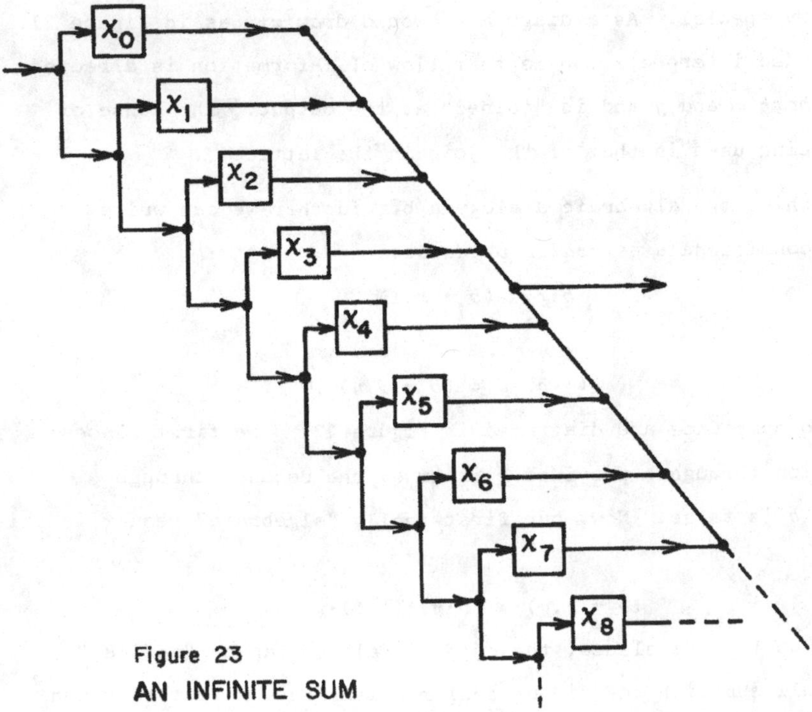

Figure 23
AN INFINITE SUM

be easily appreciated from the diagrams. This would not be very interesting if we did not have further algebraic equivalences, but note the following:

$$b\cdot(x\dotplus y) \approx b\cdot x\dotplus b\cdot y$$

$$b\cdot(b\cdot x) \approx b\cdot x$$

$$b\cdot(c\cdot x) \approx c\cdot(b\cdot x)$$

$$b\cdot x \ \underset{\sim}{\sqsubseteq}\ x$$

(If we had introduced some algebra into B, These results would be even more regular. But we chose here not to algebracize B.) One must take care to remember that T is not a Boolean algebra; thus, while it is correct that:

$$b\cdot x\dotplus(1-b)\cdot x \approx x \ ,$$

it is *not* correct that:

$$b\cdot(1-b)\cdot x \approx \bot \ .$$

The reason being that $\mathcal{B}(b)(\sigma) = \top$ is possible.

Having sufficient illustration of the properties of scalar multiples, we turn now to products. First, there is one distributive law:

$$x;(y\dotplus z) \cong (x;y)\dotplus(x;z)$$

that is correct; but the opposite

$$(y\dotplus z);x \cong (y;x)\dotplus(z;x)$$

is *not* correct. The reader may carry out the semantical analysis of these two proposed laws. The correctness of the first turns on the fact that for $f, f' \in [S \to S]$ and $\sigma \in S$ we have

$$(f \sqcup f')(\sigma) = f(\sigma) \sqcup f'(\sigma) \ .$$

The incorrectness of the second is a consequence of the failure in general of the equation

$$f(\sigma \sqcup \sigma') = f(\sigma) \sqcup f(\sigma')$$

for $f \in [S \to S]$ and $\sigma, \sigma' \in S$. Similarly, one must take care to note that *neither* of the following are correct:

$$(b\cdot x);y \approx b\cdot(x;y),$$

$$x;(b\cdot y) \approx b\cdot(x;y)$$

However, the *associative law* for ; is valid:

$$(x;y);z \approx x;(y;z) .$$

Returning to consideration of the operation $\dagger$, we remark that it is associative up to equivalence also; and since it is a join operation, we can prove

$$x\dagger y \sqsubseteq z \textit{ if and only if } x \sqsubseteq z \textit{ and } y \sqsubseteq z .$$

This means that $E/\approx$ is algebraically a *semi-lattice* with $\dagger$ as the join. Whether $E/\approx$ is a lattice, the author does not know at the moment of writing. However, we can define *countably infinite* joins in the partially ordered set $E/\approx$ as follows: Given a sequence x_n of elements of E, there is a unique element we shall call

$$\sum_{n=0}^{\infty} x_n ,$$

which is characterized by the equation

$$\sum_{n=0}^{\infty} x_n = x_0\dagger\sum_{n=1}^{\infty} x_n$$

In pictorial form the diagram is illustrated in Figure 23. This type of combination is clearly only of theoretical interest, but it does show why $E/\approx$ is countably complete. It may be possible that $E/\approx$ is a complete lattice, but the author doubts it.

As examples of equivalences involving infinite sums we have:

$$(x;\sum_{n=0}^{\infty} y_n) \approx \sum_{n=0}^{\infty}(x;y_n) .$$

In case $y_n \sqsubseteq y_{n+1}$ holds for all n, we would also have:

$$(\sum_{n=0}^{\infty} y_n);x \approx \sum_{n=0}^{\infty}(y_n;x)$$

as a consequence of *continuity*. There are many other similar laws.

9. <u>CONCLUSION</u>. Starting with very simple-minded ideas about flow diagrams as actual diagrams, we introduced the idea of *approximation* which led to a *partially ordered* set of diagrams. A rigorous, mathematical construction of this set produced what proved to be a *complete lattice* -- the lattice of flow diagrams. Defined on this lattice were several *algebraic operations* and the lattice as a whole satisfied an equation that connected it with the approach of *analytic syntax*. But the *limits* available in a complete lattice introduced something new into the picture: *infinite diagrams*. In particular, these infinite diagrams provided solutions to algebraic equations (the solutions were *algebraic elements*) which could be identified with the intuitive concepts of *loops* and other "*recursive*" diagrams (i.e. with feedback). So much for *syntax*.

Semantics of flow diagrams entered when the mapping of *evaluation* was defined from the algebra of diagrams into the *algebra of functions* on a *state space* (which was also a lattice). A bit of argument was required to see that evaluation captured the intuitive idea of *flow* in a diagram, but it became clearer in the example of a while-loop. From there, it was safe to introduce the notion of *equivalence* of diagrams and to study the resulting algebra. The reason for working out the equivalence algebra is, of course, to *formalize* some general facts about semantics of flow diagrams.

Much remains to be done before we have a perfect understanding even of this elementary area of the theory of computation. For one thing, only a start on the systematization of the algebra under equivalence was made in Section 8. It may be that equivalence is *not* at all the most important notion, for there may be *too few* equations between diagrams holding as equivalences. A more useful notion is the *conditional equation*. That is, we might write

$$x = x' \vdash y = y'$$

for x, y, x', $y' \in E$ to mean that for all S and all $\mathfrak{F}$, $\mathfrak{B}$, and all $\sigma \in S$, *if* it is the case that

$$\mathcal{V}_S(\mathfrak{F})(\mathfrak{B})(x)(\sigma) = \mathcal{V}_S(\mathfrak{F})(\mathfrak{B})(x')(\sigma),$$

then (as a *consequence*) we have

$$\mathcal{V}_S(\mathfrak{F})(\mathfrak{B})(y)(\sigma) = \mathcal{V}_S(\mathfrak{F})(\mathfrak{B})(y')(\sigma).$$

Note that this is not an implication between two equivalences, but from each *instance* of one equation to the corresponding instance of the other. Similarly, we could write:

$$x \sqsubseteq x' \vdash y \sqsubseteq y'.$$

Also it is useful to be able to write:

$$x_0 \sqsubseteq x_0', \; x_1 = x_1', \ldots \vdash \; y = y'$$

to mean that in each instance in which all the *hypotheses* on the left are true, the conclusion on the right follows. Many important algebraic laws can be given such a form.

Thus, in order to have a really systematic and useful algebra of flow diagrams, one should study the consequence relation $\vdash$ and attempt to axiomatize all of its laws. An effective axiomatization may only be possible for equations involving a *restricted* portion of E, because E contains so many infinite diagrams. But it is an interesting question to ponder.

JOINT BIBLIOGRAPHY

Ashcroft, E. A. (1970). "Mathematical Logic Applied to the Semantics of Computer Programs", Ph.D. thesis, Imperial College, London.

Ashcroft, E. A. and Manna, Z. (1970). "Formalization of Properties of Parallel Programs", Computer Science Dept., Stanford Univ., Artificial Intelligence Memo AIM-110.

Arbib, M. (1969). Theories of Abstract Automata, Prentice-Hall Series in Automatic Computation.

Bakker, J. W. de (1968). "Axiomatics of Simple Assignment Statements", Report MR 94, Mathematisch Centrum, Amsterdam.

Bakker, J. W. de (1969). "Semantics of Programming Languages", Advances in Information Systems Science (Tou, J. T., ed.), 2. Plenum Press, pp. 173-227.

Bekić, H. (1970). "Definable Operations in General Algebra, and the Theory of Automata and Flowcharts", to appear.

Birkhoff, G. (1967). Lattice Theory (third edition), Amer. Math. Soc. Colloquium Pub., vol. 25.

Brice, C. and Derksen, J. (1970). "The QA3 Implementation of E-Resolution", Stanford Research Institute, Artificial Intelligence Group, (to appear).

Burstall, R. M. (1969). "Proving Properties of Programs by Structural Induction", Comp. Journal 12, pp. 41-48.

Burstall, R. M. (1970). "Formal Description of Program Structure and Semantics in First-Order Logic", Machine Intelligence (Meltzer and Michie, eds.), 5. Edinburgh Univ. Press, pp. 79-98.

Caracciolo di Forino, A. (1964). "N-any Selection Functions and Formal Selective Systems", Calcolo 1, pp. 49-81.

Chroust, G. (1970). Compilation of Expressions--A Bibliography, IBM Laboratory Vienna, Lab. Report, LR 25.3.061, (in preparation).

Cooper, D. C. (1969a). "Program Scheme Equivalences and Second-Order Logic", Machine Intelligence (Meltzer and Michie, eds.) 4. Edinburgh Univ. Press, pp. 3-15.

Cooper, D. C. (1969b). "Program Schemes, Programs and Logic", Computer Science Dept., University College of Swansea, Memo No. 6, (see this Symposium).

Dijkstra, E. W. (1960). "Recursive Programming", Numerische Mathematik 2, No. 5.

Eilenberg, S. and Wright, J. B. (1967). "Automata in General Algebra", Information and Control 11, No. 4.

Elgot, C. C. and Robinson, A. (1964). "Random-Access, Stored Program Machines, An Approach to Programming Languages", Journal ACM 11, pp. 365-399.

Elgot, C. C. (1968). "Abstract Algorithms and Diagram Closure", Programming Languages (Genuys, ed.), Academic Press, pp. 1-42.

Engeler, E. (1967). "Algorithmic Properties of Structures", Math. Systems Theory 1, pp. 183-195.

Engeler, E. (1968). Formal Languages: Automata and Structures, Markham, 81 pp.

Engeler, E. (1969). "Algorithmic Approximations", to appear in Jour. Comp. Syst. Sciences.

Engeler, E. (1970). "Proof Theory and the Accuracy of Computations", Symposium on Automatic Demonstration (Laudet, Lacombe, and Schützenberger, eds.). Lecture Notes in Mathematics 125, Springer Verlag, pp. 62-72.

Floyd, R. W. (1967a). "Assigning Meanings to Programs", in Proc. Sym. in Applied Math. 19, Mathematical Aspects of Computer Science (Schwartz, J. T., ed.), Amer. Math. Soc., pp. 19-32.

Floyd, R. W. (1967b). "Non-deterministic Algorithms", Journal ACM 14, pp. 636-644.

Floyd, R. W. (1967c). "The Verifying Compiler", Computer Science Research Review, Carnegie-Mellon Univ. Annual Report, pp. 18-19.

Foley, M. (1969). "Proof of the Recursive Procedure Quicksort: A Comparison of Two Methods", Master's dissertation, Queen's Univ. of Belfast.

Good, D. I. (1970). "Toward a Man-Machine System for Proving Program Correctness", Ph.D. thesis, Univ. of Wisconsin.

Good, D. I. and London, R. L. (1968). "Interval Arithmetic for the Burroughs B5500: Four Algol Procedures and Proofs of their Correctness", Computer Sciences Technical Report No. 26, Univ. of Wisconsin.

Good, D. I. and London, R. L. (1970). "Computer Interval Arithmetic: Definition and Proof of Correct Implementation", Journal ACM, (to appear).

Green, C. (1969a). "Application of Theorem Proving to Problem Solving", Proc. of the Intern. Joint Conference on Artificial Intelligence, Washington, D. C.

Green, C. (1969b). "The Application of Theorem Proving to Question-Answering Systems", Ph.D. thesis, Stanford University.

Green, C. and Raphael, B. (1968). "The Use of Theorem-Proving Techniques in Question-Answering Systems", Proc. 23rd Nat. Conf. ACM, Thompson Book Co.

Hayes, P. J. (1969). "A Machine-Oriented Formulation of the Extended Functional Calculus", Stanford Artificial Intelligence Project, Memo 62.

Henhapl, W. (1969). "A Storage Model Derived from Axioms", IBM Lab. Vienna, Tech. Report TR 25.100.

Henhapl W. and Jones, C. B. (1970). "The Block Concept and Some Possible Implementations, with Proofs of Equivalence", IBM Lab. Vienna, Tech. Report TR 25.104.

Hoare, C. A. R. (1969). "An Axiomatic Basis for Computer Programming", Comm. ACM 12, pp. 576-583.

Ianov, I. I. (1958). (See Yanov (1958).)

Igarashi, S. (1963). "On the Logical Schemes of Algorithms", Information Processing in Japan, 3, pp. 12-18.

Igarashi, S. (1964). "An Axiomatic Approach to the Equivalence Problems of Algorithms with Applications", Ph.D. thesis, Univ. of Tokyo (also: Report of the Computer Centre, Univ. of Tokyo 1(1969), pp. 1-101; and: Publications of the Research Institute for Mathematical Sciences, Kyoto Univ. B, No. 33 (1969)).

Igarashi, S. (1968). "On the Equivalence of Programs Represented by Algol-like Statements", Report of the Computer Centre, Univ. of Tokyo 1, pp. 103-118 (also: Publications of the Research Institute for Mathematical Science, Kyoto Univ. B, No. 33 (1969)).

Kaplan, D. M. (1968a). "The Formal Theoretic Analysis of Strong Equivalence for Elemental Programs", Ph.D. thesis, Stanford Univ.

Kaplan, D. M. (1968b). "Some Completeness Results in the Mathematical Theory of Computation", Journal ACM 15, pp. 124-134.

King, J. C. (1969). "A Program Verifier", Ph.D. thesis, Carnegie-Mellon Univ.

King, J. C. and Floyd, R. W. (1970). "Interpretation-Oriented Theorem Prover Over Integers", Second Annual ACM Symposium on Theory of Computing, Northampton, Mass., pp. 169-179.

Knuth, D. E. (1968a). "Semantics of Context-Free Languages", Mathematical Systems Theory 2, pp. 127-145.

Knuth, D. E. (1968b). The Art of Computer Programming, vol. 1: Fundamental Algorithms, Addison-Wesley, 632 pp.

Lauer, P. (1968). "Formal Definition of ALGOL 60", IBM Lab. Vienna, Tech. Report TR 25.088.

Lawvere, F. W. (1963). "Functorial Semantics of Algebraic Theories", Proc. Nat. Acad. Sciences 50, pp. 869-872.

London, R. L. (1970a). "Bibliography on Proving the Correctness of Computer Programs", Machine Intelligence (Meltzer and Michie, eds.), 5, Edinburgh Univ. Press, pp. 569-580.

London, R. L. (1970b). "Computer Programs can be Proved Correct", Theoretical Approaches to Non-numerical Problem Solving (Banerji and Mesarovic, eds.), Lecture Notes in Operations Research and Mathematical Systems 28, Springer Verlag, pp. 281-302.

London, R. L. (1970c). "Proof of Algorithms: A New Kind of Certification (Certification of Algorithm 245 TREESORT 3)", Comm. ACM 13, pp. 371-373.

London, R. L. (1970d). "Proving Programs Correct: Some Techniques and Examples", BIT 10, pp. 168-182.

London, R. L. and Halton, J. H. (1969). "Proofs of Algorithms for Asymptotic Series", Computer Sciences Technical Report No. 54A, Univ. of Wisconsin.

Lucas, P. (1968). "Two Constructive Realizations of the Block Concept and their Equivalence", IBM Lab. Vienna, Tech. Report TR 25.085.

Lucas, P., and Walk, K. (1969). "On the Formal Description of PL/I", Annual Review in Automatic Programming, Vol. 6, Part 3, Pergamon Press.

Luckham, D. C., Park, D. M. R., and Paterson, M. S. (1970). "On Formalized Computer Programs", J. Comp. Syst. Sciences 4, pp. 220-249.

Luckham, D. C. and Nilsson, N. J. (1970). "Extracting Information from Resolution Proof Trees", Stanford Research Institute, Artificial Intelligence Groups, Technical Note 32.

Lukasiewicz, J. (1941). "Die Logic und das Grundlagenproblem", Les entretiens de Zurich (F. Gonseth, ed.), pp. 82-108.

Manna, Z. (1968a). "Termination of Algorithms", Ph.D. thesis, Carnegie-Mellon Univ.

Manna, Z. (1968b). "Formalization of Properties of Programs", Stanford Artificial Intelligence Report, Memo No AI-64, Stanford, California.

Manna, Z. (1969a). "The Correctness of Programs", J. of Computer and Systems Sciences 3, pp. 119-127.

Manna, Z. (1969b). "Properties of Programs and the First-Order Predicate Calculus", Journal ACM 16, pp. 244-255.

Manna, Z. (1970). "The Correctness of Nondeterministic Programs", Artificial Intelligence Journal 1, pp. 1-26.

Manna, Z. and McCarthy, T. (1970). "Properties of Programs and Partial Function Logic", Machine Intelligence (Meltzer and Michie, eds.), 5. Edinburgh Univ. Press, pp. 27-37.

Manna, Z. and Pnueli, A. (1970). "Formalization of Properties of Functional Programs", Journal ACM 17, pp. 555-569.

MacLane and Birkhoff (1967). Algebra, Macmillan.

McCarthy, J. (1962). "Towards a Mathematical Science of Computation", Proc. IFIP Congress 62, North-Holland, Amsterdam, pp. 21-28.

McCarthy, J. (1963a). "A Basis for a Mathematical Theory of Computation", Computer Programming and Formal Systems (Braffort and Hirshberg, eds.). North-Holland, Amsterdam, pp. 33-69.

McCarthy, J. (1963b). "Predicate Calculus with 'Undefined' as a Truth-Value", Stanford Artificial Intelligence Project, Memo No. 1.

McCarthy, J. and Painter, J. (1967). "Correctness of a Compiler for Arithmetic Expressions", Proceedings of Symposia in Applied Mathematics 19, Mathematical Aspects of Computer Science (Schwartz, J. T., ed.), Amer. Math. Soc., pp. 33-41.

Mendelson, E. (1964). Introduction to Mathematical Logic, Van Nostrand.

Milner, R. (1969). "Equivalences on Program Schemes", Journal Comp. Syst. Sci. 4, pp. 205-219.

Morris, J. B. (1969). "E-Resolution: Extension of Resolution to Include the Equality Relation", Proc. of the International Joint Conference on Artificial Intelligence, Washington, D. C.

Naur, P. et al (1960). "Report on the Algorithmic Language ALGOL 60", Comm. ACM 3, pp. 299-314.

Naur, P. (1966). "Proof of Algorithms by General Snapshots", BIT 6, pp. 310-316.

Naur, P. (ed.) (1962). "Revised Report on Algorithmic Language ALGOL 60", Comm. ACM 6, No. 1, pp. 1-23.

Painter, J. A. (1967). "Semantic Correctness of a Compiler for an ALGOL-like Language", Ph.D. thesis, Stanford Univ.

Park, D. (1970). "Fixpoint Induction and Proofs of Program Properties", Machine Intelligence (Meltzer and Michie, eds.), 5, Edinburgh Univ. Press, pp. 59-78.

Paterson, M. S. (1967). "Equivalence Problems in a Model of Computation", Ph.D. thesis, Trinity College, Cambridge.

Paterson, M. S. (1968). "Program Schemata", Machine Intelligence (Michie, ed.), 3, Edinburgh Univ. Press, pp. 19-31.

Paterson, M.S. and Hewitt, C. E. (1970). "Comparative Schematology", Unpublished notes.

PL/I Language Specifications (1966). IBM Systems Reference Library, Form No. C 28-6571-4.

Randell, B. and Russell, L. J. (1964). ALGOL 60 Implementation, Academic Press.

Robinson, J. (1965). "A Machine-Oriented Logic Based on the Resolution Principle", Journal ACM, 12, pp. 23-41.

Rutledge, J. D. (1964). "On Ianov's Program Schemata", Journal ACM 11, pp. 1-9.

Rutledge, J. D. (1970). "The Problem of Correct Programming", IBM Research Report RC 2746.

Scott, D. (1967). "Some Definitional Suggestions for Automata Theory", Journal of Comp. Syst. Sci. 1, pp. 187-212.

Scott, D. (1969). "A Type-Theoretical Alternative to CUCH, ISWIM, OWHY", Unpublished memo.

Scott, D. (1970). "Outline of a Mathematical Theory of Computation", Proceedings of the Fourth Annual Princeton Conference on Information Sciences and Systems.

Shepherdson, J. C. and Sturgis, H. E. (1963). "The Computatibility of Partial Recursive Functions", Journal ACM 10, pp. 217-255.

Simon, H. (1963). "Experiments with a Heuristic Compiler", Journal ACM 10, pp. 493-506.

Slagle, J. (1965). "Experiments with a Deductive Question-Answering Program", Comm. ACM 8.

Strachey, C. (1966). "Towards a Formal Semantics", Formal Language Description Languages (Steel, ed.), North-Holland.

Strong, H. R. (1970). "Translating Recursion Equations into Flow Charts", Second Annual ACM Symposium on Theory of Computing, Northampton, Mass., pp. 184-197.

Van der Mey, G. (1962). "Process for an ALGOL Translator", Report 164, Dr. Neher Laboratorium, Leidshendam.

Van Wijngaarden, A. (ed.), Mailloux, B. J., Peck, J. E. L., and Koster, C. H. A. (1969). Report on the Algorithmic Language ALGOL 68, Mathematisch Centrum, Amsterdam (second printing).

Waldinger, R. J. (1969). "Constructing Programs Automatically Using Theorem Proving", Ph.D. thesis, Carnegie-Mellon Univ.

Waldinger, R. J. and Lee, R. C. T. (1969). "PROW: A Step Toward Automatic Program Writing", Proc. of the International Joint Conference on Artificial Intelligence, Washington, D. C.

Walk, K. et al (1969). "Abstract Syntax and Interpretation of PL/I", IBM Lab. Vienna, Tech. Report TR 25.098.

Wegner, P. (1968). Programming Languages, Information Structures, and Machine Organization, McGraw-Hill, 401 pp.

Yanov, I. I. (1958). "The Logical Schemes of Algorithms", (English translation in: Problems of Cybernetics 1, Pergamon Press, pp. 82-140, publ. 1960).

Zimmermann, K. (1970). "Loop Optimization: Representation and Proof of Correctness", IBM Lab. Vienna, Tech. Report TR 25.108.